The Hybrid Diaries (On Electric Wheels)

By: Andy Barrientos

The Hybrid Diaries
On Electric Wheels
Acknowledgements

First and foremost,

I would like to thank my wife Rhonda for allowing me the freedom

to indulge myself in all of my creative endeavors.

No matter how silly or off the wall they may be.

Thank you Rhonda.

I love you Baby,

You have always been my rock.

I love you very much.

And to everyone who offered support.

And sometimes materials and supplies to my project.

It was those special people who made my dream a reality.

And without their assistance and help.

None of this would have been possible.

I'd like to thank Mark Henderson Welding for their great work building the

prototype frame work and all the fabrication.

And, to The New Paluxy Breweries,

"Home of RepTex Beer."

The Unofficial beer of the Republic Of Texas" For all of their support.

And of course for all the many cases of "The delicious and refreshing taste of

RepTex Beer.

In Dark rich and blonde light.

Look around. you just can't find a better tasting beer.

It's in your grocer's cooler.

God love you guys. especially you.

My friend Richard Wiggler.

Without your support we'd never have gotten the project off the ground..

And I'd like to give a big shout out to Global Composites of Grand Prairie, Texas

For their great job, on the body for our racer.

It was a work of art.

And helped us cut the weight down and made for a stronger safer and more aerodynamic body.

Thanks Bill..

Thank you, Thunder Chicken Paint and Body.

And you too Max Powers for your custom Texas flag paint job.

We were truly representing the Great State Of Texas.

The flaming bull skull on the nose looked awesome.

And of course Unidynamic Motors of Dallas, Texas.

For their faith and support.

And a big shout out to Toby Caliente motor designer and shop foreman.

I hope that my designs and your building expertise

make us both millionaires.

To the Tri Solar Confed Solar Power Panel Corporation.

For their wonderful solar panels.

And all of their help with the wiring and energy management systems.

They did an outstanding job with integrating the inverters and energy control modules.

Thanks again, Rich Richardson.

And also a big thanks to "All Nationwide Batteries" for their support and the use of their prototype batteries.

We couldn't have raced without them. Thanks again Earl.

Thank you Mizzenmast Cruises, for providing us with transportation all the way home From Australia.

The trip alone was an adventure in itself.

And we enjoyed everything about it immensely.

Thank you Captain Stewart for making us feel at home.

And to the crew of the good ship " SS Idle wild".

We miss you guys.

And a big Hi there to Big Jon D's Auto Salvage. For allowing me to rummage and scrounge

And not charging me full price to all the "junk" you said I was taking.

By the way, thank you for your advice.

And giving me a place to keep working on my projects,

I know it was just a shed, but, it was my emergency "port in a storm"

And also for offering to let me use your welding rigs, rods. and cutting saws.

To "On The Shore Restaurant and Bar" and Manager Sonny Maldonaldo for your support and for everything.

This was perhaps the greatest undertaking I have ever tried to organize.

I know I am forgetting way more people that I am thanking…

But, I want all of you to know,

that we love all of you guys and we wanted to thank all of you.

For being a part of our lives and for being there for us when we needed you.

Thank you, Jesus.

You can keep the tube socks.

I hope we meet again. In the very distant future.

Love to all of you. Bless you all.

For this is a story that all of you took part in,

And I would like to dedicate this story to all of you.

Enjoy the story

The Hybrid Diaries
Chapter One
In the Beginning…

Ok, here is the plan….

I'll design and build a Human Electric Hybrid vehicle.

That also has solar cells and a small gasoline powered generator to provide additional energy to recharge the batteries

And to provide the initial power to get the whole thing moving.

Then we'll go to Australia and compete in the 1991 International Energy Challenge in Woolloomooloo, New South Wales.

And beat the best Universities in the world.

Then we'll win the 25,000.00 purse and prove to the world that an underdog can win.

I've always wanted to fight in a struggle and triumph over overwhelming odds.

And this would certainly end up being a huge struggle to get this thing to work.

All I have to do is design and build a vehicle and then obtain financing.

So, we can actually afford to go on the trip and compete. No problem…right?

I first saw the story in the Dallas Morning News.

There it was, plain as day.

A story about "The International World Energy Challenge in Woolloomooloo, New South Wales, Australia."

I thought all about the 1909 British Air race Across the English Channel.

And I thought all about how cool it would be to show the World.

That a little guy with a little support can beat the best funded projects,

From all of the finest Universities in the world.

Hey, maybe they'll offer me a honorary degree after

I beat them to the finish line?

So, I cut the article out of the newspaper.

And the picture of a solar racer that also accompanied the article.

Then I sent a letter to the address of the race officials for the Challenge in Australia.

To request a copy of the rules and entry forms.

And forgot all about it.

And then a few weeks later I get this large business envelope in the mail.

It came all the way from Australia.

Sent by the guy who is organizing the energy challenge.

They were the entry forms and rules and categories of competing vehicles.

The date is April 4th, 1990.

At this time I am working for a Thrift Store that was run for the benefit of The Vietnam Veteran's Foundation.

They were in the process of trying to raise funds to build the Vietnam Veterans Memorial at Fair Park in Dallas, Texas.

I was their store manager.

And I ran the store and the entire collections operation.

I arranged for pickups and hired drivers, cashiers and handled the day to day business.

It was a very happy time for my wife and I.

We'd been married for a little over four years

We had just moved into our first rent house.

It was the first rent house that we had ever lived in together.

Thanks to her sister's dog,

We have just acquired our first puppy, "Thor,"

He was the smartest dog I had ever owned.

He never had any accidents and was always happy and ready to please.

He'd even catch Frisbees in his mouth.

I believe that Labrador Retrievers are the best "All around breed of dog".

And "Thor" was the best of the best.

In spite of making very little money working for a Nonprofit organization like The Vietnam Veterans Foundation, we were happy.

Then I saw that News article about the Energy Challenge in the paper.

And something clicked inside of me.

Ideas for my own entry started to fill my brain and invade my slumber.

My mind was filled with concepts.

I decided to investigate the possibility of me actually designing and then building the vehicle that I would eventually call "The Slingshot."

I came up with the name because the vehicle would generate potential energy.

And use it to produce electricity.

Then It would supply the energy to storage batteries.

The batteries would run electric motors.

That in turn would drive the wheels and provide forward motion.

Stored energy is converted to kinetic energy.

Just like a slingshot went you pull back on the sling.

You create potential energy and when you let go…

You convert it to kinetic energy.

That in turn, hurls the rock or projectile into the direction you were aiming at.

Simple enough, right?.

I originally came up with the idea way back in 1980.

While I was dating a girl who lived A long way from where I was living at the time.

She never got her driver's license and so, She couldn't drive, or own a car

Since my car was broken down at the time.

And I wasn't working,

If I wanted to see her I would have to ride my ten speed bike to her town to visit her.

And believe me, there were hills.

A whole lot of hills.

Some seemed like they were almost vertical.

And I really hate riding uphill .

On those long rides, I would think about old world war two movies.

Where the platoon is pinned down by enemy fire.

They needed to call in an air strike or artillery support.

To keep them from being overrun by "the enemy".

The field commander would call for the radio man.

Here he'd come in running with a big field radio on his back.

And another guy would come with him.

His job was to turn a hand crank generator.

That would in turn power the radio.

And I thought, what if I had a crank generator and I would turn with my feet.

Surely, that would be much easier than using my hands

It would have to be easier using my legs to turn a generator at a constant speed than using my legs riding up hills.

I also, knew that trains have been using a similar system to power diesel Locomotives

The diesel motor doesn't drive the train at all.

It turns a electric generator that powers a huge electric motor.

It's the electric motor that actually drives the train.

So this was my original thinking way back in 1980.

A easier and faster way to travel to my girlfriend's house.

On a machine that would redirect my physical strength to powering a generator.

That in turn would power the electric motor that would drive me and my invention and me down the road.

and to her front door.

If I could make it light enough .it would take less energy to drive it down the road,

It'd have to be aerodynamic to decrease wind resistance and to improve efficiency

Now, ten years have come and gone.

So has the girlfriend that I was seeing at the time.

We had both moved on with our lives.

I met the woman that I would soon call my wife.

And unfortunately she too had lived many miles away from me.

And there were many freeways between us.

So, although I never tried to ride to her house on my bike.

I did think about my invention now and again,

and how much easier it'd be than riding a conventional bike.

Or driving a conventional car.

Now, being a retail manager working for a nonprofit organization like the VVF.

That meant that I was almost poor and not having much working capital available.

I was going to have to do some creative rummaging around for supplies and building materials.

But, I would still have to hire someone with a welding rig to put the prototype test frame together.

And that is how I met Mark Henderson. of "Mark Henderson Welding."

Now, Mark had a shop on the same street as my thrift store and I actually hired him to weld a steel box around my store safe.

Then we filled the box with concrete and welded a steel rod onto the top of the box that passed through eyelets on the side of the safe strongbox.

I could run a padlock through to keep the bad guys out of the money.

(the store was in a kind of rough part of town and break ins were frequent.)

So, one day I approached Mark and told him all about my ideas and described to him what I was planning on building, and he said to me.

"Well, you draw up the plans and the measurements and If I can read them, then I'll weld it up for you."

" You just buy the materials and I'll only charge you labor."

It sounds good so far. Right?

The Hybrid Diaries
Chapter Two
Building the framework of a vision

With the welding shop lined up and the welder agreeing to do the frame .

I was now time to start the hunting / gathering process.

You may not know this, but working at a thrift store is really a great source for crap and junk.

A lot of times people would give what they have for a good cause, other times they would use us for a garbage service.

Our policy was,

if someone put something out and put a sign on it

with the letters VVF on it,

Then, we'd HAVE to take it.

No matter what it was,

That could include broken bicycles that were rusted solid or missing parts.

And these bikes were usually from the late sixties or the early seventies.

Old heavy and made of 1020 gauge steel..

No lightweight composites, no aluminum parts.

While these bikes were usually one toss away from the scrap pile.

They were almost gold to me,

After all it was much easier and cheaper to weld steel than aluminum.

And since I was only building a test bed prototype proof of concept vehicle.

Weight wasn't going to be a real consideration at the time.

Broken chairs, busted exercise bikes, busted bed frames with metal bed rails and scrap metal angle iron became the building materials of choice.

They were trash,

And since they were in plentiful supply and no one wanted it.

It was an obvious choice.

No one wanted these materials and since I was supposed to throw away anything that would not sell,

I would buy the junk materials that I could use from the store and take them down to the welding shop.

At night, I worked on designing the frame.

I used graph paper and came up with a scale of one

block on the graph paper would equal two inches square.

And so I drew the plans according to this scale.

At last, I finished the plans and I took them over to the welder's shop so he could review them and then get started on cutting and welding..

Mark took a look at the plans and said,

"Yes" he could work from my plans

As soon as I gave him a deposit.

He'd get started on the project.

I said alright.

And the following week I gave him the first payment.

We were off and running.

The next couple of weeks were very busy for me.

With getting materials and making trips to and from the welding shop.

To check on his progress.

Mark wasn't too busy so progress progressed fairly quickly.

I checked the calendar and I had less than a year to build the entire vehicle.

Then get financing and all the parts.

I'd need a way to make the trip.

We would need to be there by January

That would allow us some testing time before the actual race.

In the interim,

I had gone by a couple of bike stores searching for some way to stop

"The Slingshot".,

Instead of using "Fred Flintstone feet through the floorboard brakes."

I found some 26 inch bicycle rims that had disc brakes attached to the hubs.

I though these would be ideal for my project .

So I bought two rear wheels

And two front wheels with their calipers and all the hardware.

I would need to mount the calipers to the frame.

I took them over to the welder's shop so we could put them on.

By now the shop was filling up with tubing and parts and other junk.

Mark was starting to get annoyed with the clutter.

It made it hard on him when he had to weld on trailers or over head truck

Racks and other things like that.

In spite of the clutter the frame was almost finished.

And soon I'd be out of his hair.

I'd have to take all my junk with me.

I was sure that Mark would be overjoyed when I left.

because he was tired of trying to explain to his customers "what the heck that thing was"

And what it's supposed to do?"

Finally the big day came.

It took us almost three weeks to weld the frame.

Mark (with a flair for the dramatic), had thrown a

couple of blankets over the freshly completed frame.

And before I went by, he had attached the wheels and the chair

(which was nothing more than a metal and plastic chair attached to a metal frame mount. that was mounted to the frame with ¼ inch bolts.)

Mark jerked the covers away and there it was.

The frame that I spent days and weeks.

Designing and redesigning and building.

In 3-D and finally a reality.

It was unbelievable.

I stood there in a state of amazement, and joy.

I was very pleased with Mark's work and he followed my plans to the letter.

I asked Mark if those bolts were strong enough to hold the chair to the frame.

Mark said "If it's good enough for NASCAR then it'll be good enough for you too."

I never knew that NASCAR rules require you to use ¼ in bolts to hold the seat in the car.

But, Mark was a BIG NASCAR fan.

And he had race car posters all over the shop.

I looked over at Mark who was watching me to see my reaction.

After a few minutes of silence,

Mark looked over to me and said,

"I'm not really sure what you are trying to do here.

But, You Sir, are one hell of an engineer."

I was never more pleased or surprised in my life.

No one had ever said anything like that to me.

And I was flattered and almost embarrassed by the praises.

We just stared at it for a while and my mind was racing to the future.

I was on my way to challenge the rest of the world…

Me, And my greatest creation. " The slingshot."

Now all I need are … batteries, motors, alternators,

wiring, inverters, solar panels,

A body, windshield, etc… everything else… this will all be so easy… Right?

I went back home and told Rhonda about the frame being finished.

God love her, she was patient, and very understanding with me.

I know she didn't understand why this was so important to me.

But, that didn't seem to matter to her.

All the hours and money and materials lying in the yard,

And up against the side of the garage.

I'm sure it bugged the hell out of her.

She never let on.

When I brought the frame home for the first time.

She too was amazed.

She couldn't imagine that I designed and had this built.

It really was unlike anything any one else had ever seen before.

She just smiled.

Shook her head and said..

"It looks nice, Honey"

I had spent the last few weeks trying to explain my motivations and my plans to her

And I promised her that I would take her with me on the trip to Australia.

I promised her that she could be my team manager.

And she could coordinate everything for us.

"That would take a huge load off of my shoulders.

And allow me to concentrate on finishing and designing and building the Slingshot.

I still had tons of things to do.

And now that I got the frame home, it was time for me to get started.

The Hybrid Diaries
Chapter Three

Now there are thousands of things that need to be decided before you build your electric hybrid vehicle.

And there are very few how to books written on the subject.

And please, try to keep this in mind,

It is 1990, and there was no internet to speak of at the time.

So, no free input or advice from other people who have done this before.

So it was to be expected that I would make quite a few mistakes along the way.

And one of the first things that I'd have to figure out,

Will it be A/C or D/C ?

The pros and cons of the current choices are many.

At first I was going to go with the D/C system.

Thinking it was more direct and required fewer working parts to make work.

I started searching for a generating system that might get me and the project running.

I went to an Auto parts supply house and started looking for the largest generator that I could find.

And I finally found one.

It was for a dump truck and it put out 100 amps

I'd need the amperage to power the drive motor that I found in an electrical supply catalog.

Now, I had to order the generator,

That would take three days to arrive.

So, I had no idea about the size or weight of this generator.

Imagine my surprise and alarm when the generator was delivered.

The generator was about three feet long.

It weighed about 80 pounds.

And the shaft was almost impossible to turn by hand.

I was stunned…

That was way too heavy and too hard to turn either by hand or by leg power.

This was NOT what I would need.

Besides, D/C current is hard to regulate.

It a bit tricky to control the flow of the current.

Especially, with a rheostat, or some other speed control system.

I decided then, that maybe I could pursue an A/C system…

Nicolea Tesla would have been proud of me for choosing his alternating current setup.

So, I sent the generator back.

It's weight and size and stiffness of the bearings

were making it a poor choice.

I now set my sights on automobile alternators.

And I finally found what I was looking for.

The solution came from General Motors.

I found out through looking through stacks of catalogs and parts suppliers books.

What I needed were two alternators from a couple of Cadillac Limos.

Each one would crank out 100 amps.

I could build a flywheel friction engine just like the kind in toy cars.

To power a belt that would in turn spin the alternators.

Thus producing electric alternating current.

I'd need batteries to supply what is called a carrier current.

To get the flow going and then the motors could run off the electric current.

And here is how I met Toby Caliente.

Toby works for a company called Unidynamic Motors of Dallas, Texas.

Toby is the shop foreman and he is a master electrical motor rebuilder.

Toby has been rebuilding electric motors for many years,

And is a very pleasant person to work with.

I told Toby about my project and showed him pictures of my frame.

I let him know what I had decided about the choice of A/C current.

Toby said that A/C current would be a better choice

due to the speed controllers that I was looking at in a motor supply catalog.

He liked the speed controllers and thought that they might be able to handle

the loads and the strains that a vehicle like mine would put on them.

With all the stopping and starting up etc…

I told Toby that while there were motors out there that would work.

I was really wanting to build the motors directly into the wheels.

That would cut down on weight and make it a direct drive system.

And when the motors were not running,

the motion of the spinning wheels would produce an electric current that can

be redirected back to the batteries in a process called regenerative braking.

I called my concept "electric wheels" and this would be a major headache

until Toby had worked all the bugs out of the system.

It was like something straight out of the Bible.

"A wheel within a wheel"

The wheels have a fixed inner wheel with another wheel that spun around

outside of the smaller inner wheel

When the outer wheel spins it creates a magnetic field which will produce

electricity

Copper brushes strip the electrons from the wheel / armature and direct it to

the batteries

And when I would pass an electric current through the wheels,

they would produce rotational motion,

just like a regular electric motor.

With the use of a speed controller I could control the R.P.M.'s and speed of the vehicle.

Toby and I got started on designing and building the wheels from scratch using 26 inch rims spinning inside of 27 inch rims giving us about ¾ of an inch to place the magnets and the brushes.

The copper coils would wrap around the spokes and down towards the center of the wheel axle hubs.

These electric wheels were to be my secret weapons.

They provided me another way to generate more electricity and also propel the vehicle.

"I'm sure that they will cost a fortune" I said to Toby,

He just smiled at me, and said

"Don't worry, I always wanted to experiment with new motor designs.'

'And I've never seen anything like what you are proposing before.."

"You are either a genius, or else you're completely mental". then he laughed.

And I knew that I found someone who shared my vision.

And I also found another place to store some of the materials we'd need to build the Electric Wheels / motors.

OK, I have told you guys way more about the wheels than I was supposed to.

For the sake of this story, and in the interest of everyone from myself to you the reader

And especially for Toby,

I am not going to elaborate any more on any technical specifications of the wheels.

Toby did a great deal of research with me in developing the designs and manufacturing, the first pair of these wholly remarkable drive units.

And Toby wants his check for a million dollars before he spills the beans on anything.

Toby is a brilliant, but private person.

He doesn't like being spotlighted for his kindness or good deeds done at others behest.

So, in all fairness, I'll just give a few more brief statements on Toby's behalf.

Toby says that he did no work for me on company time.

And this was after all just an experiment.

Done on his own time with his own materials (and mine of course)

And, if you wish to contact Toby.

He requests that your forward any and all Communication, correspondence and requests to me care of this publisher. Thank you,

And, now, with that being said, back to the story…

When I told Rhonda about all the progress that I had made.

It made her happy to see the light in my eyes and the expression of complete and total excitement on my face.

I'm not a great poker player.

I have a very expressive face.

Just look in my eyes,

you can see that most people can read me almost instantly.

She knows when I am holding a full house.

It was this a character fault that she loved about me.

She knows that I'm not a good liar.

So, I make it my business to avoid lying at all cost.

Because, I am not that good at it.

And besides, it is a whole lot easier not to lie.

Because you don't have to remember what lies

you've already told other people.

Now that I have the frame and the electric wheels are in production.

I will need to focus on other parts and materials

that I would need to complete the project.

It was coming together,

But time is always against you when you are trying to do something that's

never been done before.

I always try to expect delays.

Now I need something to generate electricity and to resupply the batteries and

provide the electricity to drive the motors when the electric wheels are finally

finished.

Anyhow... - I went to the wrecking yard to find my alternators..

I used the phone book and started calling wrecking yards to try to find the

special Cadillac alternators, that only come on Limos.

I called a wrecking yard called "Big Jon D's Auto Salvage"

Jon answered the phone.

"Big Jon D's Auto Salvage, This is Jon, can I help

you?" he said into the phone.

I said, "Hi Jon, my name is Andy and I am looking for Cadillac limos."

" do you have any there?"

there was a pause, and then I heard him cough and

then he answered me " a couple"

So, I got their address and went there to pull a couple of alternators.

Now, Big Jon D's Auto Salvage has employees who actually pull the parts for

you.

They charge more than most wrecking yards.

But they usually DO have what you are looking for.

I walked around the yard with a guy named "Clovis" he said that was his real name

He used to get ribbed about it when he was in the army.

So, he got used to calling himself "NM" (for New Mexico, where the City of Clovis is).

He said that "his friends still call him Clovis"

I just thought that was interesting.

So, there we were.

Me and Clovis, wandering around the wrecking yard in the GM section looking for Cadillac limos.

Amid the ruins of other, much less fortunate Caddy's.

As luck would have it, we found two of them.

(alternators that is.)

I bought both of them and took them over To the auto parts store.

Where they could be tested to make sure they work properly.

They both checked out, so I now had the alternators.

And now I needed to build the crank assembly and the friction motor to turn them.

If you've ever rode an stationary exercise bike.

Then you probably know that there are several models available on the market..

One has a regular bicycle wheel, the other has a fan instead of a bicycle wheel.

That type is called a wind rider, wind rower or a wind machine.

Because of the breeze kicked up by the fan as it's spinning around .

The third type of exercise bike is the kind with a metal flywheel.

That is kind of like a spin cycle in that it continues to spin even after you stop pedaling.

A flywheel naturally stores up energy as it spins around.

It takes a lot of energy to stop it once it gets going.

And that is what makes it perfect for the application I was considering.

I don't know if there will be pictures of the Slingshot here.

so please forgive me if I can't give a perfect explanation about the layout.

Please, bear with me.

The Slingshot. Is a three wheeled vehicle.

The two wheels in the front are drive wheels

The rear wheel is also a drive wheel.

So technically, it's a three wheel drive system.

The driver sits up front between the two front wheels.

The front wheels are steerable.

While the rear drive wheel is fixed and only rolls forward or backward.

It doesn't turn.

There is a metal boom with a bicycle crank on the front most part of the vehicle.

That crank has a long bike chain that passes under the seat and behind the rider.

That Is where I put the flywheel and the drive wheel to the crank.

And that in turn was hooked up to the twin alternators via a drive belt.

And that provides the electrical current to recharge

the batteries and also provide electricity to run the electric wheels.

Now all I need is a power management system that will transfer power from

the alternators directly to one of the batteries

And after it is fully recharged.

It automatically switches the power to another battery.

One that has just been depleted.

And then transfer the power to the other depleted

battery when that battery has been fully recharged and so on,

Until all the batteries have been fully charged.

I also thought that Solar Cells would be able to charge the batteries even when the system is not in use.

I needed a special power inverter system to manage the electric flow

And to replenish all of the batteries and manage the solar cells output and

power the electric wheels and hopefully take me somewhere.

So, What I have created is (to the very best of my knowledge) is the World's first and only Human, Electric, Solar, Gas Powered Hybrid vehicle.

The Hybrid Diaries
Chapter Four

I used to have a friend who was a restaurant manager. Sonny Maldonaldo.

Sonny worked for a Seafood Restaurant called "On The Shore."

It was a great little dive that my wife and I loved to frequent.

I loved the casual eating experience and the dollar beer served in plastic glasses.

They looked like the same glasses you'd get if you were going to a kegger or a rave or a Frat party or some other place where mass amounts of alcohol are served.

Sonny was a good friend who loved his customers.

He had a "Pay as you go" policy where you ate and drank,

And then you'd tell him what you had consumed

and he'd charge you accordingly.

I'm sure some people would always understate their purchases.

And I am certain that Sonny knew that they might

be undervaluing their meals or drinks,

But, Sonny didn't seem to mind He felt that there were enough honest customers to offset the few scammers.

He even fed us a complementary meal when we were having a tough time a while back..

Now, One day I was eating at the restaurant like we frequently did at the time.

And when I came in I noticed that there was a tall professional looking businessman at the counter talking to Sonny.

He told Sonny.

"I know that your store will double it's beer sales

once you start selling RepTex Beer."

" It's from a small brewery not far from the Texas Hill Country."

The guy turned around and looked at me and said,

"So, Do you Like Texas?"

I was caught off guard and I blurted out "Of course

I do. Why?"

"How about the taste of a great tasting beer?,

"Served cold in a frosty glass?"

"Sure, I like to drink beer, especially if it's cold and

taste good. Why?" I asked.

"Because, I'm about to give you a six pack of "RepTex Beer"

" It's the Unofficial Beer of the Republic of Texas!"

I stared into his bespectacled eyes and I gazed at his semi muscular build,

His graying temples with his short cut hair.

Which reminded me of a military style haircut.

His eyes had a sparkle when he started talking Beer.

It seemed to be his passion.

His demeanor kind of reminded me of professor Harold Hill from

"The Music Man".

"Who are you?" I asked with more than a little bit of suspiciousness.

"MY name is Richard Wiggler, and I'm the district distributor for

The New Paluxy Brewery.

I'm looking for ways to increase the face value of our fine product."

He said proudly

"So, How about sponsoring some crazy stunt to draw attention to your

brewery?"

I asked him. (I couldn't believe that I said that. It just blurted out of my

mouth.)

26

Richard thought about it for a minute and then he

Asked me point blank…

"So, what do you have in mind?" he said. looking

more than a little puzzled.

At first I though he was joking, but, the almost somber look on his face let me

know that he was as serious as a heart attack.

I cleared my throat, and asked him, if he'd like to sit down,

because my story would take a few minutes.

So, he pulled up a stool and I began to tell him my story.

After I finished my story,

He looked at me like I had a snake on my head.

And then he asked me "you are doing WHAT?"

"You must be crazy." He said.

So, I went out to my car and pulled out a package of pictures and started

showing him the Slingshot frame,

The shop where it was welded together, and a picture of the electric

motor supply company where I planned to buy my motors

And a picture of me with my new friend Toby Caliente.

Richard rubbed his eyes, and looked at the pictures again.

Then he said, "You really are crazy, do you know that?"

"So, I have heard" was my response.

I had to smile at his almost shocked expression on his face.

Richard leaned back in his stood and said,

"You outta know, you're never going get anyone to

sponsor you or your project, unless someone steps up

and offers to help you first"

"It's like the old foot in the door".

No one wants to be the first to sponsor anybody."

He took a sip of water and then he continued

"Companies look for a project or teams that are already established to sponsor,

because someone has already done the legwork."

"And this cuts down the risk that is involved in sponsoring a major undertaking."

"Only an idiot would shell out large sums of cash on an untried and untested anything."

" I don't mean to sound harsh, I just wanted you to know what you are up against."

I sat there listening to the words Richard said to me,

And I knew that everything he said to me was the truth.

I knew this because I had already sent out proposals

to all the major beverage companies and distributers and they all had

responded to my inquiries with letters in which they all politely turned me down.

Some were very nice about their rejections. Others not so much.

A major brewery to the north of Texas said that

"They would have to wait until October for the start

of their fiscal year before they could sponsor anything new."

I didn't have the time to wait until October for them to reject me again.

And with January looming in the not too distant future.

I was starting to get a little nervous about finding sponsors.

So far I had zero.

I couldn't count the guys who have already promised to help me because they

were either doing this out of kindness or else I paid their company for their services.

So, I couldn't say they were actually sponsors.

I was hopeful that I could get the local media interested in doing a story about my project,

and maybe that would help me get financing from an interested corporation.

But, so far, no media outlet had returned any of my calls.

And I was starting to think

"Maybe I AM crazy."

And maybe I didn't stand a chance of making it to Australia,

Much less competing.

Maybe I should just chalk this up to a stupid idea or obsession.

So far, I've dumped a little 2500.00 in parts and welding and all I have to show for it was a neat looking frame and not much else.

I sat there in silence, probably looking like someone just ran over my dog.

Finally Richard spoke up.

"I am sorry, I didn't mean to crush your spirit,

"I just know how hard it is to get people to listen or to persuade them to listen to your line of thinking or get the word out."

"In my business, visibility is key to everything.

If people see you, then people will remember you later on". He said.

"And maybe the next time they see you, or your product they might think,

"Yeah, I remember seeing something about that."

"That is called product knowledge. "

"A product will never sell until someone has acquired some product knowledge about that particular product. In this case the product is beer. "

"Now, In your case, it's the same with race car teams and the like."

" They slap Kellogg's or BUD, or Texaco, all over their cars."

" Those names, are from the major sponsors who pay hundreds of thousands of dollars to have their mark or brand name on that particular car.

"With that particular driver, and that particular builder

And that particular PIT crew."

" They usually sponsor that team for the entire year and at the end of the season,"

"They have meetings to determine if they will choose to sponsor that team or driver in the upcoming year."

"Based upon his performance or his winnings or his standings with the race patrons."

"How much do you think you would need to actually compete in this race?"

He asked me directly.

"I was expecting the trip itself would be the most expensive part of the entire project,

And that was going to be around twenty five thousand dollars just to ship the slingshot and myself."

I responded.

Honestly, I had not considered a team before.

Because, no one seemed to be as interested in the project as I am. And none of my friends wanted to or else were unable to take that

much time off their jobs to even take a short trip to California.

Or to help me get my vehicle to Long Beach,

So, we could send it out by ship to Australia.

"Listen" Richard said

"I've got to go and check on a couple of

distributors, can you give me your phone number,

and, I'll call you later, and we can talk more about this?"

I said, "Sure"

And so, I gave him my phone number.

Completely convinced that I'd never hear another blessed word from him ever again.

I said "Thank you" to Richard for all the time and the advice he had shared with me.

And I told him that he had given me a lot of information
that I would have to digest over time."

As, I was walking out of "On The Shore" I said bye to Sonny,

and was heading out the door,

when I heard Richard calling after me.

"Hey" he shouted "what about your six pack?"

"Huh?" I answered, (I had forgotten that he offered me a free six pack.

 I was too busy mulling over all the information that he had just given me.

And I was thinking about what else I was going to need to get the project up and running.)

"What? " I said

"Do you think I'd turn down free beer?"

We both shared a laugh, and we walked outside to Richard's car.

He had a very nice car of European heritage.

And we went back to the trunk and he opened it

The trunk was crammed with beer and flyers, literature and some banners.

Richard reached in and pulled out a six pack,

"You'll have to chill them in the fridge before you can really enjoy them"

He said,

"Thank you so much, I actually had forgotten all about the beer."

I said.

He said, "Yeah, well I almost forgot it too."

"I am a man of my word. And I always try to do right by others"

"But, I say I am going to do something, I always try to do it."

I said good bye to him again,

As I was getting into my beat up old crap-mobile

I heard him yell at me one more time.

"I'll call you."

And that was the last time I expected to hear from him…

After I got the beer home and it sat in the fridge for about an hour

My wife and I drank a couple of bottles each,

It was a really nice tasting beer.

It had a good bite and it was smooth.

Very hoppy with a hint of malt It was a very good beer.

The Hybrid Diaries
Chapter Five

After my talk with Richard,

I realized that If I didn't do something soon,

I'd run out of time and I'd never get the project any further than what I could accomplish on my own.

I was going to have to do something pretty dramatic

to generate some interest.

So, I started looking for a small Briggs and Stratton edger type motor.

I had an idea to make the frame move.

And maybe get the TV crews out to view my project.

But, it'd have to be amazing.

So, I decided that maybe a test drive would be in order.

I still had the bicycle wheels with the disc brakes.

They were there to keep the frame off the ground,

And served as place holders until the electric wheels were finished.

It was going to be several weeks before they would even be ready to test.

So they were not going to be used for this demonstration.

I still needed a rear drive wheel that I could use for the demonstration run.

I looked at all In all of my junk bike parts and I found a twenty inch rear bike wheel and hub assembly that I could adapt to suit my needs.

Next I went to a lawn mower repair place and I purchased a go kart clutch and a drive pulley.

Then I went to a appliance repair place and bought a used dryer belt

And I proceeded to mount the gas edger engine

towards the back of the frame next to the new rear wheel.

I used the throttle cable that came with the edger

and I mounted the throttle lever to the side of the seat.

Next I mounted the clutch and the pulley to the drive shaft of the edger engine.

And then I ran a drive chain from the clutch to the rear wheel.

That would provide the torque that would get the thing moving.

And when the clutch was not engaged,

The edger engine would turn one of the alternators and produce electricity.

Then I hooked up a couple of electric motors and bolted them to the frame

Where they belong and ran bike chains around the front wheel hubs so they can at least look like they were driving the front wheels.

The bike chain around the hub cassettes spun the electric motors around when the wheels were moving.

I was hoping that it would work.

So, I borrowed my father in law's pickup truck and got out my video camera

And I took the frame up to the local community college parking lot on a Sunday.

When there wouldn't be anyone around.

I set up the camera, and had my wife videotape me in front of the Slingshot.

And I began my monologue.

Just like all of those wackos in the early 1900's

That would stand proudly in front of their aircraft that they just finished building in their garage or in some out building.

And they started spouting off about how they were going to

"Sail like the birds through the air and fly high in the sky around the field .
Before making a perfect three point landing right in front of the cameras. "

Only to crash about two feet front the spot where they were taking off from.

Now, it was my turn to look stupid…

"Hello, My name is Andy B. and this is my invention.

I call it the Slingshot,

This is my entry for The International World Energy Challenge in

Woolloomooloo New South Wales, Australia..

"The Slingshot is the world's first Human powered electric, solar, gas/ethanol Hybrid vehicle".

"It runs on human power, solar panels and electric motors.

It's a first step into a more energy efficient future.

And I believe it to be an alternative to the traditional automobile which most of us already own."

And now for the airplane crash.

I got onto the seat and hit the electric start, and the engine actually came on,

And on the first try too.

I considered this to be a positive sign.

So, I increased the revs and the clutch started to engage the rear drive wheel.

And I started rolling.

Slowly at first and then it got faster as I started building up more and more momentum.

I was looking good rolling around in the parking lot,

making sure I was keeping the Slingshot on a downward grade.

It looked like I was running on electric power instead of the gas engine

I was actually running on.

It was all an illusion,

But, the vehicle was rolling under power and I was moving.

Everything was rolling on the frame that I had made.

And it looked like It was rolling on it's own power.

And it looked cool to me.

My wife shot about thirty full minutes of film footage.

Mostly showing me riding around the parking lot on the Slingshot,

looking every bit as competent as those guys in the aforementioned old timey airplane failure movie footage.

But, at least it was moving.

After the ride,

I packed up everything and we took it all home and I put the Slingshot in the garage and shut the door.

Then dropped the truck back off at the in laws.

After I got back home,

I went and hooked up a couple of VCR's to the TV and made several VHS copies of the tape.

And then I mailed them out to all the local and major network news affiliates.

So, with the footage mailed out,

I sat back and waited to hear back from someone.

Finally I got a call from a reporter for Channel 7 news.

He saw the tape and was very interested in doing a human interest story about Me, the Slingshot and the Challenge.

I told him that I was still working on the "Car" and it wasn't running "right now"

but, he was more than welcome to come out and film it.

And I could give him my story.

The following Wednesday, the news crew showed up at the house.

So, I took off early to meet with the reporter and film crew.

I had just enough time to pull it out of the garage,

hook up a 12 volt battery and used the electric start to get the engine running.

I hoisted the frame onto a few cinder blocks and gassed up the engine.

So I got the rear wheel spinning and I did my best to look like an engineer.

I looked greasy and dirty.

Like I had been slaving over a hot engine compartment all day long.

"Hello? Are you Andy?" came a voice from outside the gate.

"Hello, I am John Procktor with Channel 7 News."

" And this is Mike Heller My cameraman."

The sounds of unfamiliar voices at immediately caught my dog Thor's attention.

His hackles went up when he noticed those strangers standing at his gate.

And his hackles bristled when he saw those two strangers try to enter into his yard and talk to his master without his permission.

He whirled around and started barking at the strangers.

Gnashing his teeth at them like a hound from hell.

Most impressive.

I yelled loudly and forcefully.

"No Thor!" I said,

And then I took him to the back door by his collar

And let him in the house.

Then I invited the crew and the reporter into the back yard.

And then we began the interview.

They asked the basic questions first,

like why was I building this and for what purpose and so on.

And what was my inspiration for building the vehicle

So, I came up with …

"For the benefit of my fellow brothers and sisters

and to help clean up the environment and to make some kind of

personal contribution to the rest of mankind."

"Alright then", John said,

" Who do you consider to be your inspiration?"

"Well, I've always been inspired by people like the Wright Brothers,

Glen Curtiss, and Paul Macready."

" He's the guy who built the world's first successful human powered airplane

to cross the English Channel".

"Oh and then there's Dick and Burt Rutan. "

"They were the ones who designed and built the Voyager airplane that flew

around the world on a single tank of gas."

" Dick Rutan and a lady aviator named Jeana Yeager took off in The Voyager

on December 14, 1986".

And they were the first to fly around the world nonstop."

"In about nine days I think."

And my favorite inspirational person would have to be

Clarence "Kelly" Johnson.

He was the chief designer for Lockheed.

Some of the planes that he designed are the World War II fighter plane,

The P- 38 Lightning ,The P-80 Shooting Star, The F-104 Star fighter,

the U2 and the SR-71 Blackbird."

And then we got into the more technical questions.

I didn't want to give too much info so, I clammed up about everything

Except for the race in Australia.

And that we were looking for sponsors to help us get to there,

So we could get our chance to compete and win.

We talked for about 20 minutes with the cameras rolling the entire time.

They filmed me and my responses to their questions.

Before too long, they were finished with the interview,

and they thanked me for the opportunity to interview me.

I told them that they were more than welcome,

And that they were welcome to come back anytime they wished.

They said to be watching the five o' clock news tonight.

I told them that, "I would not want to miss it"..

I ran back inside to an excited Rhonda who was watching the interview

process through the window.

"What did they say?" she asked me

The second I came inside the back door.

"The story should be on at 5:00 O'clock." I replied

"So, what did they ask you" She asked.

"Just wait, and see, I don't remember everything they asked me,

'or how I answered their questions."

" Let's just wait and we can watch the story together."

Rhonda, got on the phone and called her Mom and Dad.

And then she called her sister.

So, five o'clock came around and it was like the second to the last story.

It showed a clip from the tape I sent the TV station.

And I couple of minutes of me talking to the reporter.

And that was it.

I was excited about seeing it on the television.

So much so, that I forgot to record the news segment on my VCR.

"Darnn it.

I could have used that news story to help me get some more sponsorship."

Rhonda just laughed. and, said that maybe someone was watching.

I smiled back at her

And, all I could say was

"maybe"

so much for my three minutes of fame.

And so much for my time in front of a state wide audience.

It was a good experience for me.

I was delighted for the opportunity to be interviewed by Channel 7 news.

I hoped that maybe someone out there, with deep pockets would see the news

clip and call the station for my phone number

And that calls would be flooding the station.

With companies lining up to sponsor my project.

But, I cannot dwell on my latest accomplishments

or the hopes that it brought to me and my cause.

I've got too many things on my plate to have to worry about something like this,

and I needed to focus my attentions on other problems.

The Hybrid Diaries
Chapter Six

Well the day started out OK,

I was at my office in the Thrift store.

I got a little bit of ribbing from a couple of employees who saw me on the television last night.

The truck driver was out making pick ups,

When I get this call from a very distraught woman on the other end of the line.

"Hello?. My name is Janet DuBois and I need your help!"

I told her my name, and I asked her "How could I help her?"

That's when she started telling me in a sobbing voice

that her husband had died recently and that she had just received her insurance settlement and she wanted to hide it in a safe place.

So the safest place she could think of was her late husband's coat.

Unfortunately,

Her sister had come over and helped her try to clean up her house,

And one of the things that her sister did was clean out the hall closet.

And, when she did,

She stuffed all of his old shirts and coats into a garbage bag

And left it on the porch for the Vietnam Veteran's Foundation.

And that the driver had just come by and made a pick up.

I asked her "how much money are we talking about?"

She screamed into my ear…"Twenty five thousand dollars, in cash!"

"It's everything I own!"

" Can you PLEASE, help me???"

I got a description of the coat and what color the trash bags were and I told her that I would have to wait for the truck to get back before I could contact her.

So, she left her number and I promised her I'd call her back.

So the day passes and finally, the truck driver returns.

Now my driver is named David, He was from the Midwest and nothing can ruffle this guy.

he's got a handlebar moustache and the kind of poker face that I wish I had.

I took him to the side.

I told him that we were looking for a couple of trash bags of donations.

So instead of piling the bags on the pallet dolly,

we piled them on the floor,

I went through each bag until I found the coat.

I looked in all the pockets, and yes, it was there.

I pulled my hand out of the pocket and I was holding twenty five thousand dollars in my hand.

Dave did raise an eyebrow when I showed him that wad of cash,

And he said,

"Wow, If I only knew that I was driving around with all that money…"

"And what would you have done IF you HAD known?" I asked him.

"Call in, and tell you where you could find your truck!"

He said

I knew he was only joking,

He was the best driver I had ever hired.

He was always on time and I seldom got a call about missed pickups when he was on the job,

He always brought back a full truck on his pick up runs.

I submitted him for a raise.

And I walked back to my office.

Wow, this is exactly what I need to finish the Slingshot and book a trip to

Australia I thought to myself.

I could use the money to buy a new car, or maybe make a down payment on a house.

Or buy my wife something nice, she deserves that..

All I have to do is say, "Sorry Ma am, but, I couldn't find your missing money…

" And It would be mine. Her word against mine, no one would ever know.

So, I walked over to the phone and I called her.

"Hello Mrs. Dubois? This is Andy at the thrift store."

Can you tell me how much money you lost again please?

She said "twenty five thousand dollars all in hundreds."

"Could you come up to the store please?" I asked her

She said, "I'll have to get my daughter to drive me,

I can't see too well"

I said, "That would be alright."

And she said that "she'd be there in about thirty minutes."

I know, if I lost that kind of money I would be there in five minutes.

Thirty five minutes later this old woman with a walker came into the store with a younger woman holding her by the arm for support.

They came walking through my store's front door at a snail's pace.

They asked my cashier if I was there, and Cindy said,

Yes. He is in back"

So, I came out of the back area, and I met the two ladies.

And I asked Mrs. Dubois if I could see her ID.

I took it back into my office and I copied several of the hundreds and her driver's license.

Her license was old, and had recently expired.

So, I know she hasn't been using it in quite a while.

Then I went out to the floor and asked them if they could come back to my office.

They came back, and I closed my door.

And then I unloaded on the old lady, telling her that

"she should not have that kind of money laying around in a coat or stuffed in a mattress".

"It should be in a bank where she can keep track of it."

And she agreed,

Then I pulled out the money and gave it back to her.

I told her that she was very lucky to have met someone like me,

because, someone else might have kept the money.

And done god knows what with it..

They both thanked me again for my honesty,

and then they left.

No reward, just a thank you, and that was it.

They were gone.

That night, I thought about her and I wondered

What I would have done with the money.

I dreamed about the trip and then the Challenge I dreamed about crossing the finish line ahead of everyone else.

And I dreamed about accepting the trophy and waving to the adoring crowd.

And then I woke up.

I was still hurting for money,

I was running out of time, and I still didn't have any real sponsors.

All I did have was just the spirit to compete and I needed everything else.

I still had to design a body for the project and I needed to get it painted and solar cells etc.

I went over to Toby's shop and spent some time there working on the wheels and assembling them.

"Soon, it will be time to test them." Toby told me.

He was starting to get excited too.

"I was thinking…"Toby continued "If you reverse the electrical field, you just might be able to stop the wheels using just the electric fields already in the motors without having to install external brakes."

He continued "You'd lose regenerative braking, but you would not need additional brakes to stop the vehicle."

I laughed out loud, and I told Toby."

I would rather have a back up braking system.

That is independent of the electric drive units just in case."

" Besides, I think the rules require us to have emergency brakes."

I wasn't sure, But, I believe that I read that in the rulebook.

I also discovered that because our vehicle is unique, then we would not be racing with the purely solar vehicles.

We'd be racing against any and all hybrid vehicles.

And while there was not any listing of competing teams in each category I can assume that there would be more purely solar powered vehicles than hybrid vehicles.

That would improve my chances of winning in my category.

So, I could be a winner just by showing up.

As opposed to competing against all of the solar racers.

They might call foul if they figured out that my entry was a hybrid instead of a pure solar car.

But, when I sent the entry forms in there was a place to describe the vehicle,

They said that they would notify all ineligible entries within 48 hours of receiving their entry forms back.

It has been a little over two months now

The actual date is July 6th, 1990, and I have not heard back from the racing board.

So, I was determined to press on with the project and hope for the best.

Toby had said that he had found some inverters and control boxes that we could use to regulate the flow of electricity.

And make sure it goes where it is needed the most.

I said "Cool!"

With all the voltage and current changes we'd need something pretty robust to handle the system that we are building,

Stepping down current, inverting current from DC to AC and vice versa.

It would take a really stout system that could handle our power management requirements.

Of course, I never asked him how much the system would cost.

I just keep an open mind

I just hope that somehow, if this is meant to be,

then somehow the money will come in.

no matter how long it takes,

It will be provided for.

So, I kept my "Pollyanna" type attitude and pressed on with my mission.

I had been pawing through a couple of aerospace magazines,

And I found out about a Company in Grand Prairie, Texas,

That is not too far from my house. They work in carbon fiber,

Manufacturing aircraft parts for General Dynamics.

Now General Dynamics manufactures the US F–16 Fighting Falcon Fighter plane and many other aerospace projects for the U.S. Air Force

And the Air National Guard

I thought " carbon fiber is stronger and lighter than steel or

aluminum. and is great for projects

requiring light weight and great strength. "

So, It seems that carbon fiber would be the way to go.

The Company "Global Composites of Grand Prairie, Texas" the article says,

has been a leader in the manufacture and fabrication of high quality high

stress carbon fiber products for the aircraft and Armed Forces applications."

So I thought , "I'll give them a try".

Surely, they would be interested in a project like mine.

So, I looked them up in the good old phone book

and I gave them a call.

I spoke to the receptionist, and she transferred me to some guy in promotions.

I was wondering "Why would they transfer me to promotions?"

I wasn't asking them to promote anything..

I just wanted to meet with them and maybe tour the facility..

THEN, I would show them the project and the portfolio that I was compiling.

With all the news clippings and the video that my wife shot

And some pictures of me and Toby.

I was hoping that I could speak with the company

CEO or someone in special projects.

Instead, I got to talk to "Bill."

Now, Bill, was a really nice, likeable guy,

He had a good sense of humor. And I had a feeling

that he was the type of guy that would

hear your proposal and then politely inform you

that, "they were not interested, in any special

projects at this time."

But., since I had nothing to lose.

I went ahead and told him about my project .

I told him about the blueprints that I had drawn up and the frame that I had made and the size and my dimensions.

Bill was amazed by what I was able to accomplish and his voice

smiled at my proposal,

Instead of telling me

"No thank you"

He said that he would discuss my proposal with his management staff.

Of course, I took this to mean,

"No Thank you"

And I hung up the phone feeling like I just wasted my time.

I kept telling myself

"Why did I just waste my time talking to some

flunkey who was only there to blow me off?"

As I hung up the phone, I was wondering what else

I could try to obtain more parts?

I still needed solar cells an energy control system for the solar panels

I needed to figure out how to tie everything together into a cohesive system.

There was so much to do, and time is always running faster than the project.

And I was still without transport or sponsorship.

And I was starting to feel the strain.

I felt alone, because, no one seemed to share my vision.

I was starting to feel more foolish than I ever did.

Maybe, I am just a nut. Why would I want to do this anyway?.

The next day, I went to work at the Thrift Store.

It was a typical day mot much happened.

When I got home, Rhonda, told me that

someone had called to speak with me.

Some guy "I think is name is "Dick"

I was wondering who the heck is "Dick?"

I asked her if "he left a number for me to call him back?"

My wife had written it down on a pad and she handed it over to me,

So I picked up the phone and I dialed the number

and I was shocked almost out of my mind when the voice on the other end of

the phone said

"Dick Wiggler, Can I help you?"

"Richard???" I asked.. "IS that you???"

"Yes, this is Dick Wiggler, How can I help you?"

"Well, this is Andy B. and I was returning your call."

"Andy! So, How have you been doing?"

"I saw you on the news a while back, So, how is the project coming?"

"Have you got any sponsors lined up yet?"

"No, not yet, I have some companies that are offering me assistance, and

support, but, no major sponsors as of yet." I replied.

Richard went on

"I was wondering if you and your wife would like

to come down here to

Austin and spend the weekend as our guests?

"Candace and I would love to have you down."

I said, "I'm sure that we do not have anything going

on this weekend. So, Yeah, We'd like to come down."

"When would you like us to show up?" I asked

I was still in disbelief. I kept wondering

"Why would he even bother to call me?"

I guess, that Candace must be his wife, or girlfriend

or some significant other.

Richard replied

"I know it's short notice, but, do you think you

could come down this weekend?"

I looked at the calendar and then I said,

"Hold on for just one minute please."

I looked Over at Rhonda hopefully and with a

"Oh Brother" look on her face,

she smiled and said

"Alright, let's go"

"Hello, Dick?, Yes, We can come down this

weekend. Do you need us to bring anything?"

"I don't suppose you want me to bring beer right?"

"Oh my no, I think we have the beer thing covered."

Richard Laughed.

"You might want to bring your swim suit,

We have a pool and a hot tub."

"OK", I said, "I'll let Rhonda know."

What time do you want us there? I asked again.

Around 6:00 PM, that way we can have a few drinks before we eat dinner.

"Well, Alright then, We'll shoot for as close to six as we can come,

without getting lost of course."

So, for the next twenty minutes I painstakingly wrrote down the directions to his house and I got his phone number and wrote it

on the directions just in case we did get lost on our way down there.

I hung up the phone and I looked over at Rhonda

and said

"So, It looks like We are going down to Austin this weekend."

Rhonda said, "It looks like"

So, I felt a little better.

At least We can have a beer or two at the house of a beer distributor,

who has a pool and a hot tub.

And I can only assume he might have beer there as well.

I was looking forward to just getting "the hell out of Dodge"

and, just getting away from the store and my ever demanding employees

for a couple of days.

It all sounded so good to me.

So, I called my boss and told him that I needed to take the weekend off,

And that he needs to have the someone cover the store for me

during the time I would be away.

It all went downhill after that. So, now my boss was peeved at me.

But, I had the time coming,

and It was only two days.

Surely he could spare me for just two days. Right?

Wrong. It seems that my boss had decided that I didn't need to take the time

off after all.

And since he was after all my boss, he knew best.

And his fat assed oversized word is supposed to be law.

And, if it didn't mean so much to me,

I would have sucked it up and kept working without so much as a complaint.

But, this WAS important to me. I called my boss back

and let him know that I would not be there that

Saturday and Sunday and I would be back on Monday.

That is if I still have a job to come back to.

That being said.. I hung up the phone, finished the day, and started to pack.

Well, the rest of the week went by slowly and I did as much around the store

as possible,

trying to get everything in order before I left.

I promoted Diane my head sorter and Pricer

to be my assistant manager while I was gone,

I gave her the keys to the door and to the safe

I did my weekly reports up to Saturday.

So, all I would have to do when I got back to the store would be add

Saturday's and Sunday's sales to the report, total them out and then

submit the figures and deposit slips to my bosses house.

I have to admit it here, that I was scared.

I had never had to "walk off of a job" before. (even

though I wasn't.)

And, I was so worried that I didn't even tell my wife about it.

The last thing I wanted to do was worry her.

And since, Randall never did say "You are fired!"

I technically still had the job.

I was just going on a very short vacation with my wife to visit some friends

in Austin, Texas. Nothing more.

With all the work that I could do done,

I made sure that Diane watched the other employees.

And make sure that they have plenty to do while I was gone.

Like clean the floors, dust and restock

Purge the old clothes to make room for the new ones that had just come in,

Or else, mark them down so they will sell faster.

I gave Diane the phone number for where we would be in Austin,

and told her not to call us unless it was an emergency.

Then Rhonda and I climbed into our yellow, un-air

conditioned Datsun B-210 hatchback

Crap mobile with the rust holes in the floor board.

Loaded up the dog and drove Thor over to his grandparents house

(the In-Laws) and headed down I-35.

On our way to Austin and the Texas Hill Country.

It'd be an uneventful three hour drive

if we didn't stop to eat along the way.

But, I'm not the kind of guy who tries to barrel

through traffic to get to the destination as soon as possible.

I like to look or stop along the way, and just enjoy

my freedom for a little while before continuing on my way.

Rhonda and I like to play silly games on road trips.

One of our favorite is the classic game "Windmill, "

You play "Windmill" like you play counting Blue cars.

Whoever sees the windmill first shouts out

"Windmill" and points to where it was sighted.

And then you keep score.

Now being the DRIVER and having to focus my attention on DRIVING,

I usually lose this game.

But, Rhonda likes it and It does help to pass the time,

Especially since the radio will not work once we leave the broadcast area that surrounds the City of Dallas

We passed through Waco, a little more than an hour into our road trip,

that's just about the time the last Dallas radio station faded out of range.

As the sounds of Classic rock faded into static.

We saw a sign for "The Starving Road bum Restaurant" five miles ahead"

And since I missed breakfast, and it was close to lunch time,

I looked over at Rhonda and asked her

"Have you ever eaten there before?"

Now, Rhonda and I are both from Dallas, and we

both had parents that would load up the

family into the good old station wagon

And head out to the coast or to Mexico,

Or to the Hill country,

For a good old fashioned car vacation.

And on these trips we would pass the Rattlesnake Farm and Prairie Dog own,

or the Natural Bridge Caverns or the Candle Factory or Nickerson Farms,

Or Stuckey's or even a Kip's Big Boy.

So, It was only natural that I would do this as well,

seeing as it was ingrained into my

personae by my earlier family experiences.

So, we decided to have lunch at the Starving Road bum.

It was a lot nicer than the name suggests, and they

had decent hamburgers.

The fries were good so were our drinks.

Then, we were back on the road again.

A couple more hours later, we were pulling up to Richard's house.

Rhonda won the windmill game again,

this time she saw nine windmills to my six.

Now, Richard and Candace live in a part of the Hill Country called the Million Dollar View area.

They have a very nice home. It was very large and very comfortable.

OK, in all honesty, the house was immaculate.

We pulled up to the front door and got out and walked up to the front door.

I rang the bell and a very pretty woman answered the door.

She smiled at us and said

"Hello, you must be Andy, Hi, I'm Candace, Richard's wife."

We both smiled and shook her hand and I said, that

"It is a pleasure to meet you. This is my wife Rhonda, and Rhonda smiled and reached out to shake her hand as well.

And she said "Likewise, A pleasure to meet both of you."

That said, Candace took a couple of steps back opened the door and she showed us inside.

"Richard will be back in a little while, she said,

"Please, do make yourselves at home".

So, Rhonda and I sat down on the very expensive leather sofa in the den and waited for Richard to come back home.

The Hybrid Diaries
Chapter Seven

Richard came home about twenty minutes after we got there.

I was surprised that Candace didn't bother to offer us anything to drink

or anything to nibble on while we were waiting.

The reason was soon to become very evident.

"Andy!!" Richard called out with a big smile on his face.

"How are you Richard?" I asked.

"We met your wife Candace, and she is lovely."

"Thank you so much, I think she is lovely as well.

How was your trip?

"It was uneventful, but, I'm glad to be off the road now."

But, I am kind of dry." I said, I am normally not this

forward with other people

But, I was thirsty and my throat was getting dry as well.

"No Problem" Richard said, "I've got something

special I want you to try."

We followed Richard to the kitchen and he opened

up the fridge and he pulled out a

Fancy looking water type bottle and poured us both

a glass of watery looking liquid.

I took a long sip, and to my surprise, it was water.

Except, It really tasted good.

Some people think that water is colorless and

flavorless, but, it is not the case.

Water takes on the flavor of the region where it is pumped from.

And this water was clear and refreshing and it had a good taste.

Richard asked us to sit down at the table, and he started to tell us his story.

"Water, "Richard began

" Is the single most important ingredient in great tasting beer."

"Let me tell you a story about where you are.

"We are sitting on top of the largest limestone and sandstone deposit in Texas."

" And trapped inside these deposits is the Edwards aquifer."

" These hills you see out the window channel the rainwater down through the sediments and into the aquifer."

" Which acts like a natural filter and it purifies and flavors the water with rich minerals."

"Now, north of here, is a small town called Paluxy, Texas."

" Way backing the 1900's there once was a brewery called The Paluxy brewery."

"It made the very best tasting beer in Texas until over usage of the local water resources caused a dry up of the Edwards Aquifer".

No water, no beer."

" So, the brewery went out of business."

" And the people stopped draining the aquifer."

"So, many years later there is an abundance of water in the aquifer again And there are strict controls in place to conserve the water levels in the aquifer."

" It's monitored by the Army Corps of Engineers and The University of Texas at Austin."

"I come from an old brewing family that's been making beer for at least the last one hundred and fifty years."

" I have my grandfather's recipe and I started brewing beer in college."

" That made me very popular at the frat house."

" It almost got me thrown out of school being a freshman minor in possession of alcohol

but, fortunately , there was a loop hole at the time that let me off.

"Because the rules stated no open containers and since home brewed beer doesn't have bottles."

" I was able to stay If I promised to stop brewing beer in the frat house."

" So, I started brewing beer at some frat brothers houses.

They in turn supplied me with empty beer bottles and they helped me through my chemistry classes."

"And that is where I met Candace."

" She was in my English Literature class, and we kind of hit it off."

" I was studying to be an Chemistry major and she was a geology major."

"She taught me all about the aquifer and the local geology of the area."

"And I started brewing at home after we were married."

" So, I am not actually a beer distributor I am a brewer and I own the brewery."

" And I make RepTex Beer."

" I've been brewing RepTex Beer for going on four years now."

"And I have contracts to supply beer to several of the local restaurants in Austin and in the surrounding area.   "

" We have recently bought some land outside of Granbury, Texas
that according to "Candace's research should have plenty of water underneath the soil for our brewery. "

"Aren't you worried about pumping all the water out of the aquifer
like they did way back into the early 1900's?" I asked

"Not at all, I will just use the local water from the taps

And pass it through a special filter that I designed back in college."

" I analyzed the water and added minerals to the filtering process to improve the flavor and color of the local water."

" This process should make the water taste almost as good as the pure aquifer water."

" And it should make one heck of a great tasting beer."

As he finished his story he reached into the fridge

and pulled out a couple of beer bottles

with no label on them and handed them to Rhonda and me.

I opened the bottles and took a long sip of the bubbly refreshing brew…

"Wow, this is wonderful," I said,

After I swallowed the first drink of my frosty beer.

"Thank you, this is from my private reserve." Dick

added.

We sat and talked for a long time, before I noticed

that Rhonda was nowhere to be found.

I heard this shriek of what sounded something like laughter.

And I bolted to my feet and called after Rhonda.

When, I got up Richard and I went downstairs to the

lower level where the pool and hot tub are located.

And I found the girls sitting down at a patio table,

drinking a couple of mixed Drinks that looked like screw drivers or else it

could have been tequila and orange juice.

It wasn't a sunrise, because there was no grenadine in the drinks.\

Rhonda and Candace were getting along swimmingly.

And they were laughing and cutting up and having a good time

Sitting out by the pool and next to the hot tub.

After discovering that the girls were both alive and

still in good health we men folk

retired back to the kitchen and we drank a couple more beers.

Then Richard dropped a bombshell on me.

"I want to tell you how excited I was when I saw

that story about you on the news a while Back".

" I was surprised, and I started screaming at

Candace to come in here and watch this news clip."

"Then I told her about how we met and about your

project and that I would like to sponsor you."

" Do you know what she said???

" I couldn't believe it. All she said was,"

"Well, I've always wanted to go to Australia."

" Maybe do some shopping…Is he married?

I hate shopping alone."

"I couldn't believe that."

" Completely missed the purpose of going."

" I am thinking about competition and she wants to go shopping…"

I sat there like the statue of the thinker.

Remember that I am a lousy poker player

But, this time, after all the color drained out of my face.

And there was almost no expression on my face.

Just this stone cold stare into oblivion as my mind

tried to desperately get a grip on what was happening.

All I could say was,

"Could I have another beer please?"

"Oh, sure thing."

" But, before We do anything else, I want you to promise me something."

"Would you please stop calling me Richard?"

"That's what my Mom calls me when I am in trouble."

" Just call me Dick".

"OK then, from now on you are Dick" I said.

I was always kind of uncertain about what to call

Mr. Wiggler.

I didn't want to insult him or hurt his feelings by calling him Dick.

I mean, I grew up with the name Andre.

'I hated it.

So, I changed my name to Andy instead.

Not that much better, but, at least it was better than Andre` in my opinion.

Dick called me back into reality when he asked me

if I was feeling alright.

"I'm just a little bit flushed. That's all."

"Dick? Do you know how much money it would cost to sponsor me? "

"I am flattered that you want to help me,

but, you've got to understand that this venture would cost more money than

we would receive even if we won the race."

" we'd only get the twenty five thousand dollar purse."

" And I am not even sure if it will be paid in US or Australian dollars."

" If it's Australian dollars, then we'd really be boned."

Dick didn't even flinch.

"You heard what my wife said, She wants to go shopping in Australia."

" So, now we have to go."

" And since we now HAVE to go, and you need to get there anyway, then why not kill two birds with one stone?"

"But, Dick this is going to be a HUGE sum of money."

" No one I know can afford to even buy me a well wish."

" And I can't even afford to pay attention." I sighed

"Look", Dick said, Remember the race cars? and their sponsors? "

"The odds are statistically against all of the team sponsors."

" Only one car can win one race at a time."

"The sponsors aren't buying winners."

"They are buying billboards that travel around the

track at almost two hundred miles per hour".

They are purchasing big rolling displays that

showcase their name and they are buying

the imaginations of thousands of loyal race fans."

I'll be doing the same thing."

" I want to sponsor your entire project and I want to

slap my beer labels all over your vehicle."

" I want the world to know about my beer and I

want you to be our Corporate face to a brave new world."

"Besides, my wife hates to shop with me almost

As much as she hates to shop alone. "

"You know, for a salesman you sure like to back the

reverse psychology approach."

Dick said about me trying to sheepishly refuse his help.

Refuse his help??? What was I thinking?

I can't afford to buy myself a decent car and I am

throwing away a little over three

thousand dollars of my own money on a dream.

I have already raided my bank accounts and now."

A stranger who I met in a bar wants to give me

money to do what I have been working so hard to accomplish.

And I am refusing it?

What the heck is wrong with me?

I know the old adage "never look a gift horse in the mouth.

And don't get me wrong,

I am extremely grateful for the offer of financial support just to

slap a few beer decals on the slingshot.

I would have done it for free anyway because I like

the guy and I admire his pioneering spirit.

"Dick, I can't take your money."

(my brain screamed out "I can't believe I said that!")

"In all good conscious I can't take your money."

(In God's name what am I doing???)

"I just can't imagine you sinking your hard earned money into my project

because, it offers no return for your investment."

"Besides, What if I fail? Or don't even place? "

Then Dick spoke up and said

"Rolling billboard…Hello?"

" All you have to do is show up on the dock with the car.

We'll slap some labels on the body and put the name of the brewery on it

and that'll be all you have to do."

"People will flock to see your vehicle and they will see the name of the

brewery and that's how we'll hook them."

" Once you have a sponsor other people will flock to you just so they too can

be part of the experience."

"But, what if you can't afford it?" I asked him point blank.

I really didn't want to offend his sensibilities

but, I had to know.

Dick took me outside and we sat on the porch and me told me about

Candace's family.

Dick started talking

" Candace" he started "comes from Old Texas Oil money".

"Her family is rich, and I am not exactly hurting either."

"Just look around you,"

" I worked for a large chemical company for many years

before I retired to pursue my dream of owning a brewery

and she supported me without question."

"And I love her more than anything because she believed in me."

" And she has always been there for me."

" And she knows how much I want to sponsor you."

" So, I know you are having trouble wrapping your mind around all this."

but, both Candace and I want to sponsor you."

"And with our help and your hard work,

I think that we can make a great team."

"That is, if you choose to let us help you."

"And besides, I want to go to Australia. " laughed

Candace from just outside the kitchen door.

Rhonda was right behind her .

"How long have you been there listening in?"

asked Dick.

"Oh, long enough to know that you love me!"

Candace and Rhonda were both smiling at us.

64

I couldn't help but feel embarrassed.

"So, what's for dinner? Asked Candace...

"You know, I haven't gotten that far yet."

replied Dick.

"What would you guys like to have for dinner?"

" There's a lovely restaurant in town that we go to."

"If you don't mind eating out do you? I'll buy!".

Rhonda and I both said

"Where ever you want to go would be fine with us."

So we went out to one of the restaurants that sells RepTex beer

The food was wonderful.

And the atmosphere was casual and very laid back.

And I hate to say it

but, I ate like a horse.

Rhonda wasn't as big an eater as I can be,

but, she did her part to put away her fair share.

I just wanted to interject, that the ride to therestaurant was also wonderful.

Candace also has a vehicle of fine German descent

It was nice to ride in a car that has a working air conditioner.

And it rode like it was on rails.

Smooth and .precise.

It made me want to have something that has a working air conditioner.

Yes, I did feel ashamed of my little crappy Datsun, B210.

But, It DID get us down to the Hill Country with no problems whatsoever.

So, I can't complain.

But, It sure was nice to ride in something other than my subcompact..

We went back to the house and we got into the hot tub.

Candace brought us some glasses of wine.

It was so relaxing that we just zoned out,

Dick and Candace got out and went upstairs

And they got the guest room ready.

I helped Rhonda get out of the hot tub and up the stairs.

When they called down to us that the room was ready

We came up to the guest room.

It was decorated in antique Duncan Phyfe style furniture.

Very early American style with a wheat straw headboard and king sized

pillow top mattress and actual box springs.

We found robes on the inside of the bedroom door

And all of our stuff was in the room already.

There was an attached bathroom

With all the toiletries that we could possibly need.

We showered and washed off all the chlorine from the hot tub and got ready

for bed.

Then there was a knock on the door, It was Dick.

"Find everything alright ?" he shouted through the door.

I made sure Rhonda was decent and then I opened the door and said

"We found everything Fine"

"Thank you again for inviting us out here again,"

We both chimed.

"You are very welcome." Dick replied

"We are a bunch of late risers so, just get up whenever."

" And we'll see you in the morning."

I said, "Good, so are we."

"Tomorrow, we'll go down and visit the brewery".

"You guys are going to have a busy day tomorrow"

"See you in the "morning"

" Good Night, Guys, Sleep good."

We both wished them a pleasant night's sleep and

we laid down and our heads hit the pillows

And that was all she wrote.

Maybe it was the long drive,

or, maybe, it was all the good food and alcohol,

Or maybe, it was the hot tub relaxing us so much,

Or, maybe it was because the bed was so big and comfortable

I just don't remember.

We closed our eyes and we sleep like babies.

I don't even remember dreaming,

Tomorrow would be a different day indeed.

The Hybrid Diaries
Chapter Eight

The sun light shone brightly into our room in the morning.

I got up and closed the curtains.

Then, I climbed back into bed and slept for another thirty minutes

Rhonda was awakened by the smell of freshly brewing coffee.

Nothing can keep Rhonda in bed if there's fresh coffee being made somewhere.

The most important thing to remember is to stay out of her way

until she can get her first cup down.

So I let her stumble down the stairs first.

And I followed up the rear as she made her way to the kitchen.

Dick and Candace were already awake and sitting

around the kitchen table drinking coffee.

Rhonda asked where the cups were and Candace

said that they already had a couple of coffee cups and saucers already on the table.

All Rhonda had to do was stumble to the table and

pick up the coffee cup and take it over

to the coffee maker and pour herself a nice hot cup of morning thunder.

I was a different story all together.

You see, I do not drink coffee. I am more of an Iced tea drinker.

That's my coffee.

So, I apologized for being a bother,

"Do you have any iced tea in the fridge? " I asked hopefully.

"Oh, I'm sorry, We don't have any made right now,

but, would you like a soda?"

Candace asked me.

"Do you have any Dr. Pepper?" I asked

"Sure we do"

"You must be from Texas too". Candace said with a smile.

"Yes, Ma am, Dallas, Born and bred."

Richard got up to go to the fridge in their garage to bring up a twelve pack of

Dr. Pepper.

Candace was getting curious about me…

"So, what school did you go to?" she asked.

"Well, I went to Mountain View College,

and also to the University of Texas in Arlington as a

"Talented and gifted student. " I replied

"Go Short Horns!" she said.

"Yes, that's right, They are called the short horns" I responded.

(Now, I had been to The University of Texas, for a U.I.L. Debate, prose and

impromptu speaking Tournament when I was still in High School.

But, that had been a while ago.

And I was always going to the Texas/OU football game in the Cotton bowl

every year

I was in High School at the time and because a good friend of

mine was in ROTC and they worked as ushers during the game

The ROTC rules would not let them march

wearing jackets or coats.

So, I got to carry their coats and jackets into the cotton bowl and for doing

That.

I got to stay and watch the games for free.

And those were some good games.

Texas didn't always win, but, they were fun to watch.)

Candace seemed suitably impressed.

And she smiled as I looked around the kitchen and into the den

I just happened to notice the furniture colors and made the connection.

The den had an awful lot of orange highlights in the upholstery.

With earth tone colors it's easy to do and very subtle.

"You know, I just noticed that you guys have a lot

of orange highlights in your earth tone furniture."

I said.

"Ahhh, you noticed that.

"Yes, we support our alma matter.

"But, we really don't like decorating the house in a

western motif with Bull skulls all over everything.

" Or else solid orange and white walls and orange furniture."

"That's a little bit on the obsessive side,

if you ask me" said Candace.

"It would kind of reminded me of visiting and eating at an "Orange Julius"

I joked

I wasn't sure if she had ever been to an Orange Julius,

but, we seem to be close in age,

I went to Orange Julius when I was a kid.

So, maybe she too had been to one once upon a time,

either way.

"You know, I am only joking, Right?"

I asked Candace just to make sure.

"Yes, I know, that, " She smiled when she said that.

Then Candace looked at Rhonda and asked her

"Is he always like that?"

"I'm afraid so," Rhonda chimed

"He's usually a lot worse, you caught him at a slow period."

"And, I love you too, Sweetheart."

I gave Rhonda a big hug and then acted like

I was going to give her nuggies.

And I said to her as I wrapped my arm around her head..

"Why I oughta…" in my best Moe Howard tone of voice

And we all cracked up. Then I heard Richard call up

to me from the garage.

"Andy, Can you come down here for a minute,

I want to show you something."

"Sure, I'll be right down.".

And Candace pointed to the stairs I would need to

go down and get to the garage.

And as I opened the door to the garage I stared

down the stairs and into the garage.

but, instead of seeing their cars, the room was almost empty.

Except for a big table in the center of the room,

And on that table was a diorama.

A scale model of the Alamo battlefield complete

with the model of the mission and Fortifications.

The long barracks, barricades and cannons and

miniature lead castings of all the Texians and the

Tejano volunteers, and their little lead cast horses.

And the Centralists Army of Santa Ana,

All of them brightly painted in authentic colors done in great detail.

The table was about eight feet by eight feet and the

model covered most of the table top

with most of the troops concentrated along the north wall.

With the remainder of the Mexican forces dispersed

along the three remaining walls.

I stared at the model and I could tell by the way the

Mexican soldiers were dispersed

along the model walls that this was Day Thirteen of the siege.

And judging by the model soldiers lying on the

ground and others walking on top of their

"dead bodies" that this was shortly after the start of

the final day of battle.

And very soon the volunteers would be overrun and slaughtered.

"How cool is that?" I asked myself out loud.

"Yeah, Dick said,

"The Mexican Army buglers have just blown "Deguello"

(for anyone who doesn't know what Deguello is,)

Please, let me explain.

"Deguello" is a trumpet signal / command blown by the Mexican Army buglers.

To signal to the Army that there will be no quarter

asked for and none given…

Now for those who do not know what "quarter" means,

Please, let me explain. "Quarter means "Mercy"

So, basically, If you are surrounded by a

hostile Spanish or Mexican Army and they start

blowing their bugles as their Army is

marching up to you and firing their guns or trying to

stick you with their bayonets,

or slashing at you with swords, they are probably

playing "Deguello" on their bugles.

I said, "So this is March 6th around 5:30AM?"

"Yep, the final battle."

There is no doubt about what the battle has come to symbolize.

People worldwide continue to remember the Alamo

as a heroic struggle against impossible odds,

That seems to be the underlining reason for my existence."

"Ever since I was a little kid people were always

trying to tell me what I can or cannot do."

"And quite honestly, I never listened to them."

" Sure, sometimes it turned around and bit me in the ass."

"But, I never stopped trying to do the impossible."

said Dick

"You know, you and I seem to have that in common."

Richard smiled at me.

"C'mon, let's go upstairs and have some breakfast.

"After that, I'll take you to the brewery."

"Oops" said Dick as he stopped in his tracks.

"Forgot the sodas." He went back over to the garage

fridge and got out a twelve pack of Dr. Pepper.

And we both went back up stairs and turned the

light off on the battle scene.

"What took you so long?" Rhonda asked us when

we walked back into the kitchen.

"I am not from here, and, I didn't know the fastest

way up the stairs." I replied.

"Dick probably showed him that stupid model of "The Alamo".

He's always showing people that model.

He's been working on it since he was a child."

chided Candace

"I actually understand where he's coming from." I replied,

"I am a bit of a history buff myself."

"It's true, Rhonda said,

"He's got military books and books of famous battles all over the house."

"Does he have toy soldiers lying all over the place?"

Candace asked with a little Ire in her voice.

"No, but, he does have ship models of the German,

Japanese, American and British fleets, from World War II."

Rhonda replied .

"So, I know exactly what you are probably going through."

They both sighed.

"Really?" said Dick.

"Wow, what's your favorite battle? Dick asked me.

"Well, I have several, but, My favorite U.S. victory over the Japanese Navy

was in Leyte Gulf

"The Battle of Surigao Strait."

"Which one was that?" dick asked

"Oh that was the battle when The Japanese were trying to send a battle group

through the southern strait of Leyte Gulf to attack the US transports "

"As the Southern Force approached Surigao Strait,

It ran head long into a trap set by the U.S. 7th Fleet bombardment and Support

74

Force.

Under the command of Rear Admiral Jesse Oldendorf."

"He had six battleships: The USS West Virginia, USS Mississippi,

The USS Tennessee, The California and The Pennsylvania."

" All but the *Mississippi* having been sunk or damaged in the Attack on Pearl Harbor."

"They had been raised from the mud and had been completely repaired.

"There were also the 8- and 6-inch guns of four Heavy Cruisers and four light cruisers."

" There were the smaller guns and torpedoes of 28 destroyers and about 39 PT boats "

"The Japanese were sailing up the strait, when they had their "T" crossed by the Pearl Harbor salvaged U.S. Battleships. "

"And they blew the Japanese battleships and cruisers out of the water."

"I love history."
\

"OK then what was your favorite Japanese victory?"

Dick was quizzing me.

"The Battle of Savo Island" in the battle for Guadalcanal.
"Admiral Mikawa sailed down the "Slot" with a force of four cruisers.
He took them by surprise and he beat the combined
Australian / US fleet of four cruisers in a fair night action battle."
"I liked that one because It was a "fair fight."

75

"Four Japanese cruisers vs. Four ABDA cruisers"

I added.

"Wow, that is pretty impressive."

You see Candace? I'm not the ONLY one."

Laughed Dick.

"Well, I love you in spite of your many faults"

Candace fired back.

"And, I love you too, Sweetheart." Dick said as he

blew Candace a kiss.

"Don't you go a changing" added Dick.

"So, who's up for some breakfast?" Dick asked

"Well, Rhonda and I have decided that we're not

going to clean up after you destroy the kitchen making breakfast."

"So, We're going out for breakfast. Maybe at La Mammoselles" or Cheeta's,

or even Big Dick's Restaurant."

Where do you want to go Rhonda? Asked Candace.

"Oh, anywhere, they all sound good to me".

Rhonda replied.

"You mean you girls would just leave us here to starve?"

I asked in my weakest sounding voice

Doing my best trying to sound pitiful.

"You guys know what to do. "Candace interjected

" If you are hungry then you can meet us at the restaurant."

" Besides, I need to take my car anyway because

Rhonda and I have some shopping to do

after breakfast anyway."

Alright." Dick said

So, we all climbed into two cars and drove over to

"Frenchy's" for breakfast.

We ate breakfast there . I had an Omelette du Fromage (a cheese omelette).

And then we parted company.

Rhonda and Candace went to God knows where and

Dick and I headed over to the Brewery.

The Hybrid Diaries
Chapter Nine

Now, I have never been in a brewery before, So, I had no idea what to expect.

Some part of my brain was thinking that it would be

Like a scene in "Willy Wonka and the Chocolate Factory.

Where he'd throw open the doors

And I'd run in

drinking everything in sight.

And that everything would be made out of beer.

We pulled up in front of an old warehouse.

It was very plain outside and there was nothing

outside like a sign or a banner or anything

that would identify the building as being a brewery..

Just this drab looking old building.

We parked in front and Dick got out his keys.

He opened the door and we went inside.

OK, when we went inside everything was nice,

clean and very modern.

Nothing outside would betray what was kept on the inside.

Then we walked past the reception area and we

walked through two thick wooden doors

and then out to the actual brewery area.

We walked past by the four big brewing kettles.

They were large copper and very clean.

Then we went to the hopping room and the malt storage bins.

He showed me the coolers where they keep the

brewer's yeast and then on to the bottling facility.

There were several people working at the bottling

machines filling and capping beer after

beer and placing the freshly filled bottles in carton

after carton of six pack holders and twelve pack boxes.

And then placing the packs on their case boards and

wrapping them up in plastic wrap for shipment.

So, Where do you fill the kegs? I asked

"Oh, we fill them in the next room over."

"All this is our first micro brewery."

explained Dick.

" We will be building our second brewery up in the

Glen rose lake Granbury area just as soon as we find water on our land."

"And when we get our environmental impact study

we should be ready to build".

"Dick " I asked, "Who buys the beer from you?

"Restaurants mostly."

"Some restaurants have what they call a micro

brewery in their restaurants."

"But, they don't actually brew the beer there."

" they buy the beer from us and they fill their onsite

tanks with our beer."

"And then they sell it as their own product."

" Their equipment and stuff in their restaurants

looks authentic, but they are really just

using their equipment for cold storage".

"They buy their beer from us."

I'm in the process of trying to break into the retail

markets with new bottles and labels

doing our best to establish a trademark or a brand

name that the public will become familiar with."

That will get us on the map. " Dick concluded.

Now I can see how my project would help get the

name of the beer and the brewery out to the masses.

This could work out well for both of us. I thought.

Now, you can't take a tour of a brewery without

getting to try some of the product for yourself.

That would be inhospitable.

So, Dick showed me to the Hospitality room which

was decked out in wood and leather furniture.

Complete with beer barrel tables and chairs and a an

antique wooden bar with a great big

picture framed mirror behind the bar.

And a sign over the bar that says there is a two

drink maximum.

"Don't worry about that sign."

"It's for the patrons only." smiled Dick.

"We are not supposed to give more than two beers

to the public for liability reasons."

That was a rule that didn't make sense to me.

Because they have to restrict each visitor to two beers.

But, Dick's brewery makes four different types of beer.

So, no one can try all of their beers without having

to buy a couple of the ones that you didn't get to try for free.

But, let me tell you…

The beer was very good.

I liked the wheat beer Dick calls it his "Blonde".

And I liked the dark beer too and the "Light beer"

has some redeeming qualities as well.

Dick said, that "All of my beers are Lagers but,

I am experimenting with pilsners."

"I just don't make any pilsners at this time."

"But, that wouldn't stop me from trying."

So, we sat down in a couple of beer barrel chairs

and finished off the first couple of beers

And then Dick got up and went behind the bar and

poured us two larger schooner type

glasses of the dark lager beer. he handed one glass

to me and he held onto the second.

And then he sat back down,

"Here, try this one, It's sweetened with molasses".

" Not enough to make an ale or mead out of it"

"Just enough sweetness for flavor and body."

Well, after four beers I wasn't able to tell which one

I liked the best.

In all honesty, I would tell any host that their

whatever was the best whatever I tasted,

just to be polite.

But, in Dick's case, I really did enjoy his beer.

It was fresh and bubbly, and it wasn't bitter or flat.

If you ever get a chance to tour a brewery, Please

do, because there is a tremendous

amount of difference in the way bottled beer taste

and the way "Fresh Beer" taste.

And believe me, the fresher the beer, the better the taste.

So, I had four more, and did my best to remain critical.

But, I still couldn't make up my mind.

So, we tried four more just to be on the safe side.

Nope, still no decision.

OK, then We tried it one last time to see if a clear winner emerged.

And Nope, they were all good.

But I could not finish the last of the last dark beer I drank.

So, I guess it was between the regular and the wheat beer.

I've always had a weakness for wheat beer.

But, now, I was getting closer to the snockered side.

And Dick was still going strong.

I used to think that I had a hollow leg and I could

drink anyone under the table.

but, Dick made me look like a panty waister.

"Don't feel bad, Andy, Dick said, "I've been

brewing and tasting beer for over twenty

Years now, so I sort of have a high tolerance for alcohol. "

I would not dare challenge a giant.

Or try to beat Nolan Ryan in a pitching match.

"So, are you still going to refuse my help?" asked

Dick point blank.

"I didn't mean to sound like I was turning you down."

" I was worried that you might have had the wrong

idea about the project".

" Like I said, if you were interested in backing a

proven winner, then, you would most

likely be disappointed."

" All I have is a dream and desire to do something

that even fewer people have tried to do".

And there would most likely be no return for your investment"

" The purse is only twenty five thousand and that

wouldn't even cover the cost of sending

a team over to Australia, by boat. Much less by air."

I tried my best to explain all of this to Dick in as

gloomy and depressing terms as humanly possible.

It didn't seem to deter Dick one bit. and after

discussing this back and forth,

We drank a couple more beers and I caved. I could

not believe my own ears.

I said "OK", to let Dick be my first sponsor.

And I don't know why, I was screaming in my

head, (Noooo!!!)

I guess more than anything.

I didn't want to be a burden on anyone, and I never

thought about having a team

because, I didn't think anyone was all that

interested in my project.

And if no one was interested, then why would

someone want to back me?

So, I tried to put all myself doubts aside.

And now I had to try and figure out how we were

going to get back home.

I was thinking in deep thought, when I heard the door open.

"Hey you, is anybody there?"

Hey there, is somebody there?

I distinctly heard Rhonda's voice.

But, I could barely respond to her inquiries.

"Well, There you boys are!"

I heard Dick mumble,

"Now that's Candace."

I could smell the distinct scent of hair care products

and the faint smell of fingernail polish.

I knew what that meant.

I did what any smart husband would do, I looked at

Rhonda and Candace, and I said,

"Hey Dick, Don't they look beautiful?"

"Why, huh, Yes, they do look beautiful" Dick said smartly

"Well while you boys were getting snockered,"

" Rhonda and I went out to the spa for some long

overdue repair work." Candace said.

"And look, we got mannies and peddies". Rhonda

held her hands out to show me her

finger nails." She always looks so cute when she gets to do girl stuff.

"I had my hair done too, do you like it?" Rhonda inquired.

"I strained to focus my eyes in her direction and I

could see that she had the hair cut,

highlighted, and styled. And she looked really hot.

Candace looked very pretty too. Her eyelashes

looked very full.

"Yes, Honey, You look beautiful.

" I love the hair. Nice nails too".

Dick opened one eye and looked at Candace and said,

"My, how pretty you look, Sugah!"

I also noticed that Rhonda was wearing different clothes too.

So, I guess they really did do some major shopping.

" I love you, Honey!" I blurted out then I laid my head back down.

After the girls looked us over and decided that the

men folk were not in a good enough condition to drive.

They had to pull a plan together so that we could

get both cars back to the house.

"Rhonda, could you help me?"

" I need to drive Dick's car back to the house."

"or if you prefer my car would you drive it home for us?"

" do you like my car better?"

"You can drive either one"

I'll get Willy to come and help us carry them down to the cars."

Now, Willy, besides being the name of my nephew,

is also the name of a little Filipino

janitor who has worked for the Wiggler's for many years.

Willy never got a driver's license.

So, he rides a bicycle to and from work every day.

 He lives outside of Austin and has to ride about

forty miles to work each day and forty miles back home again.

So, He's in really good condition for a little guy.

And he's strong for a little guy as well.

He hoisted me up out of the chair so fast that

It woke me up enough to walk outside to the car under my own power.

And Dick soon followed me.

Don't ask me how many beers we had between us.

More than I can remember and more than my head wanted to think about.

So, We all convoyed back to the Wiggler house.

I slept most of the way while Rhonda drove to the

Wiggler's home with a white knuckled

death grip on the steering wheel fully realizing that

the car she was driving was worth

more that I make in about two and a half years at my old job..

Even though she was terrified on the inside,

I could tell that she really enjoyed driving it.

She had this huge, worried, almost crazy grin on her face.

I had to listen to her berate me about being drunk

without out her all the way back to the Wiggler's house.

"Well Dick's drunk too. I said weakly.

Trying not to hurt my head anymore that it already was.

"He wants to sponsor us, Baby, he really wants to

sponsor us." I said almost under my breath to Rhonda..

"Yes, I know, Candace told me, and she says that

she's OK with it."

"Really???" Good, because I am so worried about

everything now."

We went back to the Wiggler's house and I had to

take a shower to freshen me up.

followed by a couple of laps in the pool.

Then it was off for Dinner with our new sponsors.

We're supposed to leave tonight,

So, I can make it back to work in the morning.

but, to be quite honest.

I really was still a little too hung over to drive all the way home.

Besides I don't know HOW I am going to explain all of this to Randall.

I probably don't have a job to go back to anyway.

So, I'll just let the evening go and see what develops.

And maybe we can wake up extra early and drive

back to Dallas in time for me to open at 10:00AM.

But, I seriously doubt it.

I'll need to start talking about finances with Dick.

I have no idea if he'll just sponsor the project, or

will I still need my job.

.

.

The Hybrid Diaries
Chapter Ten

OK, dinner was great as usual.

I will not go into details about our business arrangements.

Let's just say that I was not as worried about the drive back home after dinner

as I would normally be.

We had a wonderful time at the Wiggler's and they have a lovely home.

And they are just about two of the nicest people

I have ever had the pleasure of meeting.

Now, I did the adult thing and I decided that we

needed to go back home and I needed to

go back to work and have it out with my old boss.

And in spite of Dick wanting to hire me as a brewery employee.

I had too many loose ends to tie.

I needed to go back and check in with Toby and

meet with some people Toby wanted to introduce me to.

I never wanted to leave Austin,

I wish that I could have moved the entire project there.

Believe me, if there were a decent job waiting for me in Austin.

I would have so wanted to move there.

But, my roots and my contacts were all there in Dallas.

So, I HAD to go back home,

Besides my in-laws would only keep Thor for the weekend.

We bade our host farewell and thanked them again for their hospitality.

Then we headed back home.

I don't know why it seems like it takes less time to

get home than it does to get to your destination.

But, We made it home with plenty of time to spare.

Being at home seems rather drab and boring now by comparison.

But, I was home and I missed being where all my stuff is.

And I missed my dog.

Thor was really excited to see us when we arrived at the In-laws house.

He ran around us in circles and he was wagging his tail so hard,

I thought it would break off of his body.

He barked and gave us a happy whine and then

rolled over on his back and gave us his belly to rub.

We thanked the grandparents for watching their "grandson".

I didn't give them any information about the

meeting or what was discussed.

I was very tight lipped about the whole deal.

No sense getting people excited about nothing.

So, we loaded Thor and all of his stuff into the back

seat of the crap mobile and went home.

I took a thirty minute nap before I had to head back to the store.

I thought I would have to wait for Diane to show up

so she can let me back into the store.

But, when I got there the store was already opened

and I could see Randall's pickup truck in the parking lot.

My stomach tightened.

OK, here goes nothing.

I took a deep breath and walked up to the door and walked inside.

So, you decided to show up eh? Randall said

In a sarcastic tone of voice.

"I wasn't sure that I would even be welcomed here."

" Or am I mistaken?

Randall didn't answer.

"So, Do I still work here or should I just gather up my things and go? I asked him directly

He still didn't answer me,

So, I went back into my office and I started pouring over the weekend's receipts.

And I noticed that Randall had come up and HE watched the store

instead of letting Diane do the job I gave her.

That was his choice since he was the Company owner.

And They would have to do whatever he said.

I didn't feel angry about Randall being up here.

I was more peeved that Randall didn't trust my judgment anymore than he did,

Heck, I've been working for him for over a year and I was turning a profit.

So I felt that he should have trusted me more.

So here I am in my so called office and I have my boss outside the door and I still didn't

know if I was even supposed to be there.

And I sure didn't feel welcome at the moment.

"So, Randall, how was the weekend?" I asked him.

"Lousy, I was supposed to be in San Antonio for the weekend but, I had to come up here because,

"You had something to do."

"I'm not going to apologize because I had to take a couple of days off Randall.

"So, just do us both a favor and just say what's on your mind."

I was starting to get upset by the way he was acting towards me.

If I didn't think I needed this job,

I would have told Randall off and leave the entire store in his lap.

And just wash my hands of him.

But, I still needed the job and I was resigned to the fact that I would have to

follow Randall around like a puppy and kiss his ass as long as he was up here.

Believe me when I said that

I hoped he had to leave and go back to Austin right then and there.

I kept hoping that someone would call him and tell

him his dog had died or someone had

run over his great grandmother, or that an airplane

crashed into the bank where he keeps his money.

ANYTHING, to get him out of my store.

But, he stayed all day so, I had to stay all day.

And I resented him walking around my store and

pointing out things that needed to be fixed.

These were things that I had been after him for months to order replacement equipment

or else authorize me to spend the money to get it fixed.

So now, he can see what I was dealing with

But, instead of offering some support.

He was bitching at the problems that I have been begging him to address.

Like the leaking roof and the broken A/C unit and

the holes in the parking lot..

Well the day dragged on into the evening.

The sun was going down before Randall finally went home.

I was amazed that I didn't lose it and just walk off the job.

But, I didn't. I stayed because, I wanted to work the

job at least through Christmas.

And now that Randall has left the building,

I was able to call Toby and set up a time to meet with him at his house.

And I was excited about bringing the Slingshot over

to his house and maybe get started

on installing the wheels and the wiring and the fuse

block and the inverters, and the alternators.

We set up the time for the following morning at 8:30 AM.

And I got my keys back from Diane, and I went home to Rhonda.

Who still looks hot with her new hair style and her

freshly painted and manicured nails.

I still had some money left. So, I asked Rhonda if

she wanted to go out to dinner.

But, she was too wiped out by the trip to want to go anywhere.

I went down to a hamburger place down the street

and bought us a couple of burgers, drinks, and some French fries.

I took the food home and we ate and watched the TV for a little while.

Then she went to sleep.

I've got a busy day tomorrow and I'll have to meet Toby at his house.

because, he said he had something to show me.

So, I went back up to the store and I borrowed the

delivery truck and drove it home.

I backed it into the driveway and I opened the back

door of the truck and extended the ramp.

Then I opened garage door and I went inside the garage.

I pushed the Slingshot, out of the garage and up the ramp .

And with a heavy grunt, I pushed the slingshot into

the back of the truck.

Then put the ramp away and closed the truck door.

So all I would have to do was drive the truck over to

his house and drop off the frame.

Toby could start testing the equipment as soon as

we mounted the stuff onto the frame.

Morning came and I drove over to Toby's house.

Together we took the frame out of the truck and

wheeled it over to Toby's garage.

Inside the garage, Toby had set up a mock frame

with all of the batteries and a gas powered generator

and although the wheels were not yet finished

We were able to test out the system.

The first two batteries were not fully charged the

second bank of batteries was partially charged.

The third bank of batteries was fully charged

and the last bank of batteries was not charged at all.

Toby started up the generator and used a circuit

tester to determine the voltage that each battery bank had.

Then he hooked up the inverter and the sensors that measure the voltage levels

and automatically starts charging the bank with the least amount of electricity.

And when that bank is fully charged,

The sensors tell the inverter when to start charging the other banks of

batteries.

 And so on and so on until all the batteries are fully charged.

he even affixed lights that came on when the battery banks were at full power.

"I do not have the D/C to A/C inverters hooked up right now."

We are just running off the gas powered generator

and that produces 120 A/C current"

"That's the same type of current that the motors and speed controllers use".

Then he hooked up a power strip that had a TV and a refrigerator plugged in.

And he turned on the fridge.

After it started up, he then turned on the TV.

"These appliances simulate the load the wheels

would be putting on the generator".

"And as you can see, you still have a little voltage

left to send to the batteries in case they ran out of electricity."

" I don't think it would charge the entire bank of

batteries under a full load,

"But, it'd definitely be enough juice to run the car on just the generator."

That was great news.

So, we can run the Slingshot on the gas powered generator alone.

And I can use my leg muscles to recharge the batteries.

And when I tie in the solar cells,

they could charge the batteries as well.

So, maybe we won't need to use my legs to

recharge the batteries after all?

I'd have to wait and see.

"Hey, Andy, can I ask you a question?" Toby

looked concerned and he said

"I didn't want to offend you or make you mad or anything,

But, Why do you have a flywheel hooked up to the alternators?"

"It helps generate the electrical current to recharge the batteries". I said

"But, that is a big flywheel, right?"

Toby asked me again

"Yes, It is. Why do you ask?"

"Because. I can do something with that." Toby said.

"I can make it generate electricity and you won't NEED the alternators".

"And you can toss the alternators"

" And we won't be limited by their small power output."

I stood there with a blank stupid expression on my face.

The thought of getting rid of the alternators never occurred to me.

I had no idea that Toby could turn the flywheel into a generator too.

I never thought about that before.

The savings in weight would mean we could carry

more batteries or more fuel.

I stood there still reeling from the bomb that Toby dropped on me.

Honestly, I didn't think about that because I did not

have the expertise to build a generator out of the flywheel.

So, I became married to the alternators.

And the clockwork type motor I was building.

Suddenly, I no longer need either the clockwork

type motor or the alternators.

That gives me a simpler system and less weight.

But, WHY didn't it occur to me to ask Toby to help

me reduce the size of the power generating system?.

God, I felt so stupid.

All I could say was "Go ahead and do what you can

to get rid of the alternators."

And Toby said,

"I'll get right on it, but, you'll have to leave the frame with me."

I said that would be alright, and I headed off to work,

The store just might need the truck to make pickups

this morning.

On the way back, I just kept beating myself up for

not noticing that I could have done that from the very beginning.

The answer was right in front of me the entire time.

And I just didn't see it.

Sometimes it's far more helpful to have a second

pair of eyes looking at the project you have been working on.

A fresh prospective can see things that you completely miss .

Because, you are too close to the project to take an objective look.

Anyway, I got the truck back and David was going

to take the truck out for the days pickups.

And I got the store opened up and I started my work

day doing the usual once around the store to make

sure that the shelves were policed for trash and that

everything was hanging where it is supposed to be.

And that Diane had done her weekly mark downs on all the clothes.

Everything looked alright except for the gaping hole

in the roof and the hole in the ceiling.

and the fact that the safe had been smashed open with a pick axe.

A sledgehammer and a hack saw.

And the fact that the back door where the truck

comes to unload had been smashed open. and left that way.

I came to the conclusion that the thieves had

fled out the back, instead of the way that they came in.

through the roof and the ceiling.

It is a long fifteen feet from the ceiling to the floor

So, I hope they got hurt when they landed on the

clothes racks and the floor.

I looked at the damage done to the safe

And I discovered that the crooks could not get into the combination safe.

All they were able to take was the morning cash

that I leave for the morning crew.

It's just the change that I leave in the top of the safe

for the cash register.

It's usually about two hundred and fifty dollars in

cash and change and change rolls.

"Wow, I guess Mark Henderson does good work."

was all I could think of,

So, I picked up the phone and called Mark to come

by and bring his mobile welding rig.

Because we had to repair the safe / strong box. again.

And of course I called the cops.

They were not much help at all.

And they came to the conclusion that the store had

indeed been broken into.

And the thieves might have come in through the

roof and they used the tools that they

found in the store to try and break into the safe.

They got no finger prints and they made no arrests.

But, they sure took their time getting there.

It only took Mark about twenty minutes to get to the store,
but, his shop is right across the street from the Thrift store.
And he had to get his portable welding rig together.
so, that was to be expected.
I had the crew clean up the busted sheetrock and the
fiberglass insulation that fell out of the attic crawlspace.
And I went over to Home depot and buy some two
by fours and some two by sixes to close off the
entrance that the thieves had made.
Mark cut out a piece of metal grating and I put that
on top of the two by fours that I used to fill the hole.
And then I covered it up with two by sixes.
I made sure that they wouldn't be getting back in
the way they came in the last time.
That's for sure.
As for the back door I had Mark burn a hole
through some two inch square metal tubing.
And we welded steel bar brackets on the door frame
so we could use the square tubing
to replace the one that the thieves tore up.
Mark had cut away all the twisted steel and we
redid the safe and welded a new security bar door.
And I could run a pad lock through the hole that
Mark had made for me.
So, I can lock the bar in place with a padlock.
And then I bought lunch for everyone for doing a
good job above and beyond the call of duty.

I felt fortunate to have such good employees.

The Hybrid Diaries
Chapter Eleven
Here is where I say "Hello, to my friend Jesus."

One day I was working outside the Thrift Store

trying to patch a hole in the parking lot

With some left over asphalt that some guy dropped

off at the store.

Now as I was spreading it out with a shovel.

I noticed this guy walking up the street.

He was walking in my direction

and his form was getting bigger as he got closer to me..

I finally got a good look at the guy and I swear to you,

This guy looked just like Jesus.

He had long brownish black hair and a full beard of similar color.

And he was wearing what looked like a white robe

and was barefoot.

He was holding a pair of white tube socks with two

red stripes at the top of each sock.

At last he was finally up to where I was standing.

He stood next to me on the sidewalk.

And he never said a word.

He just extended his hand and held out the tube socks to me.

I remembered a passage in the bible…

It was a selection from the Book of Hebrews.

It sheds some light on what Jesus was teaching

"Let mutual love continue.

Do not neglect to show hospitality to

strangers,

For by doing that, some have entertained

angels without knowing it."

So, I gave Jesus the benefit of the doubt.,
And I humbly declined to take the tube socks from Jesus.
And I said unto him. "You may need these more
than I, You keep them."
Jesus smiled and he walked away.
I returned to filling in the parking lot and then
looked up and he was gone.
He just disappeared.
I know I didn't take my eyes off of him long
enough for him to run down the street and out of sight.
Then I though about The Book of Genesis.
Please, allow me paraphrase.
Abraham invited the strangers to rest and refreshment.
But Abraham really went out of his way,
offering to wash the three strangers' feet,
and to have a meal prepared for them.
The three strangers blessed Abraham and
promised the aged Abraham and
Sarah that they would bear a son within the year.
Abraham offered hospitality to strangers,
who turned out to be angels who brought him
God's blessing, in the form of Isaac."
I felt that I had a meeting with a messenger of God.
And I hoped that I did not offend anyone by
declining the gift I was offered.
 The meeting was enough for me.
And I decided to keep the feeling of continence that

he imparted to me.

I know it may not have actually been the big guy himself,

but, then again. Maybe it was him.

You never really know for sure.

Either way, If it was the real Jesus,

It was a pleasure to meet you.

And if it wasn't JC…

Then It made me think about my faith and about God.

That in itself would be considered a blessing from up high.

I went home after work and plopped down on the

sofa and I told Rhonda about my experience.

I was surprised that she didn't laugh, and she really

caught me off guard when she agreed

with me that it might have actually been Jesus or at

the very least an angelic being

delivering a message from God.

I guess the message might have been

"Keep your feet clean and dry."

I am open to other people's interpretation of the incident.

You are more than free to come up with any

conclusions you may like except for

"Under the influence". I assure you, I was sane,

sober and fully aware.

And there was no extracurricular activities that had influenced our meeting.

I am not what anyone would call a religious person.

In fact I tend to lean a little more towards the gray side.

But, I have always tried to keep an open mind.

I felt truly blessed in spite of the store getting

broken into and everything,

I had learned that the things that happened to the

store had nothing to do with my life and

that things like that can only bother you if you elect

to make it part of your existence.

I was beginning to distance myself from Randall

and the Thrift store.

And I was beginning to focus on myself and the

challenge and my vehicle.

So, now, I had to refocus my goals to getting solar

panels and batteries that can hold a lot of energy

and can be recharged over and over again very quickly.

I spoke with Toby and he said that he wanted me to

come by his house tomorrow morning,

That he had something to show me.

So, I agreed to go by his house in the morning and

see what he has to show me.

The Hybrid Diaries
Chapter Twelve

Morning came into our bedroom like gang busters.

Light shone onto my as yet unopened eyes,

and I rolled out of bed,

grumpy from not enough sleep.

I stayed up the night before and I drew what I

decided would be the body of the Slingshot.

If you take a typical paper airplane and flip it upside

down and look at it.

That is what the body's shape will be.

Like an upside down paper airplane.

No curves here.

I didn't have the tools to build the body by myself.

So, I made a mock up body out of cardboard and

wood fireplace matches for support and Elmer's glue.

The real body would be ten times the size of the model I made.

That kept me awake because my mind was racing

around inside my skull.

Trying to envision what the finished ride would look like.

It was very exciting to me having conceived and

then designed and then finally built

something that came from my fertile imagination.

This was what I always wanted to do,

And maybe, it would leave some kind of legacy in its wake.

I was hopeful that somehow everything would work out right.

I hadn't thought about the Wiggler's much.

It's been three weeks since Rhonda and I came back

from Austin and I haven't heard one

word from either of them.

So, I was thinking that they might have had second

thoughts after my dismal attempt to

paint the most bleak a picture as possible about my

ability to pull this off.

Or, maybe it was about the lack of rewards that the

project would offer or produce.

Maybe I was more persuasive than I thought.

From what I could tell about Dick.

He's a guy who likes it when things are afoot and,

when progress is being made.

He's sometimes forgetful, but, he always seems to

remember to keep his word.

He may not remember something,

but, he usually catches himself before he completely forgets.

I don't think they were backing out of our

"Gentleman's agreement."

I wouldn't blame him if he did.

But, this was HIS idea to sponsor us.

I don't know.

I was in so much conflict. I wanted and needed a Sponsor.

And I needed money to finish out the project.

I wanted their help, but, I was scared that I would let them down.

Or I'd spend too much of their money or else,

Not enough to get high quality parts that I would

need to complete the vehicle.

Instead of organically thrown together pieces of

crap that I had to adapt to make the vehicle work.

I mulled all this over while I was driving over to Toby's house.

When I got to Toby's house there was a van from a

battery company that has its Corporate offices in

Dallas, parked in front of his house.

I walked past the van and up to Toby's front door.

Then, I knocked on Toby's door and he came out

and hurried me into the house.

We went back into his garage and he introduced me

to a man named Earl Malloy.

Now Earl works for "All Nationwide Batteries" and

he has been a friend of Toby for many years.

Toby had filled Earl in on the project and Earl told

Toby about some batteries that the company has

been developing for electric postal and commercial vehicles.

They are similar to deep cycle marine batteries that

fishermen use on their boats to power their trolling motors.

Except they can hold a big charge and they can be

recharged very quickly.

They are not lead acid batteries they are much

closer to lithium Ion batteries except they

can hold a charge twice as long as lithium batteries .

And they weigh about a third less than lead acid batteries.

Earl said that these were experimental batteries.

And they are not for sale to the general public.

Earl said that the company was looking for a project

to test out the feasibility of these new batteries.

I told Earl that I would be honored if they would

consider me for their testing.

"Good" said Earl

"Then, I'll go back to the office and talk to the

product development division."

"Once I hear back from them,"

" I'll give you or Toby their answer."

"Great I said,

"It was a pleasure to meet you, and thank you for

your time."

" I'll be waiting to hear back from you, soon."

I had to go to work and get ready to open the store.

So I bade farewell to Toby and Earl and went to the store.

I had a pretty easy morning.

I got everything ready did the books and issued paychecks .

I paid my telephone solicitors and opened the doors

to my employees and then to the customers that

were already waiting outside.

It was colored tag day today where the colored tags

on the clothing would indicate the amount of

savings already on top of our already low prices.

Example: Red tags 20% off, Blue tags 30% off and white tags 50% off.

And all of the regular shoppers were lined up

outside to catch some of the bargains.

So, we were busy.

But, I still made time to call a company that manufactures solar panels.

They were a little out of the way in Grand Prairie.

But, still close enough to me so I can pick up any

panels that I would need to purchase.

The name of the Company was Tri Solar Confed. Solar Power Panel Corp

I spoke with a guy named "Rich Richardson".

I explained my project to him and I told him that I

already had several sponsors.

I needed solar panels to recharge the batteries

when the vehicle is parked.

And to supply voltage to the batteries that have already been discharged.

Mr. Richardson was reluctant to show any interest

in the project (probably thinking I was a nut job or something to that effect.)

I would need to meet with him in person if I was

going to get anywhere with him.

And I better have something to show him to prove

that I was serious about my intentions.

It was lunchtime and I headed back home to eat.

That's the biggest advantage of working close to home.

I can go home for "Nooners" and lunches.

Plus, I get to see Rhonda.

She would have a sandwich, chips and a freshly

made pitcher of tea waiting for me when I got home.

But today was different.

She greeted me at the door with the mail in her hand.

And she had a puzzled look on her face.

After a "Hello" a kiss and a hug.

Rhonda gave me the mail and she handed me two

letters that were most likely the cause of her confusion.

I looked at the envelopes and one was from the

Wigglers and the other one was from the

"New Paluxy Brewery Corporation."

Now the one from the New Paluxy Brewery

Corporation was in a large brown business

envelope and the other letter was in a regular sized

number 10 business envelope.

It was not as thick as the big brown business envelope.

"So, Which one do you want to open first?"

I asked Rhonda.

"Hey, they're both addressed to you" she said

 "You pick"

"I decided to go for the big brown envelope first.

I read the letter out loud to Rhonda.

It was not written by Dick but, by his secretary.

I assume.

Dear Mr. Barrientos:

I would like to congratulate you on your acceptance

of the position of "Director of Promotions" for the

New Paluxy Brewery Corporation.

Please fill out the rest of the application and sign

and return to us as soon as possible.

Along with the enclosed W-4 and the I-9.

And please, include a copy of your Driver's license

and Social Security card for our records.

If you have any questions please, feel free to contact

us at the phone number listed below.

Thank you,

Sincerely

Abigail Van Taubbin
Operations Administrator
New Paluxy Brewery Corporation

"Huh???" I stood there in silence.

I looked down and kept reading the letter over and

over again to myself.

Trying to convince my mind that my eyes were not lying,

and that I was really reading the words in the letter

correctly.

"Seems pretty straight forward". I said

after about two or three minutes of total silence.

"So, lets see what the second letter says."

Rhonda said.

Truth be known, I forgot about the second letter.'

This one was much more personal and I won't read it verbatim.

But, the gist of this letter was much more personal

than the acceptance letter from the brewery.

Basically, it was:

"Hi Andy and Rhonda:

Thank you for coming down and spending time

with us, we really enjoyed our time together,

And hope you guys can come back down soon

Dick said that the Brewery sent you a letter,

It's merely a formality and it was very necessary for

me to fill out the application and sign it before I can

activate and use the enclosed Platinum MasterCard
and Platinum American Express cards.
There are no limits on your expense account and I
can use them at my descression
And all I have to do is send in my receipts at the
end of the month so the Accounts Payable
Department can pay the balance each month.
Also the enclosed check is what Dick says is your
signing bonus.
If you decide to take the Position of "Director of
Promotions"
Just check with the rules and make sure that a
corporation can enter a vehicle in the race.
Just let us know if there is anything that we can do.
Talk to you soon,
Candace Wiggler
PS: Tell Rhonda that "Alphonso" wants her to come back down for a retouch
of her hair.
"What is she talking about?" Rhonda asked
"I guess 'Alphonso' was the guy who did your
hair" I said.
"I know that part" Rhonda snapped "What check is
she referring to?"
"Oh, it must be this one wrapped around the charge cards" I said.
As I was unwrapping the check from around the credit cards.
Once I got a good look at the check I felt really light
headed and had to sit down.

I was now holding a check for twenty five thousand

dollars made out to me.

Drawn on a major bank.

All I had to do was fill out a job application for a

job that I am already guaranteed to get.

sign it and send the stuff back to the Brewery

And I will get to keep the signing bonus.

No strings attached.

I picked up the phone and I called Dick at his office

and I wanted to ask him about the position.

The phone rang three times and then Dick answered.

"Dick Wiggler here"

"Hi Dick, its Andy"

"Hey, Andy! I was thinking about you,

How's Rhonda and you doing?"

"We're good here Dick,"

" I was calling about the letters"

"OH, so you got them already?

"That's good news."

" So how do you like your title?"

"Yeah Dick, about that"

You didn't say anything about a salary before."

"Much less a position,"

" What's going on?" I asked

"Well Candace and I were talking to our CPA and

he suggested that we make you an employee.

So that we could get a business account set up for you."

112

"But, In order to do that."

"You would have to be an employee first."

"So, Candace and I talked about it."

"And we decided that since we do not have anyone

to handle our promotions and public image.

" Since you were busy doing your own

"Promotions for your project."

"And since you have experience in promoting an idea."

Then you would be a perfect fit for the Promotions position."

"This way we could offer you a signing bonus as a

condition of your acceptance. "

"And an expense account."

"But, Dick, I have a job."

" When am I supposed to come to work?"

I was getting confused.

"Andy, we consider you are already on the job."

"You are out in the field working on a special project."

" All you have to do is send in your receipts, the

Accounts Payable Dept will cover all of your expenses."

"So, hurry up and fill out the application and the W-4 and your I-9."

"And don't forget to send in a copy of your driver's

license and your social security card for our records.

Abigail can be a real demanding bitch if she doesn't

get what she wants. or needs,

And she needs this stuff before we can send you down to payroll."

"OK, Dick So what's my salary?"

I had to know, so I had to ask.

After I hung up the phone, I had to lay down for a

little while, Rhonda asked me

"Well, what did Dick say?"

"I now have two jobs." I replied

"So are you going to quit the Thrift store?

"Are we going to have to be moving to Austin?

"Are they going to give you a salary too?"

"Yes,"

"Well OK, How much will you be making?"

she asked

So I told her.

After she calmed down and relaxed a bit,

she asked me,

So, what are you going to do?"

"I'm not sure;

I have got to finish the Slingshot soon."

It was all I could do to keep my composure,

God forbid that I ever win the Lottery.

I know now that I could not handle the initial shock

and keep a straight face.

I filled out the paperwork and dropped it in the mail

at the post office within the hour.

In less than forty five minutes it was on it's way back to Austin.

The Hybrid Diaries
Chapter Thirteen

Well, after I deposited the check into a special

account that I set up just for the project.

I went over to Toby's house.

When Toby was bringing me back to the garage he

seemed little distracted.

But, after a while he did perk back up when he

started to show me the that the wheels were ready to test.

They were not yet mounted on the frame.

So Toby had made a test jig on the wall of his garage.

He had hung to first wheel on the jig and he fired up

the generator and turned on the speed controller and

the wheel started rotating on the jig.

The wheel motor made a low hum

And I watched the valve stem on the tire spin around and around.

I turned up the speed on the controller and the

wheel spun faster and faster and the revolutions increased .

The hum got a little louder and the pitch of the hum got higher.

And it sounded powerful to me.

It was such a rush watching the wheel spinning

around and around. faster and faster.

"Toby, this so totally rocks!" I screamed

"You are a genius! I don't know what to say."

"You DID it!"

I was so excited.

Toby smiled, and said "Thanks"

That's when I knew that something was wrong.

So I asked Toby what was up.

I was already pretty sure what the problem was.

I had been so busy and consumed by trying to get a sponsor and getting the frame made

and getting everything all lined up that I didn't

think to wonder how Toby was going to pay for his

parts and any materials that I didn't already obtain for him.

And to be perfectly honest, I wasn't going to ask

him until I had a sponsor because I was afraid that

he wouldn't finish the wheels If I told him I was

broke or couldn't pay him right away.

"Andy?" Toby began…

"I've got a problem."

" Listen, I know that you are busy and trying to get

this project of yours started."

And I understand that you've had a lot of things on your plate lately."

"" But, I haven't seem any money from you for the

work that I have already done."

" and I've already invested a lot of my own money to get us this far, "

"because, well, frankly."

" I was fascinated by your concept and your ideas for the wheels. "

"And now that I see them spinning around like that,

I know that I am looking at the Future of automobiles. "

"And that hopefully, I will have a hand in it."

" But now, my wife is starting to complain about all

the junk cluttering up the yard and the garage."

"And I'm afraid that you'll need to move the project out of my yard."

"OK Toby, I understand,

116

"Here's what I need for you to do."

" I need you to give me your invoices for all the parts that you purchased."

"And I'll write you a check for the total amount plus what you want to charge me for labor".

"Just write me an invoice."

" And then I can cut you a check."

"Can you give me a little time to get the garage ready so I can make room for it and all the stuff that we have acquired?"

"Well, If you write me a check, and it clears."

"Then Yes, I could give you a couple of days."

"Wonderful."

"I'll get started cleaning it right away."

I was glad to now be in a position to be able to pay Toby for all of his hard work and hopefully the money will help smoothe out any conflicts I may have inadvertently caused Toby and his family.

I hope that when I pay him that he'll be able to work on turning the flywheel into a power generating system.

And that It will allow him to work on the wiring and helping me assemble the final system

When I get the new body ready.

That reminds me,

I STILL need to call "Bill" and see if the carbon fiber company can make the body for me.

I had to go and get the store opened.

And then I would need to make some more phone calls.

The morning routine ran smoothly.

I was able to call Global Composites.

I asked to speak to Bill

"Hello, Bill speaking"

"Hi Bill, this is Andy B.

"I was calling to see if you heard anything from your Management Staff yet about manufacturing the body for my solar race car?"

"Huh? Who is this? Andy?"

"Andy… Andy.. Oh yes, the guy from the TV news."

"Now, I remember."

" So, How are you doing?"

" How's your project?"

" Have you drawn up the design for the body yet?"

Bill inquired.

"Well, Yes, I've made some drawings and I made a scale model of what I think the body could look like."

"Would you like me to bring them over?"

" You can give me your feedback on what you think it would need. And then you can tell me what you'll need to fabricate it."

"Well that sounds good". Said Bill.

" Yeah, you can bring it over later on today if you can get time."

"OK, I can come by on my lunch hour if that would be OK with you," I said.

"That'd be fine." said Bill.

"We'll I'll get my team together and we'll take a look at what you got."

118

I hung up the phone with a feeling that something

was going on behind my back.

For starters,

I had no idea that Bill had even seen the news clip.

That was a huge surprise to me.

That might explain the funny looks I got from Bill

when I first met him.

It was like He knew the punch line for a inside joke,

And I was that punch line.

It kind of gave me that Twilight Zone déjà vu

feeling that creeped me out when we met.

From the time I met Bill,

It was like he was thinking

"Well It took you long enough to find us."

" we've been waiting for you to show up here."

Now, if "Bill" was going to be calling a meeting of his staff.

Then maybe I wasn't talking to some company flunkey after all.

I started wondering what I was getting into.

Well lunch time came.

I had to run to the house to pick up the plans for the

frame and the scale model of the body.

And I drove out to Grand Prairie to meet with

"Bill and his staff."

I walked into the front office.

I identified myself to the receptionist.

She, in turn made a announcement over the office

telecom/ loud speaker system.

"Bill you have someone waiting in the front office"

A few minutes later Bill came through a door and

He held out his hand for me to shake,

and He seemed very happy to see me.

"Hi Andy, It's nice to see you again"

Bill said as he grabbed hold of my hand in a firm hand shake.

That felt like it could have almost torn my shoulder out of socket."

Bill was a big guy about six feet four inches tall.

Probably weighed about two hundred and fifty pounds.

His big hand engulfed my hand when we shook.

"C'mon in I'd like to introduce you to my team."

He said as he ushered me into an office off the main hall.

There four designers were already gathered around

a table in the center of the room with a overhead

projector and a screen of one of the walls..

"This is Barney, Christopher, Sam, and Our chief

designer Andrew Dylan"

"Pleased to meet you all", I said.

As they gave me the nod of acknowledgement and

we sat down around the table

I sat down and I bought out my plans and showed

the design team the basic model that I had made several nights before.

They studied the plans and started laughing.

And making comments like

"Hey, we weren't that far off after all."

"And See, it does so, go all the way around the frame"

and

"OK, we might have a clearance issue here"

I was wondering what they were talking about until Dylan spoke up.

And flipped on the overhead projector where there was an image of me taken from the news clip from the Channel 7 news story and then he flipped a button and another image of me riding the Slingshot around the parking lot.

Dylan Began to speak.

"Alright now, after extrapolating the approximate size of the Frame from the images that we got off the news story we were able to "guestimate" the size and width of the frame and the height of the vehicle and the approximate width based upon the estimated size of Mr. Barrientos here. And the size of the wheels.

"So we were able to calculate how much carbon fiber we would need to make a Monoque body that would only require metal to build a light internal skeletal framework to support the carbon fiber panels and to offer a ground for all the electrical components".

"OK guys, now we have the actual blue prints and the actual measurements.

We can start building the sub frame"

"Sam, you can start building the sub frame assembly pieces for the frame."

"I'll have Barney help you out with getting the work bay cleared out."

"So we can start cutting carbon."

"We'll need about three weeks before we'll be ready to bring in the actual frame to get more specialized data."

"What we need from you Mr. Barrientos, is the size and width and location of the solar panels. And what kind of panels you will be using."

"Christopher, can you run the plans over to the copier and make us each a copy So Mr. Barrientos can take his original plans with him. Please?" Dylan asked

"No problem "

Said Christopher

As he took the plans and left the room.

"Look" I said

"I cannot tell you how impressed I am at all the hard work you guys have done on your own".

"It shows great resourcefulness and ingenuity. I am completely surprised. And just a little bit frightened by the amount of data you guys were able to glean from just a picture or two of the actual vehicle".

"I shall do my very best to provide you with any information you might require."

Now, My question to you is."

"Do I still have to wait three weeks before I can bring the vehicle over?"

"Yes, I am afraid so,"

"It's just because we are still finishing up some final assembly work on some parts for an F-16. Over there at Hensley Field."

"And we have to get that out of the way before we will have enough room to work freely."

Dylan said.

"Mr. Barrientos here are your originals back."

Christopher said

As he handed the plans back to me.

"Thank you, Christopher"

I said as I took the originals back.

"Alright then, I'll call you guys before I bring everything over."

And I shook everyone's hands and bade everyone farewell."

Bill walked me out of the office towards my car..

When we got outside.

I had a couple of questions for Bill.

"Bill, what exactly do you do here?" I asked

"Oh, we make carbon fiber aircraft parts for the Military and commercial applications."

"I thought you knew that." Bill said coyly

"No, I mean… What do you do here specifically?"

I asked again

"Oh, I am the owner" Bill replied

"Why didn't you tell me this before?" I asked.

"Would it have made any difference?" he asked me

I thought about his response for a few minutes before I responded.

"I guess not."

"So, bill asked, "Do you have a sponsor yet?"

"I am always trying to find sponsors". I said.

"Good!" Then maybe we can work something out together." He added

"I would like that very much."

"We can discuss it more when I bring the project

over here for her first fitting."

"So, how long would it take for your team to make the body?" I inquired.

(I needed to know so I can start budgeting my time

for the next few months before I will have to be ready to leave).

And there were was still so much to do.

"Probably a couple of weeks to get everything right

and to make it all look good."

"I can tell you one thing,"

"It's going to look slicker than dog shit when it's all said and done."

" I think you will be very pleased."

"It outta be to coolest looking thing to roll out of

our shop in a long time."

"And we work on airplanes, race cars, and speed boats here!"

"I bet you guys have done a lot of really cool looking projects here."
I said.

"Oh, you would really be surprised if you saw some

of them" Bill added

My lunch time is over and I have to get back to the store ASAP.

So, I had to say good bye to Bill, and his design team.

I could not believe that they had done all that work

without even seeing the actual vehicle.

That was truly amazing.

They were able to figure how much material they would need.

And they could estimate the weight of the frame and body.

And they can greatly reduce the weight of the

vehicle by building the body out of aluminum

reinforced carbon fiber panels.

"If they are going to design curves into the body

they have the tools and equipment to do almost any

shape or any type of curves they need.

And they even have a huge oven to make any and

all appropriate bends in lexan or plexiglas.

Maybe they can get be a used canopy off of an F-16

for a windscreen.

That'd be way cool.

Alright, I had made some pretty amazing progress.

And except for having to move the Slingshot back

to my house and having to move it back in my garage.

Everything was progressing better that I could have

ever hoped it would.

Now, the only problem I was having was trying to

drive home in the rain.

It was overcast when I left for lunch,

And now, it was turning into a gully washer and

water was coming out of the sky in buckets.

And the roar of all that heavy rain was only

drowned out by the flashes of lightning,

And the reverberating sounds of thunder.

There were high water spots on the road that

threatened to stall out the little crap mobile,

And water was pouring into the holes in the

floorboard, flooding the inside of the car.

But, somehow,

It managed to get me through the extremely dangerous high water.

And return me to the store safe, sound and soggy.

My garage wasn't quite so lucky.

When I got to the store.

"There was a frantic call for me from Rhonda.

She was scared and screaming into the phone.

Because lightning had struck the garage roof and

the whole garage was ablaze.

I jumped back into the crap mobile and drove home as fast as possible.

It only took me five minutes to make the drive.

And, When I pulled up to the front of the house the

garage was fully engulfed by the flames.

I pulled up to the house at about the same time the fire department showed up.

And the Firefighters leapt from their truck axes and

hoses in hand and rushed into the backyard .

They started spraying water all over where the garage was.

And started hosing down the still burning garage.

This was a old single car garage that was detached from the rest of the house.

So, fortunately, there was no structural damage to the house.

But, the garage was a total write off.

And would have to be totally rebuilt.

Now, I really have a problem,

I wasn't worried about the garage fire because this was a rental house.

And I didn't own the house or the garage.

And the landlord would have to fix everything.

But, I did have a major problem now.

Where was I supposed to put the Slingshot for three weeks while I was

waiting for the carbon fiber body people to be ready to receive the frame?

I was lucky that the frame was over at Toby's house

instead of inside of my fully consumed garage.

I lost a lot of books and some tools and some

electric motors and stuff of a personal nature that we were storing.

But, no one was harmed and in spite of the fire,

we could still live in our house.

So, in that respect we were extremely lucky that the

fire didn't spread to the house.

And that lightning only struck the garage.

If it had hit the house,

We would have lost everything that we owned.

"Rhonda ran out of the house and she threw her arms around my neck.

She hugged me for dear life.

She was so happy to see me,

I was very happy to see that she was alright and unharmed.

I hugged her and kissed her and then I asked her if

Thor was alright.

Thor is an inside dog for the most part.

He was safe inside the house at the

time of the lightning strike.

The fire was preceded by a huge thunderclap that

sounded like it was right outside the back door.

And it scared the heck out of poor Rhonda.

When she pulled the curtains away from the back door window.

She saw the flames licking up the roof and sides of the garage.

As if to add insult to injury,

The Lightning also hit a neighbor's tree next door

And part of that tree had fallen over onto the top of

the garage roof and in turn helped feed the flames.

That in turn made the fire burn faster.

It eventually took two fire trucks about two hours to

finally put out the blaze and extinguish all the hot spots

We had called our landlord Mr. Butthead.

He showed up as the firefighters were starting to

roll up their hoses and clearing out their

equipment and climbing back into their trucks for

their drive back to the station..

Mr. Butthead was trying to see if we had burned

down his garage instead of lightning

causing all the damage.

The firefighters confirmed the cause of the fire.

They packed up their trucks and rolled

their way back to the station.

We stood outside the garage with Mr. Butthead for

a short time before he spoke up.

"OH well,"

"At least you didn't have a meth lab in there."

What a classic asshole you are"

I thought to myself.

"So, when do you think you can have the garage

rebuilt?' I asked him hopefully.

"Oh, I don't know."

"I'll have to see".

"Maybe I can hire some-"

He stopped his sentence short.

I knew what he was going to say and it was

somewhat offensive to me.

He was going to say

"Maybe I can get some "wetbacks" to rebuild the

garage quickly and cheaply."

But, since I was standing next to him.

And since I was a good 3 inches taller than he was.

And since I am half Hispanic and half native American.

He had to reconsider his thoughts and choose his

words a little more carefully.

"I'll get someone on it as soon as possible"

he finally said,.

I let the "Meth lab" remark slide.

In spite of the fact that the garage would not be finished

Or probably not even started in the next three weeks or so.

I now had to find someplace else to put the Slingshot.

There is no room for it at the store.

I didn't feel safe leaving it in a mini warehouse.

So I needed to come up with an alternative place to

park the Slingshot until the three weeks are up.

I didn't want to send it down to Austin and have

Dick hold it for me for the same reason

I didn't want to move the entire project down to Austin.

All of my contacts are here in Dallas.

If I moved everything down to Austin.

It would take me several weeks just to find a place.

And I'd still have to move in and get settled

And get a new phone number, utilities, etc…

And I would also have to find new sponsors and

new sources for parts.

I know almost nothing about the greater Austin metro area.

So, now I am in an almost desperate situation.

And to make matters worse.

I needed to go back to the wrecking yard to find some "new" (well new to me)

replacement parts for the Crap mobile.

I'll have to go there when I can work it in.

When it rains it pours.

and, as I have just learned… "Lightning crashes"

The Hybrid Diaries
Chapter Fourteen

The next day was a complete contrast to the day before.

The sun was up and the clouds have moved on.

The air now hung heavy with the sickly scent of a burned down building.

I could smell melted plastic and burned insulation

and scorched wooden beams.

Whenever Thor went outside into the back yard.

He kept sniffing the ground and the air.

Then he'd make snorting sounds like he was trying

to clear out his nose and throat.

He would try to run out into the yard as close to the back fence as possible.

to take care of his personal business.

And then, he'd make a bee line to the back door

and start whining wanting to get back into the house

and away from the concentrated burned smell.

September was starting to be a real pain in the butt for me.

Not only did I have the fire to contend with.

But, because of the fire I now need to find a place to

put the Slingshot and get other things together.

Besides.

September is one of my busiest months when it

comes to my life and spouse.

Rhonda's birthday falls on Labor Day.

And our anniversary falls on the first day of Autumn.

So, those are my most important dates in September.

And since it's September 2nd, 1990.

I've got two days to buy her something special for her birthday.

Now, ever since I met Rhonda way back in 1980,

I have always celebrated Rhonda's birthday with a

weeklong celebration that I have

called "The Celebration of Rhondaness"

Starting on her actual birthday and closing at the end of the week.

I celebrate her birthday a lot like some people

celebrate other holidays.

With small little gifts all through the week and

culminating in a big present or something

to round out the week of "Rhondaness."

Please, allow me to regress.

I will explain how I met Rhonda and what we have shared together.

I met Rhonda back in 1980 at a place that is no

longer in business in Texas called "The Best Merchants Warehouse".

It was a lot like a Service Merchandise or Best Products.

They were a discount retail outlet that sold jewelry

and electronics and home items for the house.

And toys and sporting goods for the kiddos.

Rhonda worked on the main floor in house wares

and also at the pickup window.

Occasionally, she'd work on the intercom

Answering incoming calls.

And paging the store departments by saying

"Sporting Goods, you have a call on line one.Sporting Goods Line one"

I worked in the Electronics Department selling

Stereos, clock radios, calculators and cameras.

At the time I was working there.

I had just bought a snazzy orange,

1973 American Motors Gremlin.

And I was going to Mountain View College s

Studying Psychology and Computer Science,

I was learning how to program in Fortran and

Paschal and Basic during the day.

I worked at night to pay for my school,

The car, gas and insurance.

I worked in the evenings and all day on Saturday.

Rhonda was taking voice lessons so she could

become a singer.

Rhonda has a lovely voice.

And she sings like a bird.

One time at a funeral for a relative,

 Rhonda was asked to sing,

When she sung at the funeral she was behind a screen.

No one could see who was singing,

It sounded like we were listening to a record,

I could not believe that she sounded like that.

She is really a good singer with a wonderful voice,

And a four octave range.

So, our schedules kept us apart for most of the time.

Occasionally we would have overlapping work schedules

I would hear her voice over the intercom,.

 Or else, she would hear me blasting out the latest

albums or music on the display radios

"to test out the range" of the speakers on the latest

and greatest stereos from the best

manufacturers from around the world.

And on the very best speakers that the store had to offer.

But, we never met.

I saw her once dusting the shelves in house wares

and I thought that she looked really hot.

I wondered if I could ever work up enough courage

to ask her out on a date.

But, when I would try to find her after work or

when I would be on break she was never around.

It was like she was a ghost or something.

We played this game all summer long and then

I finally saw her.

I was going on break, and she was coming off break.

Now, the break room was on the second floor of the store building.

So if you went on break,

You would have to walk up a narrow flight of stairs

before you could get to the break room.

And if someone was coming down the stairs at the same time.

You would have to hug the wall to let them pass

before you could continue up the stairs to the break room.

One day, I was going on break and Rhonda was

coming off her of her break.

I saw her walking down the stairs with one of her

co workers in tow.

I had to grab wall and let the ladies walk past me

and back down to the first floor

We passed upon the stairs.

And I looked at her and she looked at me, and I said

"Hi" to her.

She smiled at me and proceeded down the stairs and back to work.

And that was it.

She didn't reply,

We never spoke.

And that was all she wrote.

Flash forward to about a month later.

And it was the Labor Day Company Picnic at a

"local park," in Arlington. Texas.

It was about as "Local" as in another City could possibly be.

And I was wrangled into playing softball.

So, I took right field.

Because, I am left handed and it's easier for a left handed person to

play right field than say short stop or left field.

So, Here I am in right field, shagging ground balls

and throwing them back to home base,

or to the pitcher or to first base.

At last, a left handed batter came up to the plate and

drove a high pop up hit right into right field,

So, I am hollering

"I got it, I got it" and my overweight manager

named Raul came running into the backfield all the

way from the pitcher's mound.

Then he crashed directly into me as I went up to get the ball,

WHAM, his hefty bulk slammed into my body like

a defensive lineman into a rookie quarterback.

And everything went black.

A few moments later (It seemed like an eternity)

I woke up and saw Rhonda and some other girls

huddled over me.

"Is he alright?" one of the girls asked Rhonda,

"I don't know, he took a really hard hit."

"How's Raul doing?"

"He's laid out like a beached whale.

" One of the other girls said,

while the other girls giggled.

I looked up at Rhonda,

All I could say to Rhonda was,

"Hi!"

She just looked down at me and smiled that same

smile she gave me when we were

walking on the stairs.

 And then she looked away at the bleachers and then she yelled

"He's alive!"

When I came to.

I was still holding onto the ball.

And because I was able to hold onto it even after the big crash.

The batter was ruled out. by the Umpire.

I rolled over onto my stomach and then I brought

my knees in and then I slowly dragged myself upright.

And then stood on my feet.

.I was still wobbly and tried to shake it off.

Raul was still out like a light and he had some blood on his lip.

Rhonda gave me her hand and she took me off the field.

 and the other girls helped her.

And we walked off the field to a loud round of
applause from the crowd in the stands.
And I couldn't believe it.
They left Raul lying on the field.
He finally came to and his wife went off onto the
field and helped him to his feet.
All is well that ends well.
Our side won the game.
And Raul still got to be the hero amongst the other
department managers.
Albeit, a sore and disheveled manager,
But, a manager none the less.
And I got to meet the girl I was interested in at work.
(Even though the store closed up several months later.
I still have Rhonda to this very day.
And she still tries to take care of me.
even though I am not a very good patient.)
Shortly afterwards,
We started dating and we eventually moved in together.
And you would never believe this,
But, my girlfriend was still a virgin when we met.
I was her first and only.
And I have always been there for her whenever she
needed me to be there.
And, uh, no, I wasn't a virgin. Not by a long shot.
But, I had never dated a virgin before,
So what we shared was unforgettable.

I never wanted to share what we have with anyone else.
And we have been together for ten years now and
married for five.
We've had some rough times and some very good times.
But, we have always been together through all of
the bad crap as well as all of the good times.
I think that is our greatest strength.
I know that I can always rely on her to be there for me,
And she knows that I would move heaven and Earth
for her if she only asked me to do it.
I have done my best to never take her for granted.
And, now back to our story…
I was thinking about what I could do for Rhonda for
her birthday.
I had decided on something special for her birthday.
I contacted all of Rhonda's friends and family
members and my friends and our mutual friends.
And I told them that I was planning a surprise party
for Rhonda's birthday for Saturday night.
Now I needed something for her actual birthday.
And I decided on something that would show her
that I listen to her and I care about her feelings.
Rhonda is one of those rare people who doesn't
really like birthday cake.
But, she tolerates it
Because, what else are you supposed to have for a birthday?
Pudding doesn't cut it, neither does jell-o or

ice cream by itself.

So, I decided on something very special for Rhonda.

A White Chocolate cheese cake with white

chocolate shavings on top from "Cheesecakes USA.".

A new pair of jeans, and a couple of music cd's.

That should hold her until her party.

Now, keeping it all secret will be a huge challenge.

Some of Rhonda's friends are experts at speaking

before thinking and I am afraid that one of them will slip up.

And blow everything that I have been working on.

So, without too much fan fare.

I had invited her family and my family over to the

house for the celebration Rhonda's

Birthday on Labor Day.

I had begun the preparations the night before by preparing a couple of briskets

for smoking.

I used a dry rub method to season the briskets,

Then I wrapped them in foil and let them sit in the fridge while I was building

a fire in my homemade smoker.

I had designed and built the smoker out of a fifty five gallon oil drum and a

smaller twenty five gallon gear oil barrel which was cut in half and welded to

the big barrel to be used as a firebox.

The top of the fire box was modified to also function as a grill

I made the legs out of chain link fence posts and the shelves out of solid core

doors.

That I cut into shelves about two feet by eight feet long.

And I made a shelf underneath the big barrel to hold about a quarter cord of firewood.

I use hickory and pecan to give the meat that smoky almost bacon like flavor of hickory

smoke.

The scent of hickory wood burning is a heavenly smell, unless it combines with the smell of burned insulation and melted roofing tiles.

Fortunately, if you burn enough wood it can overpower the burned smell of a garage fire.

At least for a short time.

I have been smoking briskets for a while now.

And I have entered several Chili cookoffs In the past and have done reasonably well in competitions.

I usually would cook a brisket and a couple of racks of Danish baby back ribs and chili.

My wife would make beans.

We would load the smoker into the back of a pickup truck.

Then load the truck down with camping gear and tents and a screened in canopy and lots

of ice chests full of food and beer and hard liquor.

Then we'd load all of the cooking utensils and lots of aluminum foil.

And then we would drive down to the cook off.

When we got there we would check in with the Cook Off officials and get our camp site,

Then we would set up our camp site at the cook off site then go to the Cooks dinner with the other cooks and their team members.

We would eat and dance and participate in auctions to raise money usually for the Leukemia society or some other worthy cause.

When that party was over we would retire back to our campsite and start cooking .

I smoked the briskets and drank beer and did tequila shots from our own private reserves, and listened to our cd's or the radio.

I would usually burn one log an hour.

I had to babysit the smoker so the heat would stay high and the fire would not go out.

I had to be vigilant and keep feeding the smoker.

It would develop into a routine.

Start with one log into the fire box, and one beer for me.

And so on and so on…

If we had "guests" or other curious cooks, trying to glean any tips for cooking the best brisket drop by.

We would break out the tequila, slice up some limes, break out the salt shaker and do shots with beer chasers for half the night or until the bottle was gone.

And when the sun came up you could see the aftermath of a night of hard partying and hard cooking and even harder drinking.

Spend beer bottles every where and empty plastic glasses and the occasional empty bottle of Jack Daniels or Tequila or an empty rum bottle here and there.

The typical drunk passed out in the middle of the aisles or else crashed out in a port-a-can.

The funny thing about these cook offs.

It would seem that almost no one would remember to bring their own breakfast supplies.

So, the Fast food restaurant down the street would be swamped with semi sober, hung

over, hungry people who are almost desperate for food.

And something other than beer or alcohol to drink.

Like a soda, juice or even water.

Since we are seasoned veterans, I always remember to bring everything you

would need for a nice big breakfast.

Sausage, bacon, eggs, hash browns, biscuits or pancakes and of course,

Plenty of coffee and iced tea and of course lots of water.

Stay away from Rhonda until she has her first couple of mugs though.

After we had our breakfast we would check on the briskets and start preparing

the baby back ribs for their turn on the grill.

Rhonda would do the beans. I would be pretty tired after being up all night drinking and

cooking. But, I still have much to cook.

Alright, that's enough background on the Chili Cook Offs there may be more info later,

But, we need to get back to Rhonda's birthday and then her surprise party.

Alright, the briskets are in the smoker and the logs are smoking the meats with

the smell of hickory and pecan wood hanging heavily in the air.

Rhonda's sister came by to help with the decorations for the family birthday

party.

(which is a separate event from the surprise party that I am throwing for

Rhonda in about five days.)

We got the house looking presentable and I handled the cooking duties and

the drinking duties because this was going to be an all-nighter.

So, we sat around and played cards and watched music videos on TV while

we played.

And about two thirty AM, Rhonda's sister Maggie, was getting pretty sleepy. Since it is Labor Day weekend, and there are probably a lot of drunks on the road.

We made a bed for Maggie to crash on in the guest room and invited her to spend the night.

It didn't make sense for her to drive all the way home and then have to drive back in the morning.

Besides, it is a lot safer for her to stay because she had a few things to drink as well.

So she called her husband Harold and let him know not to wait up and that she would be

staying with us tonight, and not to worry.

Harold said, that would be OK and that he'll see her tomorrow when we have the party.

So, the girls crashed out and I stayed up for the rest of the night watching a classic spaghetti western starring Clint Eastwood, Lee Van Cleef and Eli Wallach.

(I think you can guess which one.)

It is a long movie so, it took about three hours off my solitude.

Just as the sun came up I was starting to fall asleep sitting in my chair.

I had to go outside and check on the progress.

The briskets were a little over three quarters of the way done.

It would only take a couple of hours before they would be ready to slice.

They had to be brought inside for closer examination and a careful inspection.

For quality control purposes, of course.

In our house, we have a tradition of "Checking for bugs" Allow me to explain.

Checking for bugs goes something like this.

Someone (other than the cook) comes into the kitchen and they say.

"Look there's a bug on the brisket!"

and then, they reach over to the brisket and tear off a piece and shove it into their mouth.

Thus consuming the imaginary insect and getting a chance to pick off a sample of meat

and getting a early opportunity to taste the delicious briskets laid on the counter before them.

And I usually reply…

"Oh Thank Goodness you were here to get that bug now get out of my kitchen"

And I chase them out with a wooden spoon.

Sounds threatening, I know.

Well since there was no one around to look after the well being of these little orphaned Briskets.

I would have to take on the awesome responsibility of having to check for bugs.

I know, I was taking my life in my own hands.

but, someone had to check. Just for the sake of the safety of all concerned.

I had to make the supreme sacrifice of searching for and finding a bug on the briskets.

So, I opened the foil to reveal a terribly delicious looking brisket figuratively covered in bugs.

I steadied myself and reached down and pulled the "bug" off the brisket.

Unfortunately, a piece of brisket was "Accidently got snagged up when I got the bug.

So I had to consume the contaminated piece of brisket for the sake of pubic health.

Ohhhhhhhhhh, It was awful.

The meat was so tender that it just kept coming apart and juices flowed down the corner of my mouth, when I took a bite.

The smoky flavor had somehow managed to permeate the meat.

And I had to force myself to check for another bug that might have been lurking around on the second brisket.

Ohhhhhhhhhhhh, that one is horrible too!

It too is tender, juicy and delicious. And I had "the terrible task" of chewing up and swallowing the imaginary bug and the all too real piece of brisket that accidently came off the other brisket when I tried to snag the other bug.

Yeah, it's a dirty job, but, someone has to do it.

And since the girls were asleep, the task had fallen on my shoulders.

And I was up to the task.

So, I wrapped the briskets back up and I took them back outside one by one and placed them gently onto the racks inside the smoker so they can cook in the sweet hickory scented smoke for a couple more hours.

I'll be having to make coffee soon anyway.

If you think Rhonda is bad, without her first cups of coffee,

You really need to stay away from Maggie.

So, I did what I usually do, I got out the package of coffee beans out of the fridge and poured some beans into the coffee grinder and turned it on for a minute while it chopped

and grinded the beans into a chunky powdery mixture.

And then I put a filter in the filter basket and dumped the freshly ground coffee into the filter basket and filled the carafe up with water and poured the water into the reservoir to finish setting up the coffee maker.

I didn't turn it on,

because I expected the girls to be out for at least another two or three hours.

So I put some tea on the stove for me.

I set the heat level on low and sat back down and watched

some more TV then I drifted off to sleep "for a few minutes".

About an hour and a half later, I woke up just as Clint Eastwood was shooting the hangman's noose from around Eli Wallach's neck.

Lee Van Cleef is already dead and Clint is getting ready to ride off into the sunset.

So much for the movie I was going to watch.

Thank you, Sergio Leone. (he was the director of the movie)

I bolted to my feet as fast as a semi conscious person can do.

And I dragged my tired butt into the kitchen turned off the tea which was still on the stove.

I grabbed some pot holders and went back outside to pull the briskets out of the smoker.

The goddess of good cooking was watching over me.

because the fire had gone out and the briskets were not incinerated.

I gathered the briskets up and carried them one at a time

back into the kitchen and set them on the counter so they could cool a little.

It's no fun trying to carve burning hot meat with a carving knife and those flimsy oversized forks that come in normal carving sets.

So, instead of trying to use the flimsy meat fork, I have a big antler horn handled meat fork that has no problem holding the meat while I am carving.

I sliced up both briskets and put the meat on a large platter and covered the platter with aluminum foil.

Then took all of the scraps the fat and undesirable parts of the briskets and I gave them to Thor.

Which he scarped down with great haste and little abandon.

Then he let out a loud burp, and I let him outside.

In the interim, I heard a stirring from the bedrooms.

The girls were starting to wake up,

So, I turned on the coffee maker and I mixed up the tea and put the pitcher of tea in the fridge.

Then I called the Cheesecake place.

And placed my order for Rhonda's birthday cake.

It would be ready in about an hour.

I had to take a shower and change my smoky smelling clothes and get something ready for the girls to eat.

I cleared off all the mess and proceeded to get breakfast started.

I opened a can of biscuits and laid them out on a cookie sheet.

And then, I placed the biscuits in the oven and I took the bacon out of the fridge and took some paper towels and placed them on the microwave bacon crisper tray.

Then I opened up the bacon package and peeled off a few strips.

I placed them on top of the paper towels that I had already laid down on the microwave bacon crisper tray .

I covered the bacon with some more paper towels and placed them into the microwave oven.

I set the timer for four and a half minutes and turned it on.

The microwave started up and I could hear the humming sound that microwave ovens make when they are cooking.

I could see the light was on inside the oven and the turntable spun around slowly as the bacon began bubbling and popping as it cooked.

And soon I could smell the scent of cooking bacon blending with the smell of freshly carved briskets. with just a hint of freshly brewed coffee.

I love the smell of breakfast in the morning.

Rhonda and Maggie came into the kitchen and I left the kitchen and went into the living room until they got their coffee.

After they had drank a couple of cups of coffee I served them breakfast.

Scrambled eggs and bacon and biscuits and coffee.

I kissed Rhonda and wished her a happy birthday.

And then I broke the news that I would have to go somewhere after breakfast Rhonda wanted to know where I had to go and how long I would be gone and when would I be back.

So, I had to tell her that it wasn't any of her business and I could not provide her with that kind of information.

She gave me a silly face and said.

"Fine, I don't care where you go or what you do."

And then she pretended that she was angry at me.

I've seen this display before and I knew that I wasn't really in trouble.

And then the phone rang.

Toby was on the other end.

Rhonda answered the phone and handed it over to me. And I said

"Hello?"

"Hey, Andy!" Toby responded. "How's everything going?"

"Hey Toby, how are you doing?" I replied

"Oh, everything is fine here, I got the invoices and the bills all together for you like you asked me to do"

"When do you think you'll be coming by to pick up the stuff?"

"Oh Toby, Is there a way that I can come by to pick up the stuff tomorrow?"

The store is closed today and I won't have access to the truck until tomorrow morning. "

Now, I can come by in a little while to pick up the invoices and the bills and cut you a check for the total. amount if you like."

"Oh, that would be fine."

" Here's the total amount so you can get the check ready."

The amount was extremely large to me.

It was about what I had expected it to cost. But, it would worth every penny if they work the way we predicted they would.

No wonder Toby's wife was angry at him.

God, Rhonda would have killed me if I had spent that much.

What am I saying? I DID spend that much! and much more!!!

So I filled out the check and stuffed it in an envelope and placed the envelope in my pocket so I wouldn't forget it.

Then I headed out to the Caliente house.

When I got to Toby's house,

I pulled into his drive way and I could hear screaming from inside the house all the way out in the drive way.

I think Toby's wife must have been Listening in on their extension and she must have found out how much money Toby had sunk into the project.

Because, all I could hear her shrill voice and not much else.

I rang the doorbell hoping that no one would answer and I was afraid that somebody would.

Well, I dodged a bullet.

Toby answered the door and he took me around the house to the garage.

And we went inside.

Where he gave me his invoices and I gave him the check which I had hoped would return peace and harmony back to the Caliente house hold.

"I'm sorry that today is Labor Day, and the banks are also closed today." I said

"I am hoping that cashing the check would make everything better for you at home."

"What do you mean?" asked Toby,

"She's always like this.

"I thought you would have noticed that by now"

"Well, no not really"

" I thought she was angry about how much money you spent on the wheels and inverters and speed controllers"

. I mean, this is a lot of money." I said

"I know, but, everything costs a lot of money these days."

" especially scientific equipment ."

"speed controllers are made out of gold"

"And su**************ing magnets cost a small fortune and you have ***** in each

wheel.".

"I had no idea" I said as I scratched my head in amazement.

"The controllers and inverters are larger than what the specs called for because I wanted

to build in a little extra protection for the equipment.

"And if we decide to upgrade anything then we wouldn't need to scrap our existing "controllers." (Toby really gets excited now

when we talk about the wheels or the project.)

" Plus, if we got struck by lightning then the system won't feed back on itself and the surge would be blocked to prevent circuit damage and degradation."

"So, when do you think you can have the flywheel ready?" (I had to know so I can plan accordingly.)

"Well, once the check clears then I can get started".

And, it should take a couple of weeks before everything is all ready".

"We can start assembling everything soon".

" So, how is the garage looking?"

"Did you get a chance to clean it up?""

"Funny you should mention lightning as a hazard". I said

"Now, about that... Did I tell you about my garage yet?"

After we laughed at my tragedy,

I had to leave to go to the cheesecake place for Rhonda's birthday cheesecake.

I got there just as they were putting the cheesecake into a box for pickup.

I Paid the lady, loaded the cheesecake into my car and headed out to CD's

Howse of wares. To buy a couple of cd's for Rhonda.

And then off to the Jeanie's Jeans to pick up a couple pairs of jeans and

locked all the gifts in the trunk.

And moved the cake on the floor board of the back seat.

For safe keeping.

And I drove back to the house to wrap and finish cooking.

I still had to peel and cut some potatoes and chop some celery and onion for

potato salad.

And open a couple of cans of baked beans and top them off with brown sugar

and honey.

It's all darn good eating.

I got home and I carried the cake box inside and stuffed it in the fridge as far

back as I could to protect it from nosey, prying eyes.

and pulled an egg crate in front of the box to conceal "the point of origin" and camouflage the box.

"She'll have to wait for a while before she gets to see her birthday cake".

I thought to myself.

"This will be the best birthday week EVER!

And she is going to love her cake."

The Hybrid Diaries
Chapter Fifteen
Labor Day September 3rd 1990   "Rhonda's birthday"

The girls had done their part.

They finished straightening out the house and putting up colored crepe paper

streamers

and balloons in every corner of the dining room.

I added the finishing touch, the "Happy Birthday" banner hung from the ceiling and stretched across the room from one corner of the room to the other.

I finished up the potato salad and the baked beans were already in the oven. So, all I had to do now

Was clean up the kitchen mess and get the veggie and cold cuts and cheese Platters ready,

I had to get the bowls down for chips and dips and fill them up with potato and Tortilla chips and set out the dips usually French onion, chili cheese dip and bean dip.

The girls had spread a table cloth over the dining room table and everything looked really nice.

So much for taking it easy on Labor day.

 the day is barely halfway over and I have run my butt off.

I just remembered that I have forgotten ice cream. So, I picked up the phone and called the in-laws and asked them to bring the ice cream they agreed.

And as I hung up the phone I remembered that I also forgot ice,

SO I called Harold and asked him if he could bring the ice.

And he said "No Problem"

With all that delegated,

I was able to take Rhonda's presents out of the trunk of the car and wrap her presents.

And I bought them into the house and set them on the table.

And then I went into the fridge and got her birthday cheese cake out and put that on the table as well.

I left it in the box and I taped the box shut so she couldn't get a sneak peek before the guest arrived.

She was safely in the shower and I knew I had at least a half hour before she got out of the bathroom.

The guests should be here in about an hour..

So, I thought I would pass the time playing some music cd's from our Collection for us to listen to.

I took out some Beatles and Jethro Tull and some Led Zeppelin and programmed the songs into the cd player and waited until the guest showed up,

It didn't take long for the guests to start arriving.

friends and family members started arriving with presents in their hands.

We escorted each one into the house and placed their gifts along side of the gifts I had bought.

And then we led them to the kitchen so that the can load up on finger food and drinks

And in no time we were in full swing

We were busy passing out plates of appetizers and drinks.

Then we ate dinner and the briskets got rave reviews as usual.

After Dinner we moved on to the opening of the presents and the taking of pictures.

When Rhonda unwrapped her last gift.

It was time to have cake and ice cream.

Everybody groaned. because, they were all still too stuffed from dinner.

But, they all did their very best.

The White Chocolate Cheese cake was a tremendous hit.

Rhonda absolutely loved the cheese cake and so did everyone else.

The party was over all too soon.

Everybody wished Rhonda "Happy Birthday" as they all left our house.

When the final guests left and we locked the front door.,

Rhonda and I collapsed on the sofa, and surveyed the damages.

The house was trashed with paper plates and plastic cups flowing out of the trash can,

And the streamers were hanging down from the walls and most of the balloons were popped and their remains were scattered all over the floor and on the counters.

There was wrapping paper everywhere and ribbons from the bows were being dragged away off to some dark corner by one of the cats.

But, the overall damages were very light and could be easily cleaned up.

Rhonda looked over to me and smiled. And then she said,

"Thank you Baby, for a wonderful party and the presents.

I really enjoyed myself." Rhonda groaned

"You're welcome. Baby, let's clean up the house tomorrow, I am trashed."

I responded.

"By the way, "Happy Birthday Rhonda"

"Thank you, I love you."

It was getting late .

We went to bed, and I my head hit the pillow and I was gone.

Out like a light.

All that I remember was waking up the next morning.

I really hate nights like that, It's like I didn't have any dreams.

And I felt like I had just went to bed,

Now I was wide awake and just as exhausted as I was when I laid down.

So, I was in a grumpy mood.

I sat up in bed and rubbed my eyes and I stumbled off to the kitchen to grind some coffee beans and set up the coffee maker.

Then I dragged myself over to the fridge and took out the tea pitcher.

I poured myself a glass and went to go sit down.

Thor came up to me wagging his tail and looking expectant.

I know what that meant.

Someone has to let the dog out for a run.

And since I was the only one up and it was early,

the responsibility fell on my shoulders.

"Come on Thor" I said as he came running up to me and he danced around in what we called "The Bladder Dance"

Thor would dance around and we would open the door and let him out into the back yard.

Then Thor would go and do his business.

In all the years that we have had Thor,

He never had an accident in the house,

We never had to house train him,

He was just such a good dog.

You know, I hate it when Rhonda would give me the "Somebody's got to do it speech."

It would usually go something like this.

"Just once, I wish, that Somebody, would just…..Blah, blah, blah…"

Now there is only Rhonda and myself in the entire house (except for Thor)

So, I am not supposed to be able to figure out "WHO" this mysterious somebody is?

Just say it to my face or ask me politely.

The other way is insulting to me.

So, Long ago, I decided that our house was full of imaginary fairies who in my mind

became the mysterious "Who, Someone People" that my wife is always referring to.

So, when "Somebody" drinks all the tea.

And doesn't bother to put more on the stove to brew.

That's when I call for "Flunky".

Now "Flunky" works for the "Tea Fairy" who magically makes pitchers of tea for us.

The Tea Fairy doesn't actually make the tea.

She sends "Flunky" ahead, and It is actually "Flunky" who fills the tea pan full of water and puts it on the stove and then turns on the burner.

And then he drops the tea bags into the water.

And he boils the tea,

Then he turns the burner off when it's time to steep the tea..

After the tea has steeped and the pitcher has been rinsed out and filled halfway with water and the sugar has been added.

That is when the "Tea Fairy" swoops in and mixes the tea.

By pouring the tea from the pan into the pitcher and using a big wooden spoon.

That also serves as her wand, she mixes the contents of the pitcher

Magically into delicious ruby rich tea.

Each Fairy has their area of expertise.

There is of course the Tea fairy, and her assistant "Flunky"

And there is the Magical Jittery Coffee fairy, "Jumpy."

And the bathroom fairy named "Todd"

He puts the paper on the toilet paper rods.

He also hauls the dirty clothes out of the hamper and puts

them in the laundry room.

And the keys fairy and the Sunglass fairy who makes sure that you

don't forget to bring sunglasses to the car when we leave on a trip.

And so on and so forth. "There are many Fairies in our house."

So, I had my tea and I had to get ready for work,

So, I got up and got myself cleaned up,

Then I threw on some clothes and made my way towards the front door.

I was just about to walk out the door when I remembered that I left the dog

outside.

I had to stop and turn around and go to the kitchen and let the dog back in the

house.

I opened the back door and Thor was on the porch just wagging his tail at me.

He seemed so happy just to be alive and delighted to see my face.

I opened the door and he bolted into the house..

that being done, I closed the back door and made sure that I locked it.

I headed back to the front door and was just about to put my hand on the door

knob when the phone rang.

It was Toby.

He just wanted to tell me that his check had cleared .

And that he would be getting started on the fly wheel as soon as we got the

project out of his garage so he could have more room to work.

I told him that I would be back later on this evening as soon as the truck driver

brought

the truck back from making his pick ups.

Great, I still don't have a place to take the project.

And I need to move it today.

And, If I couldn't get access to the company truck

I would need to borrow my father in law's pick up.

All I had to drive was the Crap mobile.

She was in need of some TLC.

It was time to get the inspection sticker replaced.

And there was no way that the car would pass inspection in its current condition.

And I was really starting to feel the need for a truck of my own.

I have got to go out to the wrecking yard and find some replacement parts

So, I now, I had something else to do.

The Crap mobile was my stop gap emergency car that I had been driving for about a year and a half.

It was a yellow nineteen seventies model Datsun B-210.

it was originally painted a bright cheery lemon yellow color from the factory when it was first built.

but, the ravages of time and also years of intense Texas heat, poor treatment and just plain cheap paint.

The finish had turned into a washed out sun faded yellow .

The Left front of the car had been wrecked (not by me)

and the headlight on the drivers side pointed down to the ground immediately in front of the bumper.

Making the headlight completely useless for seeing ahead in the dark.

The car in this condition would not be able to pass inspection,

So, I had to go to the wrecking yard and get another front clip.

And replace the fender and the hood and the grill and the bumper.

And then I would have to mount the replacement parts and hook up the wiring for the lights

I had found the phone number for the wrecking yard in my desk.

I folded it up stuffed it in my pants pocket and headed towards the front door.

In just a few minutes I was out of the house.

Sitting behind the wheel and driving the crap mobile with an expired inspection sticker on her windshield to the thrift store.

So, I guess I'll have to go to the wrecking yard during my lunch break and scout them out for auto parts.

So, I did the usual morning routine.

And I got the employees on task and doing their jobs,

I got David off to make pick ups and I went back into my office to call the wrecking yard.

I pulled the phone number out of my pants pocket and sat down at my desk and I called them to see if they had the car parts that I needed.

"Hello, Big Jon D's Auto Salvage, this is John can we help you?"

"Hi, John, Do you have a 1976 Datsun B-210?"

"Yeah, sure, what do you need?"

"I need a front clip, you know, the bumper, hood, fenders, and grill." I replied

"Yeah, Yeah, I got a couple of em, come on down and you can take your pick."

"That's great, OK, I can come by around lunch time." I said.

I hung up the phone and did some work around the store and waited until lunchtime

So I could go and pick up the parts that I was needing to make the crap mobile look a little better than it does.

Not much happened while I was waiting,

just some idiot who tried to steal a pleather jacket by grabbing it and running out the door.

Fortunately one of the girls saw him and chased him out the door and she tackled him in the parking lot.

I am not kidding,

It looked beautiful.

She ran up behind him and leapt into the air and she got her arms around his neck in a horse collar type tackle that would have probably been illegal in the NFL.

It stopped the thief cold.

He went down hard on the tarmac and Wanda landed directly on top of his back.

And he just laid there with all of the breath squashed out of his body leaving him rolling around in the parking lot and gasping for air.

At least, he was lucky enough to fall on top of the coat.

Otherwise he would have had gravel and little pieces of glass crushed into his face hands and chest.

Now, Wanda is a big girl and I had no idea that she could run that fast or much less tackle some guy who was trying to run away.

The guy even had a head start on her.

Wanda says that she has 5 brothers and that she used to play football with them and they never held back with her just because she was a girl.

They helped raise her like she was a tom boy.

And she put a serious hurting on the thief.

So, I felt the punishment more than fit the crime.

And I told the guy that if he bothered to ask me instead of trying to steal from us.

We would have been more than happy to give him a coat.

to keep him warm.

It was September and October would be coming soon and with that the cold weather would not be far behind.

And he would need a coat

I told the crook that I wasn't going to call the cops on him and I was going to let him go,

but, I said, he better not come back in my store and try to rip us off again.

or the next time,

I would call the cops and they will come and arrest his butt..

We took the coat back to the store and I saw him just sit up and he remained there in that same spot for what seemed like an hour.

It was probably about ten or fifteen, minutes,

and he was still sitting out in the parking lot.

So, I went to the coat rack and I found a coat that had been on the racks for a long time.

The ticket had been marked down several times.

And although the coat was not stylish it was warm and it had no holes,

And it looked like it was the right size to fit him.

it just looked old.

So, I walked out to the parking lot and I handed the coat to the crook who looked up at me as I handed the coat to him.

His eyes welled up with tears, and he started crying.

I felt extremely uncomfortable, because, I didn't want to make the man cry,

I wanted to help him gain a little piece of his dignity back.

He started telling me that he was a Vietnam Veteran and that he was out on the street and he didn't have anything or anyone to help him.

So, I asked him, "Do you know who benefits from the sales that this thrift store generates?"

This thrift store is run for the Vietnam Veterans Foundation."

"If you want, I can call the foundation and maybe they can get you some assistance.

Would you like me to call them for you?" I asked

No, No, no one wants to help me,

Thank you for the coat, it looks like a nice one.

"I thought it looked warm, and cozy, " I said.

"I meant what I said, about helping people."

"But, If you come around and try to steal from my employees again, I'll let Wanda beat the crap out of you.

And THEN I'll call the Cops,"

"Believe me, She's a big woman and she can put a serious hurting on you."

"I don't mind someone doing a little dumpster diving as long as you put the trash back into the dumpster when you are finished."

Now, I am going back inside and I will call the foundation and maybe they can help you." I said,

So, I went back into my store and then back to the back and into my office, and I picked up the phone and I called the foundation.

They have been using the proceeds from our sales to fund the Vietnam Veterans Memorial in Fair Park In Dallas.

I wasn't sure if they had some kind of out reach program for homeless veterans.

They said that they might be able to help the man.

And they would send someone by to pick him up.

So I went back outside to tell he homeless guy the good news that the foundation was willing to help him,.

But, when I went back outside, he was no where to be found.

I had to go back inside and call the foundation back and told them thank you for their assistance.

But, the guy had left, and there was no longer any point to them sending someone out at this time.

I said that I would call them if I saw the man again.

And they said, "OK".

So, I hung up the phone and although I couldn't help the man,

I felt better because, I had stuffed a twenty dollar bill into one of the coat pockets.

And I'm sure he left once he found the money inside the pocket..

Well, lunch time was coming up fast.

and I had to go to the wrecking yard and pick up the parts.

I wondered why the guy left.

Maybe the story about being a Vietnam Veteran was all a load of crap that he used to con money out of unsuspecting people.

Or was he an actual veteran living out on the streets?

I guess I'll never really know.

Well, It was time for lunch, and everybody had made it back,

So, I was cleared to leave.

I drove down to the wrecking yard and I went into the business office and I asked for Jon.

And Jon came to the front counter I told him that I was the guy who called about the Datusn B 210 parts.

He got on the intercom and called for a yard man to come to the front counter.

In a few minutes, one of the yard men came up and he took me back to the section of the yard where they keep all the imported cars.

And we hunted for the Datsun B210 that Jon said they had.

We finally found it the car.

It was red in color and the front end looked acceptable.

So, the yard man pulled the front clip off of the red car and he used a fork lift to haul the clip up to the front office for me.

I went back inside the office and Jon made me a great deal on the parts.

He sold the entire front clip to me for 75.00. and the yard man helped me lift it to the roof of the Crap mobile.

And we tied it down with rope to the top of the car.

I mentioned that I had a problem and I needed some space to store my project for a few days.

And Jon said that I could keep it there at the wrecking yard.

He has a shed that I could lock it up in.

if I was interested.

I told Jon that I would be worried that someone would accidently crush the slingshot not knowing what it was.

Jon said that it would not happen because,

Wrecking yards have the best security.

They have cameras and guard dogs and on site security guards armed with shot guns to watch over their assets.

That sounded safe enough and since I was desperate to find a temporary place to house the Slingshot, I took him up on his offer.

It looked really funny with the front clip tied to the top of the car,

but, hopefully it would survive the trip back home.

Well, I made it home and I unloaded the clip myself and then I went back to work to finish out the rest of the day.

So, I could go home and start rebuilding the crap mobile.

And maybe, get a new inspection sticker out of the deal

After I moved the sling shot of course.

Well, my truck and driver had safely made it back to the store and we nloaded the donations into a big bin and I had David sweep out the back of the truck.

Then I told him that he could leave.

And that I would lock up the truck.

David said OK and said

"Have a good evening." And he went back home.

So, I took the truck over to Toby's house and together we loaded the project into the back of the truck and closed the box.

I took it over to the wrecking yard.

Jon had waited for me to come back, and he took me around to the back of the office, and he showed me the shed.

It looked less than desirable.

But, it was better than nothing.

We loaded it up inside the shed and Jon locked the door with a padlock.

It was locked up and semi secure.

But, now, I was losing access to the project.

I didn't like that part one bit,

What with time running short and all.

I just couldn't feel safe with it being out at a wrecking yard.

I kept feeling like I was making a mistake.

When I got the truck back to the store, I locked it up in it's pen and I went home.

I pulled up in the driveway.

I got out and opened the gate and then I drove the car up to the burned out remains of the garage door.

I closed the gate and I started to unbolt the old wrecked hood and fenders and the old bumper and then I dragged the new clip over to the car and started bolting the new parts to the car.

It didn't take very long at all.

And when I was finished, I had a yellow car with the front section painted a faded out red color.. very classy looking. NOT.

I had already purchased some yellow spray paint just for this occasion.

I sprayed the hood and fenders spray paint yellow which did not entirely match.

So when the paint dried.

I had a two tone yellow crap mobile with a decent looking bumper and two working headlights.

And a possibility that it would now pass inspection.

I was pretty trashed out when I got finished working on the car so I went inside and was getting undressed when the phone rang.

So, I answered the phone.

It was Dick on the other end.

"Hey, Andy! How's everything going?" Dick said.

"Hey there Dick! How are you doing?" I replied.

"Did you get the invitation I sent you to the event I mentioned?" I said this coyly

(because I didn't want Rhonda to find out about the surprise party.)

"Why, yes, we Did, and all we need to know is what time you would like us to arrive?"

"Oh, you can come over anytime you guys want."

" I was hoping to have some things to show you."

" but, I am afraid that I have had a couple of set backs nothing minor."

"But, I think I can overcome them". I said.

Dick, asked me

"What kind of problems was I having, "

So, I explained the garage fire and that need to move the sling shot to a wrecking yard.

And the fact I was uneasy about leaving it there.

Then Dick said that he had a couple of issues that he needed to cover with me.

My stomach immediately tightened.

I started to brace myself for the sentence I was sure would be coming.

Something like "Hey. I am pulling my sponsorship."

But, instead, he asked me.

" Hey listen, Are you receiving any mail from my offices?"

mainly any paychecks?"

"Yes, Dick, I have received three of them so far.

"But, I haven't opened any of them yet. "

Why?" I asked.

"Well, I can assure you that they are good".

" But, why aren't you cashing them?"

And why did you pay for the wheel motors,

instead of submitting the invoices for reimbursement?"

Now, I have been receiving the checks from the brewery for three weeks now,

And I have yet to work up enough courage to open the envelopes.

I am still in disbelief about landing a sponsor.

So, I have been trying to do everything with the resources I have.

Because I still do not feel comfortable about spending other people's money.

And I wanted to feel secure about spending Dick's money.

So, I had been using my signing bonus to pay Toby for the wheels.

Because I knew that money was already mine.

And I could do with it as I saw fit.

And more than anything I wanted to pay Toby for all

his hard work and I didn't want any unforeseen repercussions."

"OK, that explains why you paid for the wheels,

but, how come you haven't cashed your pay checks yet?"

"Oh, I was saving them for the boat ride." I said.

"No, Andy, Don't save them."

" Candace and I have something to show you guys this

Saturday that pertains to that aspect of the project."

" Besides you not cashing your checks is starting to mess up the payroll

department."

I told Dick that I would deposit the checks into my account in the morning.

And that I could not wait for Saturday,

How nice it would be to see them again this time at our home.

And how much I would enjoy cooking for them and the other party goers that

would show up.

Dick said that he would bring some beer to the party.

So, I said "That would be more than welcome."

Before Dick hung up the phone he added one more thing,

"Go and rent yourself a truck and a mini warehouse and get the slingshot out

of that wrecking yard as soon as possible.

And send me the invoice."

I promised Dick that I would move it as soon as I found a mini warehouse

close to my house.

Just until the carbon fiber people were ready for it.

Told him goodbye and we would see him and Candace on Saturday.

I hung up the phone and I got the checks out and I finally opened them one at a time.

And I was shocked to see that much money on one single check.

Then to see it two more times,.

It was almost overwhelming.

I should not have any problem about finding and paying for a mini warehouse tomorrow.

The next morning I got up extra early and found a mini warehouse down the street from my house.

Then I called the rental place where I usually rent trucks from.

Just in case the company truck broke down.

And I rented a truck for the day.

Then I went back to the wrecking yard and picked up the Slingshot and dropped it off at the mini warehouse before work.

The Hybrid Diaries
Chapter Sixteen
OK, today is early Friday Morning, September 7th.

I have got to run to the store to buy birthday supplies for the party.

I was hoping that Dick would remember to bring some beer.

because I focused my paycheck on purchasing wine and scotch and tequila.

(yes, even though, I now have all this disposable income,

I still only work with my income from the thrift store.

Choosing to keep most of my money from the brewery separate from my own).

Money that I earned as the manager of the thrift store.

And I would need to cook another brisket for the party as well.

That would take some explaining.

So, I decided to get Rhonda drunk and maybe she'll pass out and not even notice that I would be smoking something outside in the yard.

That sounded like a plan to me.

So, that evening I would take Rhonda out for another celebratory evening in the weeklong "Celebration of Rhondaness".

We would go out to dinner at our favorite franchised Mexican family eatery located down the street from our house.

There I would ply her with countless strawberry and Mango margaritas, chicken enchiladas and tortillas and beans and rice and a couple of shots of Patron Silver tequila.

That was all it took. Rhonda was so full and sleepy that she didn't have room for dessert.

I had to prop her up so she could make it back to the car.

Then I had to help her back out when we got back to the house.

The last margarita was a special top shelf margarita that seemed to do the trick.

All I had to do was help guide her into the house and put her down on the bed.

I didn't even bother to undress her.

I just covered her up with a blanket and took off her shoes kissed her forehead, and turned out the bedroom light.

And she was sound asleep.

So, then I drove over to the store and picked up some more brisket and some chicken breast and some flour and cooking oil.

Tomorrow night we were having brisket and my famous fried chicken strips.

This chicken is one of my many specialties it's homemade and hand dipped and double dunked in flour for extra flavor.

And It is almost always a major hit whenever I decide to cook it.

All the kids just love it.

Which means lots of hungry bug hunters that I have to contend with..

I never have to set the table when I make my chicken strips because people keep coming into my kitchen and taking pieces of chicken that they swear are covered in bugs.

Same with the French fried potatoes I usually cook to accompany the chicken.

Why set the table when they are eating the chicken a la fingers Sans plates?

Either way, I was able to go to the store and buy the brisket and chicken breast strips and all the other goodies I would need.

I also ordered a birthday cake from the bakery.

This one would be for the party,

And if she wanted, she still had a couple of pieces of cheesecake left in the fridge.

I would pick up the cake in the morning and take it to my office until I could bring it home later that evening.

Also in the morning I would have to go by La Mammoselles French bakery.

And pick up a half dozen chocolate croissants.

These are one or Rhonda's most favorite pastries in the world.

She absolutely loves them with her coffee.

So, I prepared the brisket with a wet rub this time.

I call it a honey marmalade brisket covered in orange marmalade and honey and brown sugar and smoked to perfection.

Slowly over wood smoke.

I would monitor the progress and turn it several times over the course of the night.

In the meantime I was preparing the rest of the victuals.

This time it the menu would be brisket, fried chicken strips, French fries, baked potatoes and pinto beans and Texas toast.

And of course plenty of birthday cake for dessert.

With all kinds of drinks and lots of music.

It was shaping up to be a party to remember.

I finished the brisket around 5:30 AM and I finished baking the potatoes.

so I dumped them onto a big stock pot and covered it and hid it in the pots and pans cabinet.

I placed the pinto beans in a large bowl and covered the bowl in aluminum foil and placed it in the storage bin in the stove where it would be safely hidden until it was needed.

I took the brisket out to my car so I could carve it at the store and bring it home after work.

I also took the chicken and all the ingredients to the thrift store with me so I could cook the chicken at the thrift store.

I had already purchased a used deep fryer and tested it to make sure it worked.

I'll plug it in and pour a bottle of cooking oil in it and get everything ready to start cooking, around 1:00 PM.

After my employees had gotten back from eaten lunch.

I will not tell you my recipe for my fried chicken strips.

But, let me tell you this.

It so totally rocks, I am renowned for my cooking.

And lots of people comment to me that I should open my own restaurant.

I never really entertained the idea, because, I didn't have confidence in myself.

And I wasn't sure I could be a success at it.

I started cooking and the whole thrift store smelled like a restaurant.

The smell of Recently carved brisket mingled with the scent of freshly fried chicken strips

I fried up a big platter of strips.

And I had the chicken strips laid out on a on a big old platter.

and I also placed the brisket on a large platter.

I covered them both in foil and left them on the back counter.

I asked Diane if she could watch over the meats and the birthday cake

Until at least 5:00PM.

She said she would,

So I left to pick up Rhonda and get her out of the house.

I had called Rhonda's sister Maggie earlier.

And I asked her get the food from Diane in

the afternoon after she got off work and to let in the guests as they arrived.

And show them where to hide and to get them ready to pop out and yell

"Surprise!"

I would leave the house key under the mat and she could use it when she came over while Rhonda and I were gone

"out to see a movie".

This would be my way of keeping Rhonda away from the house.

And it would allow the guests an opportunity to show up without Rhonda finding out.

The guests were instructed in their invitations not to park in front of the house.

They were asked to park down the street so that no one would notice all the cars parked in front of our house.

I worked for a half day and I got off work at 2:00PM.

That would leave me about two hours before I expected anyone arriving.

So, I called Rhonda and told her that I was taking her out to see a movie and to be ready when I got home.

I knew that she would be nowhere near ready when I got home.

so I lied and told her I was getting off at 1:30

instead of two o'clock.

That way when I showed up at two, she would be ready and waiting for me to show up.

Or at the very least, she would be in the bathroom getting her face ready to meet the world.

I was hoping for the latter.

I had already dropped off the chocolate croissants earlier in the morning so she could have some with her coffee.

Rhonda was excited about the croissants and she ate two of them in short order.

So, the appointed hour of two came and I showed up at the house.

I opened the front door with my key then I removed the key from my key ring.

I hurriedly stuffed it under the Welcome mat.

hoping that no one saw me.

Then I walked it the house to the sound of the hair dryer blowing.

That told me that even with a half hour of extra time…

 Rhonda still wasn't ready to go.

So, I let her know that I was home and that we needed to be going.

She said, "I'll be ready in a minute".

I sat down for a minute and organize my thoughts.

 Chicken on the way, check

Brisket ready and on it's way, check

Birthday cake, on it's way, check.

I had told Maggie where I hid everything.

I checked to make sure that everything was still safe and where I left it.

That being done,

I needed to get Rhonda out of the house.

But, I still haven't picked a movie for us to go see.

There is this old movie theater not far from our house.

It was called the Vogue theater

they showed second run movies for a dollar a ticket.

They had a movie that we wanted to see called "Elvira Mistress of the Dark"

starring Cassandra Petersen.

We love Elvira and her sense of humor, so it was an easy sell to get her to go.

But, when we got to the movie house,

Elvira had been replaced by a slasher movie.

We don't really like slasher movies .

So, we left and I had to scramble to find another movie to take her to.

We needed to kill time and allow the guests to arrive.

I ended up taking Rhonda to the old Texas Theater.

Yes, the same Texas Theater that Lee Harvey Oswald was hiding in when he was arrested for assassinating President Kennedy in 1963.

They also showed older second run movies, and somehow I was able to persuade Rhonda to sit down with me and together we watched

"The movie "Under the Cherry Moon." Starring "Prince"

I apologized for the lack of Elvira and the substitution movie selection.

We were able to watch almost the whole movie before I had to "Go to the bathroom"

(In reality, I was calling the house to make sure Maggie had made it inside and that Diane had given her the food.

Maggie said that Diane was there and she dropped off the food and was helping her hang decorations.

I told Maggie to make sure to invite Diane to the party and to thank her for all of her help.

And then I asked her if any guests have arrived.

Maggie said that the house was full of people.

That it would be a good time for us to come back home.

So I told Maggie that we would be home soon.

I went back into the theater and sat down next to Rhonda.

We finished out the film.

"Prince died" Jerome Benton cried "No, No, Not Chris!"

the credits rolled and we were out of there.

Rhonda asked me

"Why did we just watch that movie?"

"I mean, It wasn't on our top ten of must see movies."

I reminded her that we were supposed to be seeing Elvira instead.

She said "Oh."

I did my best to act disgruntled about the whole thing.

And then we drove home.

We pulled up in front of the house and parked in the driveway.

I acted like I lost my house key and I asked Rhonda.

if she had her keys with her.

She did, and so, I asked her to open the front door for me.

So Rhonda opened the front door and switched on the light in the living room and then all hell broke loose as at least forty people shouted out

"Surprise" at the top of their lungs.

It scared the heck out of me and I was expecting it.

Rhonda looked completely taken by surprise, and everyone rushed up to her to greet "The Birthday Girl"

They all asked if she was really surprised.

And she was completely taken in by the surprise.

Everyone I invited showed up and some of them brought friends with them and the party had swelled to around fifty guests

The food disappeared and the beer and wine was under constant attack.

Dick and Candace showed up and Toby and his wife Isabella came.

All of our friends and some of the extended family showed up.

(I did not invite the parents or the in-laws because I didn't want to offend them by drinking and partying.

Rhonda's parents were devout Fundamentalist Christians who do not tolerate drinking,

So, I thought it would be best to leave the parents off this party lists.

When Dick said he would bring the beer…

I thought maybe a case, or two.

I was not expecting him to drive up in an large Suburban with three kegs in the back.

And all the barrels and the hoses and tanks to pressurize the beer.

I kept saying "Oh my god,

Three Kegs." Fortunately there were lots of people who just "showed up".

People from my store and some of their friends

And people from the place where Rhonda volunteers at.

(The Leukemia Society of Texas") and close friends that we have known for many years.

It was neat to finally get to meet Toby's wife Isabella.

I've heard so much from her…

She was this little woman about four feet ten inches tall.

With a voice that should belong to a woman twice her size.

We did the dinner thing.

(there was NOTHING left,)

All of the chicken and brisket was consumed.

And forget trying to find a baked potato.

I had cooked about thirty pounds of potatoes.

(roughly about forty potatoes.)

And about 5 pounds of beans and they too were

eaten without any abandon.

If we were hoping for leftovers we were out of luck,

Good thing that we were not wanting to have any leftovers.

The cake was quickly destroyed.

It was large enough for our guests that I invited.

But, I hadn't taken into consideration all the friends of friends of friends

who also showed up.

After presents were unwrapped and carried into the bedroom

And placed into Rhonda's closet for safe keeping

I chased a couple out of my closet.

I know what they were doing and I wanted them to go and do it

somewhere else besides my closet.

So, they took their show on the road, and I didn't see anything in case anyone

else wanted to ask.

The guests started leaving and Rhonda was surrounded by all the girls,

Maggie, Diane, Isabella Candace and Julie, were talking about things that men folk have little understanding.

I made sure that they all had their wine glasses topped off and

I decided that maybe my attentions were needed elsewhere,.

I had decided that the keg needed to be examined for "freshness"

so I took my leave of the living room and walked out to the back yard.

I saw Toby and Dick talking to each other.

I wish I could have heard their conversation.

I guess I missed out.

I had some folding chairs out next to the remains of the garage

So I set them up for Dick, Toby and me.

I did find out what they were talking about,

How good the beer was and how sad my garage looked..

There was also a folding table there.

so I set that up as well so we had a place to rest our elbows.

We sat and drank for a good long time.

Chasing some kids away from the kegs

and making sure that only legal aged drinkers were allowed access to the kegs.

Then I remembered the tequila and the scotch were still in the house.

"Unprotected".

"The Hard Stuff!" I screamed,

And I ran back into the house and saw some teen agers

sitting down around my living room table and taking turns swilling down my tequila.

So, I snatched up the bottles and headed outside to the more responsible adults who were now laughing at me,

For rescuing the booze from under aged drinkers who I didn't even know.

I went back inside and I asked the kids who they were and who they were with,

When they didn't have a correct answer,

I asked them to leave. which they did.

After they managed to kill almost half of my bottle of tequila.

I showed them to the door and they left and walked around to the back of the house in an effort to circumvent me and get at the precious kegs resting comfortably on my driveway.

Fortunately Toby and Dick were there to provide protection to the kegs and to shoo away the already intoxicated teens.

Once they left I reached into my pocket and pulled out three shot glasses that I picked up from the kitchen counter.

I handed one to each of us.

and Dick and Toby and I began to do shots with beer chasers.

After a few rounds we started talking.

We all had questions that we wanted to ask each other.

But, no one seemed to know how to start.

Or else just like me, they were afraid to.

So I stood up and poured each man a shot of tequila and I yelled "Salute!"

Toby and Dick stood up and raised their little shot glasses and downed them.

And we all realized at about the same time

That I had forgotten to bring out the limes and salt.

So, we had to do another shot.

This time, with lime and salt. Just like you are supposed to do it.

And then we needed to clear our palates.

So, we each downed a glass of beer.

Once we were properly braced.

And we had drank up enough courage,

We started talking about everything.

Then we got down to asking the serious questions that all of us were afraid to ask.

Like why are we all here? Why are we all together?

It's the manly version of truth or dare.

If you dare you have to drink a shot,

If you tell the truth and no one believes you,

Then you have to take a shot.

If you tell the truth, then we have to take a shot.

Either way, by the end of the game you're usually telling the truth.

I was reeling when They both looked at me and said "Truth or Dare".

I said "Truth!" and I took a shot,

Did I mention that I was drunk by now and I was feeling fine with few inhibitions?

Toby and Dick Both wanted to know.

What made me want to pursue the slingshot project.

"It was one of my dreams."

" I wanted to do something that I would be remembered for."

" I wanted to create a legacy for my family name, and I wanted to make a positive contribution to the world."

"I was trying to do something that most people said was impossible."

" I invented the concept and the idea and I wanted to see it through to its completion."

"Plus, I wanted to win."

" I wanted to go all the way over there and meet the best and finest solar powered and hybrid powered vehicles in the world.",

And I wanted to show them all that an individual with the help of a talented core of designers and builders could accomplish almost anything that they can set their sights on."

Then I took another shot.

Both Toby and Dick seemed satisfied with my answer.

And they each took a shot.

So, I asked them the question that has been gnawing on my subconscious but, I was terrified to ask…

"Toby!" I said. "Truth or dare!"

Toby thought about it for a second and then he shouted "truth."

"Ok, then Mr. Caliente" I said… then I said…

"Toby, why did you agree to help me?"

"Because, I wanted to be a part of something special."

" Something bigger than myself"

"I knew that I had the skills that you needed to make your dream a reality."

"I knew that I can make this happen. "

"And I felt that with my help you might just stand a chance of winning."

"I've been rebuilding motors for over thirteen years now."

And I always wanted to build my own. motors."

" And I never had the opportunity until you walked in the front door and asked for my help."

"And besides, I believe in you, and I believe in my skills."

" So, much so, that I sank all of my savings to make the wheels a reality."

" I knew, I don't know how, I knew,"

" But, I knew that somehow you would pay me back."

" And I trusted you."

"Here, here" said Dick as he lifted his beer above his head and toasted me.

"Here's to the slingshot! "

And so, I took two shots. And I did a beer chaser.

I refilled our shot glasses, passed out some lime slices and I raised a toast to Toby.

"And, here's to Toby, without his help none of this would be possible!"

"To Toby!" Dick added

And down went the shots followed by another beer chaser.

And Toby and I both toasted Dick.

To Dick, And his wonderful tasting beer.

"Thank you for bringing the kegs" said Toby

"To Dick, thank you for being here with us.

"Thank you for everything"

"To Dick!" down went another round and we had to chase that round down with another glass of beer.

There was something that was annoying me.

It had been bugging me for some time now,

and I wanted to make a change so I made an announcement.

"Guys, I have something that has been bothering me for a while now and I wanted to run something by you." I said.

Toby raised his head from the table and he tried to focus his eyes

Dick opened his eyes and looked up from gazing down at his folded arms

 And tried to compose himself

while his brain was under the tremendous weight of tequila and beer

I was getting tired of calling the project the "Slingshot"

And I had been thinking of a new name for the project.

"Hey guys, I have a new name for the Slingshot"

I'd like to call it " The Spirits of Texas"

"to pay homage to all of the positive spirits that have brought us all together."

"I like that" said Dick

And Toby thought it was a cool name too.

So, we all drank a toast to the new name for our project.

"To: The Spirits of Texas!"

"To: The Spirits of Texas!"

"There you guys are!"

I jerked my head back and there were the girls standing on the back porch.

Shaking their heads.

like they were saying

"there they are, drunk again, as usual".

We smiled up at our better halves and they grabbed us by our arms and

dragged us back into the house.

"Wait!. What about the kegs?"

"Someone has to guard them from marauding bands of drunken teen agers."

I exclaimed

"Look around Andy, there is nobody here."

" All the guests are gone. " Rhonda sighed.,

"We, are guys, and we are still here!" I exclaimed

"Did you have a good time?" I asked

"Yes, thank you for a truly surprising surprise party."

" I was so shocked to see everybody here."

I wonder how you were able to pull it off."

" The food and the party and the cake, "

"When did you have the time to do all of this?"

"We make time for the things that we want to do." I explained

I stood up and then I looked back at the table and I realized that I was the only

one left standing!

Both Toby and Dick had their heads down and Toby was drooling on his sleeve and on the top of the table.

Dick was asleep sitting up in his chair.

almost ready to fall off the side of the table.

In spite of the fact that their wives were tugging at them trying to rouse them from a major drunken stupor.

I had out drank both of them!

"Toby, TOBY, wake your ass up Pendayoh!" shouted Isabella.

She woke up Toby and Dick with that one shout.

"Come on, I want to go home and you are crazy if you think that you are driving."Isabella added.

"Alright, woman, I am coming."

Toby fell back on his butt when he tried to stand up.

So, we had to help Toby to his feet.

Isabella walked him out of the back yard and down the street to their car.

 yelling at him all the way.

"Look at chu, you can't even walk.

I can't take you anywhere!

You're embarrassing me!"

Isabella chided.

Dick and I just laughed and we stumbled into the house.

And there was Diane and Maggie cleaning up the kitchen and trying to clear away some of the debris.

I showed Dick and Candace (who was also quite a bit on the tipsy side.) to the guest room.

And gave them some t shirts and a couple of pairs of pajamas so they would have something comfortable to sleep in.

I showed them where everything was.

Told them where the bathroom is and I bade them good night.

I remember walking into the bed room and I asked Rhonda to let Thor out to guard the kegs.

The next thing I knew I was waking up on the floor.

apparently sometime in the middle of the night I rolled out of bed.

And didn't even know it.

"Honey? What am I doing on the floor?" I asked Rhonda.

"That's where you slept."

Why didn't you wake me up and help me back into bed?" I asked

"because you never actually landed on the bed"

" You missed the bed and crashed out on the floor, so I just left you there."

"Thank you, honey. I feel so special."

I dragged my tired still drunk butt off the floor and I dragged myself into the bathroom and then to the kitchen.

I fumbled around for the coffee beans and the grinder.

I loaded the grinder with coffee beans and turned the thing on.

To my hung over brain and ears, the sound of the grinder sounded like a Boeing 747 on the tarmac, ready for take off..

The grinder was sounding as loud as a jet engine.

And it ran for a long time. Before finally turning itself off when the beans were all ground up.

Then I set up the coffee maker and turned it on.

Then I walked over to the fridge and I wanted to pour myself a glass of tea, but, the tea pitcher was empty!

There was no tea!

I looked on the stove and "Flunky" had not shown up.

There was not any tea sitting on the stove waiting for someone to come and mix it up.

So, In a near panic, I had to put a pan of water with several tea bags on the stove.

I turned on the burner and went looking for any ice chest that might still have a Dr. Pepper or a canned soda left inside.

No luck, everything was gone.

No beverages to be found anywhere.

I knew there was beer outside, but, I wasn't in the mood for beer right now.

So, I waited for the tea to boil.

And I washed out the tea pitcher and filled it halfway with water and added a scoop of sugar to the water.

And then I took the boiling tea pan off the stove and mixed the tea up with a wooden spoon.

Then I dragged out breakfast stuff from the fridge and the pantry and I got ready to make everyone breakfast.

I got out my cast iron skillets and I sliced up some sausage and started cooking that on the stove while I was mixing up the pancake batter

Then, I took out the bacon from the meat compartment of the fridge and I placed some paper towels on a microwave safe bacon cooker..

And I laid several strips on top of the paper towels.

I laid more paper towels on top of the bacon to keep it from splattering all over the inside of my microwave oven.

I set the microwave to 4:30 and turned it on and let it cook.

Then I cracked open some eggs and dropped them into a bowl and started mixing the eggs as I seasoned them with salt and pepper.

I put some hash browns in a different cast iron skillet to cook on the stove.

When the sausage was finished cooking I took the patties out of the skillet and I drained them on a folded paper towel.

And then I washed out the skillet and set it back on the stove and I used it to cook the pancakes.

The coffee was ready and I could hear stirring coming from the guest room.

And I could also hear Rhonda start moving around in bed.

In no time I had breakfast set up on top of the stove.

I had a skillet full of hash browns,

a skillet with scrambled eggs in it,

a large stack of pancakes and microwave cooked bacon,

freshly ground coffee and a pitcher of tea.

I was really hungry I didn't get to eat much last night,

so many people.

So, I called out to anyone who was listening. "Breakfast is ready!"

I heard the sounds of people stumbling around in the bedrooms

I was sure that

Rhonda would be coming out soon to get her morning cups of coffee.

But, I was surprised when Candace came out first.

"Good Morning" I said.

"Do you have any aspirin?" Candace asked

"Why, Yes I do, how many do you want?"

"I'll start with four and take more later on." She sighed.

"So, How's Dick doing?" I queried

"He's still lying down."

" Let me call him"

"Dick, get up Honey, It's time to eat breakfast!"

I heard the muffled sound of Dick calling out,

"I will be there in a minute!"

Rhonda came into the kitchen with her eyes barely opened,

So, I let her get her coffee and sit down before I said

"Good Morning" to her.

We filled out plates and sat down and ate.

And then after breakfast I asked Rhonda

"Where is Thor?"

"Thor?"

"Yeah, you know, Thor. our dog?"

"Oh my God, We left Thor out all night!" Rhonda exclaimed

So, I walked over to the back door and I opened the door and Thor was sitting on the back porch right next to the back door.

He rushed in as soon as I stepped back from the threshold.

"Oh, Poor Thor, We didn't mean to lock you out all night without any access to water." Rhonda said.

Thor was none the worse for wear,

He had done a good job of protecting the kegs.

He's a good dog.

The Hybrid Diaries
Chapter Seventeen

After breakfast, I scraped the leftovers into Thor's bowl and he ate like a pig,

In less than two minutes the entire bowl was empty.

He burped and went to go lay down on the living room floor.

Dick said that

"They had a wonderful time last night and they really enjoyed the party."

Dick wanted to know who did the catering for the meal and I said,

"That was not catered food, I cooked everything."

"Well, the brisket was delicious."

" It was very moist and tender and it had sweet tasting over tones."

" The chicken strips were out of this world." Dick and Candace agreed.

"I was surprised that you were able to get some before they were all gone."

I said.

"They usually do not last very long,"

"And there are rarely any leftovers." I added.

"well everything was very good."

" Have you ever given any thought to opening a restaurant?" Candace asked

"No, I never really entertained the idea of opening a restaurant"

" I have always cooked for just family or friends with a few exceptions."

"Maybe I should think about it."

"Hey, listen, while we are on the subject of cooking and food,"

" I was curious about something,"

" Rhonda and I do volunteer work for the leukemia society and every year."

"For the last few years, we have been competing at the annual Chili Cook off that the Chili society sponsors"

" Now, we have done reasonably well placing ninth out of two hundred and twenty cook teams. "

"Chili cook off teams traditionally have names that poke fun at themselves or their cooking and they are as creative as they are funny.

Here are some examples of names for chili cook off teams that I have seen through my many years of experiences

Bull dog Chili

Cat nose chili

Booger Red's

Three legged dog Chili

Big Moe's

It's dead already Chili!"

Dick and Candace alternated between laughing and giggling to cringing at some of the sick or gross names

"The name of our Cook off team, is "Critter Du Jour" it means "The critter of the day"

"I cook briskets, chili and Danish baby back ribs, and Rhonda cooks the beans.

"And our team raises money for the society at the event my taking donations and giving away free food, "

"We camp out all night and we dance and have fun at the Friday night Cooks party."

" And we cook all night and compete the following Saturday,

"Then we break camp and head for home."

" It's an awful lot of work, but, it is for a good cause."

"And we have a lot of fun doing it".

" Would you and Candace be interested in being on our team?" I asked hopefully.

"When and where is the cook off?" asked Dick.

"It's on the second weekend in October."

"And this year they will be having the cook off in Irving in Las Colinas." I answered.

"Rhonda has been volunteering at the Leukemia society offices. Stuffing envelopes and mailing out cooks packets and applications."

" She also helps with planning and logistics.

"They are organizing their sponsors."

"You know, who is doing what and who is supplying what and who will be responsible for whatever and who is making calls."

"Wow, Rhonda, How long have you been volunteering?" asked Candace

"Oh, about four years now,"

" I really enjoy helping people in need and I find it all very rewarding." Rhonda replied

"Yeah, I think our calendar might be opened for that weekend." said Candace

"Great, I have plenty of camping gear and an extra tent and an extra inflatable air bed.

"I just need help watching the smoker and making sure it has enough wood and help setting up and taking down the cook site". I said

Candace seemed very impressed with Rhonda.

She too does a lot of Charity work for the local Austin Charities.

It was good that they had so much in common.

"Yeah, I think we can make it." said Candace.

"That is great. I'll ask Rhonda to send you a application packet in the mail. FYI"

"You know what?"

" These events are great advertisements for the breweries who sponsor these cook offs."

" Last year, we were sponsored by "Rocky Mountain Beer" and the year before that we were sponsored by McMillians Beer."

"Do you have any promotional banners or stickers or signs from your brewery?" I asked

"Well, I have some stuff in the Suburban I can go and get it if you want. "

"I also have a few cases of beer in the back that I was going to leave here for you and Rhonda if you want them."

I said, "We would love to give your beer a good home."

Candace asked Dick where her purse was, and Dick said he hadn't seen it.

So, Dick and I threw on some shoes and went out to the Suburban.

We found it between the front seats.

Then we went back inside with our "trophies" in hand. Dick handed the purse over to Candace who looked annoyed at him for leaving her purse in the truck and not remembering to bring it in.

(She would have really been pissed off if she found out Dick also forgot to lock the doors on the Suburban.)

I took the beer to the kitchen and then I moved the promotional materials to a safe place in the guest room for safe keeping.

Candace reached into her purse and she pulled out a stack of brochures from different cruise lines.

And she started directing her questions to Rhonda and me.

"OK, I didn't know how you guys were planning on going to Australia,"

"but, I brought along some brochures from different cruise lines"

"Each line offers different amenity levels." Candace continued.

"I like Party time Cruises…"

Then again, I also like the Little Princess Cruise Lines as well

"Well, I was planning to ship the vehicle over on a container ship honestly, "

"I haven't Any idea how many people will be coming along from the other Sponsors."

I said

"Well, believe it or not,"

"There are several container ships that also offer cabins and berths for guests.

"Their accommodations are not as nice as a Cruise ship but, it would still be comfortable,"

" They do include meals and drinks, alcohol is duty free."

" And they have ship to shore phones and the team can travel with the container and keep an eye on it." Candace added.

"Do you like to sail?" Dick asked

" I love to sail, I think sailboats are very romantic." Rhonda said

"I like them too I always wanted to own a Catalina 22 foot sailboat." I replied

"Oh Dick, let me show you some of my ship models"

"I beckoned him to follow me to the game room where I keep my ship models.

I have always liked ships and I had built large scale models of

"The U.S.S. Constitution and the "Sea Witch" and the "Cutty Sark"

Plus, The Japanese Fleet and the American Fleets and the German Fleets of WWII

I was very happy to show Dick some of my favorite ship models

There was the Japanese Aircraft Carrier IJN Akagi,

The Japanese battleship IJN Yamato.

and the German Battle cruiser Scharnhorst.

While I was off showing Dick my little toy boats,

Rhonda and Candace had all but decided on how we were going to get to Australia.

They had decided that the team should travel on the container ship with the vehicle,

And Dick and Candace and Rhonda and I would be sailing on Little Princess Cruise lines.

They had been talking about the romance of the seas and the way that the people of the

late nineteenth century used to travel the seas in their finest clothing and they strolled

along the promenade deck with umbrellas in their hands on big liners like the Titanic or the Queen Mary, or the Queen Elizabeth.

The brochure said that the cabin staff were English and the dining room staff and the chef were Italian and that the rest of the crew were of different nationalities.

They offered breakfast, lunch, Tea at four o'clock and Dinner and then a midnight Supper.

Plus in room catering.

In case you didn't feel like eating in the formal dining room.

Room service, and laundry services.

And each room has its own bed turn down service with a English cabin steward.

"Oh, that sounds lovely." Rhonda gasps

I asked Candace

"what kind of accommodations would our crew have on a container ship?"

"It says here Cargo ships may not have spas with fruity facials,

Or rock-climbing walls and water slides, or spinning classes and planned shore

excursions, but freighter enthusiasts rave about their trips.

With only a handful of travelers on any voyage, there's no dress code for dinner and plenty of space."

On freighters, cabins are utilitarian.

"There are no 24-hour buffets or gourmet cooking demonstrations;

dinner is usually a single option. "

"Travel has to be booked far in advance, and while the cost is usually less per

day,"

"Usually about $100,"

"The voyages are longer"

"So the trips remain costly."

" At the same time, cargo ships aren't in port long, largely nixing extended shore visits."

"For container ship enthusiasts, walking the deck, watching the sea, or tackling crafts and puzzles are entertainment enough"

" Many bring stacks of books or start writing one of their own:"

"An autobiography or a journal of their voyage."

" A few ships have saltwater plunge pools that can be filled for passengers".

There's also visiting the ship's bridge and talking with the crew,"

"Though crew members frequently speak languages different from the travelers." Candace concluded

Rhonda and I have always loved to travel.
We are both big fans of the Thin Man movies.
We always wanted to be Nick and Nora Charles.
wining and dining our way around the world just enjoying life.
We are also huge fans of Cary Grant movies.
In some of his movies he also played a rich world traveler who traveled by trains and ship and air planes to the more interesting places on Earth.
Maybe meeting Katherine Hepburn or Carole Lombard .
We always wanted to travel on a ship and enjoy the cruise experience just like the "Kirby's" or the "Charles".

When the girls laid out their plan I asked them,

"Well, What about the sailboat cruises?"

"I thought you guys wanted to experience cruising on an actual sailing ship."

"Well, maybe we can do that on the way home" Candace offered

(Awwww, I though to myself, There goes my opportunity to talk like a pirate or walk

around with a Gilligan's Island white sailor's hat on my head like Bob Denver wore.

And driving Rhonda crazy for the entire three weeks more or less that we would be on the ship,

She would have no place to hide from my "Arrrgggghhhhh, Matey" and

"A vast there ye land lubbers" and of course, "Shiver me timbers"

But, knowing Rhonda like I do.

She would probably just do her pirate voice right back at me.

All the way back home.

But, then again cruising the high seas in POSH luxury .

And getting pampered and fed and plied with alcohol while we sail across the Pacific Ocean sounds appealing too.

And the spas and massages and work out facilities sounded good to us as well.

"Wait, We have to get our passports renewed!" I blurted out.

In all the preparations and building and planning, we completely over looked our Passports and we needed to get them renewed because we were still single the last time we needed them."

So, Rhonda and I will most likely have to go down town to the county records building

next week and get a copy of our birth certificates and a copy of our marriage license and then get new pictures made and then send all the info to the passport office.

"Dick, do you remember where we put our passports?" asked Candace with a worried look on her face.

"I think they are in our safe deposit box."

I'll have to check when we get back home." said Dick. Reassuringly

I was looking at the brochures and I found the itinerary for the cruise ship.

"Well, according to these brochures It looks like the Little Princess "Star" will be leaving

Long Beach, California on November 26th, and arriving in Sydney on December 20th,

It looks like the voyage should last about three weeks more or less. "

"We will need to see if we can find a Container ship that will be sailing to Australia at about the same time."

" So, we will not have to be waiting for the ship to arrive before we can get the vehicle through customs.

And then on to Woolloomooloo in New South Wales to meet the race officials and get the car certified.

Then we'll be off to the races!" I shouted.

I was starting to get excited.

(I was wondering if there would be enough time to test the vehicle before we would have To pack it up and get it ready to ship.)

I'd like to try and drive it from Dallas all the way down to Galveston (about a four hour drive by car.) and then pack it into a cargo container and have the container loaded onto the ship for the long voyage to Australia.

But, I am getting ahead of myself.

I still do not have the body or the solar panels or the flywheel generator ready yet.

I also need to contact the race officials to confirm that we are entered.

I will send them a letter sometime this week, just to make sure.

I would hate to go all the way to Australia and find out when we got there that the race was cancelled or that our entry was not accepted and we would not be allowed to compete.

That would really suck. And not in a good way.

Next, I'll have to contact my sponsors and find out if they are sending support Crews and if so, how many.

And I'll have to call Toby to find out how many people on his side would be coming.

While I was thinking,

Dick got up and went to the bathroom, then I heard him call my name, and he was laughing at something,

so, I got up and I went into the hallway and there was Dick standing there, with a big smile on his face.

"Hey Andy, What's going on in this picture?"

Dick was pointing to a picture of Rhonda and me at a company picnic for a company I used to work for.

They were a lousy company to work for but, they did have good parties and Company picnics.

It was a picture of Rhonda and me,

we were participating in an armadillo race .

In case you ever wondered just how do you make an armadillo race?

You have to get down on all fours, behind your racing armadillo,

And you blow on it's butt to make it run.

So, try to imagine,

There is Rhonda and myself on all fours crawling up behind an armadillo

And blowing on their butts and crawling after them towards the finish line. Rhonda's armadillo was faster than mine and I couldn't stop laughing at her blowing on that poor armadillos butt.

So she crossed the finish line first.

And she won a joke can of "Armadillo meat" (it was really a can of tomato sauce with a fake label glued on)

And to top things off in that picture I was wearing my old Australian Army Akubra hat .also known as "The Aussie Slouch Hat", it has one side of it's brim turned up against the crown,

Akubra has made these hats for the Australian Army since the early 1900's.

It was just like the one the guy wore in the TV show Rat Patrol except mine had an armadillo pin holding up the left side of the brim against the crown of the hat.

I had purchased my hat at the Army Navy store where I used to buy my blue jeans.

And I used to wear it all the time everywhere I went.

But, one day I lost the hat or misplaced it somewhere and I never bothered to replaced it.

And now that I see the picture and myself wearing that hat,

I suddenly had a funny feeling that somehow all these coincidences were somehow related, and that in some way everything was coming together in a way that I never ever would have expected it to.

"Wow, how ironic I'm wearing my old Aussie hat" I said.

I only wore that hat because I thought it looked cool with an armadillo pin on the side.

I had no idea that somehow that hat would have actually impacted my life in a way that I didn't understand..

"I wish I still had that old hat".

" Now I actually have a reason to wear it." I sighed.

"It's too bad you lost it, that was a cool looking hat" said Dick

He took the picture over to Candace and she looked at Rhonda and asked her

"What exactly are you doing in this picture?"

"I am blowing on an armadillos butt."

And beating my husband in a armadillo race."Rhonda said proudly.

The Hybrid Diaries
Chapter Eighteen
Somervell County, **Texas**, Home of Dinosaur Valley State Park
Comanche Peak Nuclear Power Plant and the future home of the New
Paluxy Brewery

Now Dick and Candace had purchased some acreage south of Lake Granbury

not far from the Brazos river.

In an area where according to the state topography and hydrology maps there

should be an adequate supply of ground water at the base of several hills.

The whole area looks like a giant water basin from the maps and they had

already had their environment impact study.

Everything has been signed off on. So, now all they need to do is drill their well.

Unfortunately for the Wigglers they could not determine the best place to drill.

The well and access to it will determine the entire layout of the brewery and

the orientation of the building and location of the pumping equipment and where the access for the municipal water supply will go.

Dick was "relaxing on the sofa" pouring over several potential drilling sites and he was not able to make a decision on where to drill the well.

I asked Dick and Candace if they had ever considered hiring a dowser to go out to their land and "witch for water."

They said that they had heard about dowsers

but, had never actually considered hiring one.

I told them about my father in-laws brother "Ronald" who is an old dowser from way back.

Ronald was raised on a farm just like my father in-law and their father showed them both the old ways that farmers used to find water and other precious minerals.

I told them, that if they like I can ask Rhonda to call him and see if he would be interested in finding the best place to drill.

They were both amiable to the idea and we all agreed that they would have nothing to lose and it might confirm their findings based on their scientific data.

I asked Rhonda and she agreed to call Ronald and ask him if he would be interested in dowsing for a well and Ronald said that he would do it for us for a small fee.

I let him know that it would be acceptable,

And we wanted to know when would be a good time for him to go out and mark the property.

He said that Wednesday would be good for him.

And since it was my day off too that would work out perfectly for us.

So, we hung up the phone and let Dick and Candace know

that we can all go out to the property and watch the dowser find water.

They were both excited and they wanted to stay with us until Wednesday.

We said that it would be alright with us.

And besides Dick said, He wanted to see what I do

And he wanted to see the project for himself.

I told him that I was waiting on the carbon fiber body people to clear out some space at their company to make room for the vehicle,

So they can begin the fabrication process.

But, I would be more than happy to take him to the mini warehouse and see the pieces that we already have.

Being Sunday and since the football game didn't start until three o'clock we had time to drop by the warehouse and see the project and then we could grab some fast food and still be back before kick off.

So Dick and I let the girls know where we were going.

And we let them and drive over to the warehouse.

We drove up to the warehouse and entered the security code in the key pad,

The gate opened and we drove through the gate and down the aisle to our storage space.

I got out and unlocked the door and opened the storage space and showed Dick the frame

I also showed him boxes with power inverters and ammeters and switches and the power generator and some boxes of miscellaneous parts. Dick was impressed.

Unfortunately, there were no wheels on the frame

So, Dick did not get a chance to sit in the cockpit area.

But, he got a good look at the progress that we were making.

I decided to take Dick by the store and I showed him where I work.

At the thrift store Dick was not that impressed.

We walked around the store.

It was clean and bright and you could really appreciate the hole in the ceiling which was covered by a piece of plywood.

And the fact that it was almost as warm in the store as it was outside.

I checked in with my employees and made sure they had enough change in the drawer and checked with the cashiers and made sure they would be alright for the rest of the day.

And Dick and I left.

We drove out to a Chinese take out restaurant not far from the store.

We ordered take out and went back home just in time to see the kick off.

We are all Dallas Cowboy fans and we watched the game and hooped and hollered at every missed pass,

or dropped football or successful reception or sack or interception.

And then we screamed and cheered at every Cowboy touchdown field goal and extra point..

Dick and I went outside to check on the damage to the kegs brought on from the partythe night before,

Only during the commercials.

And the results were impressive.

One keg had been successfully killed,

The second keg suffered a severe mauling and only had foam left in the keg.

But, the third keg was still over halfway full.

We had our work cut out for us.

We went in and brought out some plastic glasses and filled them up with the still Ice cold bubbly elixir.

We dumped the left over ice into the third keg barrel and dumped out the remaining water out of the empty trashcan like keg barrels.

And left them upside down on the drive way to dry out a bit.

Before we loaded them back up in the back of the Suburban.

We both ran back in the house when we heard a lot of screaming coming from the living room.

We rushed in the living room when we heard Candace and Rhonda yelling

"go, Go gooooo!!! Yaay!" they were both screaming like cheerleaders.

The Cowboys had scored a touchdown and the girls went wild.

Time for me to drink a beer.

Whenever the Cowboys play, I play along.

If they score a touch down, then I have to drink a beer and do a shot.

if they score a field goal no beer, but, if they score two field

goals I have to drink a beer.

If the other team scores no beer.

if we recover a fumble or snag an interception then I can do a shot.

Or a beer, dealer's choice.

Rhonda had already gone over the rules of Beer football with Candace

So, they were expecting me to bring them another beer.

which I did obediently.

And we watched the game and did shots all during the first half.

Then we did the half time stretch and got ready for the second half.

When the second half came we were all up on our feet cheering

When the Cowboys scored their second touchdown and we yelled when the

San Diego Chargers scored their second touchdown.

And the game seemed like it was going against us until we scored a

field goal and eventually won with a final score of Dallas 17 San Diego 14.

It was a great start for a new Cowboys season.

We are one and oh!

With such a low scoring game, we did not get drunk.

But, we still had fun and practically yelled ourselves hoarse.

Then the phone rang.

I got up and answered it and it was my store.

Someone had accidently locked the keys in the safe again.

So, I had to go back up to the store again.

And open the safe and get the keys out.

Dick came with me and we pulled up in from of the store and I ran in.

recovered the keys from the safe,

And gave them back to the clumsy cashier

and told her not to lock them in the safe again.

She promised she would not do it in the future, and I said,

" Alright." and left. I got back into the car.

Dick had a confused look on his face.

"Andy, Can I ask you something?"

"Sure Dick what is it?"

"Well, you have a job with the brewery "

And We've paid you four times"

"You got your signing bonus and you have an expense account"

"You probably make about ten times what you make here at this thrift store."

"Your manager treats you like crap yet, you still continue to work for him"

" How come?"

"Well, Dick, I really don't know."

" I think it's some misguided sense of loyalty to these people."

" I hired them and I feel responsible for them."

" But, also, I am more than a little scared by everything that is happening."

" I see so much potential for bad things to happen "

"And yet there are so many positive things going on around me as well."

" My whole life has been uprooted."

" Except this time the uprooting is a positive change."

" And I see the potential for my dreams to come true."

"And at the same time it is very daunting."

"I want to succeed so badly, that I can taste it, but, at the same time, I see the huge possibility of failure."

" And you know what Dick? Even though I am building the project"

"I feel like I haven't done anything to earn my salary or contribute anything to your cause" I was being as honest as I could be.

"Andy, Let me ask you something along those lines OK?"

"You accepted a position within my company as a Director of Promotions"

"You are building a hybrid racer to compete in Australia,

"which we are sponsoring."

"You have taken an assignment and you are working your butt off to get this project built

And completed in time for us to send it via a container ship to Australia."

"What part of any of those statements sounds like you do not have a full time position?"

"You have been working long hard hours."

"And honestly I think the job at the thrift store is keeping you from being able to focus all of you attention to the project at hand."

" You are burning the candle at both ends"

"And it's leaving you exhausted and almost unable to focus."

"I think that you should quit this position and start devoting more of your time to finish the project."

" We are running out of time and we need to get everything resolved before we can pack everything up for the long journey ahead."

"You know you will have to quit the thrift store very soon anyway,"

" because we will have to leave here soon."

"And we will be gone for at least two months.

"There is no way you can get off the ship in mid ocean and come back here to open the safe." Dick added.

Wow, I will not be here.

I will not even be in the city.

Wait, I will not be in the same state.

No, wait, I will not even be in the same continent.

And when it is winter and Christmas time here,

It will summer and Boxing day

Down in Australia.

And the water will flush down in the opposite direction.

And we will be gone for at least two months.

I cannot believe this.

It all still seems so unreal to me.

I am sure that anyone would be having the same misgivings accompanied by a feeling of excitement and almost blind terror.

this is a huge undertaking and so many things can still go wrong.

I have to have faith and trust in my sponsors.

And I sincerely hope that they do not to let me down.

That is something I need to make room for inside of my head.

I still have not come to terms with the fact that we are actually going somewhere.

I know, it would be hard to deny that the trip was not a necessary part of the project as a whole.

But, I was not prepared for the reality to start trying to make an impact on my psyche.

I guess I still have not fully accepted that I am really going to do this.

And that I was really going to Australia.

I had zoned out for a minute while trying to absorb the weight of the future so much so that I had stopped listening to Dick.

I quickly refocused my attention to try and take in what he was trying to say.

My mind and attention caught up with Dick as he was saying….

"I'm sure you have your reasons for wanting to hold on to your old position."

" I know it's not greed."

" They don't pay you near enough for that to be an issue." Dick laughed.

"So, I am pretty certain that it must be because you are a caring person who doesn't want to leave his employees high and dry."

"And that is an admirable characteristic."

"But, there comes a time when you will have to cut yourself loose."

" And trust that you have trained your employees well enough to get along without you.

"You do not have an "Superman's S on your chest"

" You are only one man and you are allowing yourself to be pulled in several directions all at once."

" That is not healthy"

"If Rhonda and Candace are correct then we will have to leave right after Thanksgiving.

And we will not be here during Christmas."

" So, while you have the chance, you need to be thinking about the person you would like to nominate for your successor".

" Or who you want to train to replace you."

" Because, I need you and we are not leaving without you".

"So, please, get all your ducks in a row."

"And let's move on together and focus on the project. "

"Because, I have worked a couple of jobs at once,

"And all it did for me was allow me the opportunity to fail at more than one thing at a time."

"There, that's my suggestions and my pep talk".

I know you will do what you want to, but, whatever you decide to do,

and however you decide to do it, I want you to know that I've got your back"

"And Candace and I are here for you whenever you need us."

"And that's not just for the project. That's for anything."

"Thank you Dick" I said. That helped me feel somewhat better

I knew he was right about me trying to hang onto the Thrift store job.

I wanted to originally work there through Christmas so I can give my employees a Christmas party.

And gifts. I wanted to play old Fezziwig! From Dicken's "A Christmas Carol."

Now for those of you who do not know who old Fezziwig was, just pick up a copy of Charles Dickens "A Christmas Carol" and look up the Ghost of Christmas past and there within that chapter you will find old Fezziwig.

He was Ebenezer Scrooge's boss when he was an apprentice.

Fezziwig was a good boss and a kind hearted employer and I wanted to emulate his example with my employees.

I wanted to show them that you didn't have to be a asshole to be a decent boss and that people would be more loyal and not steal from you if you were kind and treated your employees with respect.

Old man Fezziwig would do this for his employees by throwing a big Christmas party every year for his employees and family.

And I wanted to do that for my employees.

I wanted my place of business to be a happy place for my customers and my employees and I wanted them to now how much I appreciated all of their hard work and help them have a happier Christmas.

But now, it looks like I will have to forgo those plans and settle on a Halloween night

Beggars Banquet Halloween Party. And maybe a big Thanksgiving day spread. Right before we leave. For good.

It all sounds so final.

But, I guess it will have to happen before I can move on to my new position with the brewery.

I am just so worried about the position at the brewery.

I feel like everything is being done for me,

And there is nothing that I can do to contribute.

I feel like I am nothing more than a weight around the Wigglers shoulders but, then again,

they did ask to carry the weight and I shouldn't feel so nervous.

I am not usually this uptight.

I am normally a fairly laid back easy going person.

But I am scared of letting go of the life that I had built for Rhonda and myself.

And I wanted to have some sort of control over my circumstances no matter how illusionary

At least at the thrift store I knew what I could feasibly get away with and I also had some leeway with who I can fire and hire.

And I feel totally out of control with the Wigglers,

it's not that I do not trust them, I do,

I just feel like I was not helping them.

And that the ride would be over as soon as we crossed the finish line and the race was finished.

I wanted to be useful, and feel like I am helping out.

No matter how imaginary

I wanted to be able to do something helpful for the Wigglers.

and that is why I was so happy that Rhonda's uncle Ronald was going out with us to the Wigglers property and help us witch for water.

It wasn't much, but, in my mind at least it was something.

Now I have to come up with an excuse for me to miss work on Wednesday.

God, Why can't I just let go of it all and say frig it. I'm quitting?

I really think that Randall (my boss) is a moron, and an asshole

But, fortunately for his Sorry butt,

My father raised a little puritan with puritan work ethics.

And so I usually have to bleeding out of my eyeballs before I would allow myself the luxury of calling in and missing a full day of work.

I used to always joke that the epitaph on my tombstone

would say,

"I told you I was sick"

So, I was instilled with a strong work ethic that meant that I had to work hard every day

and it didn't matter if I was working for a complete ass or not.

I still had a job to do.

Kind of like the old saying "Love the game Hate the players" this is the kind of thing that is causing conflicts within me.

And it's probably holding me back with my position at the brewery.

I must try to convince myself that it would be OK for me to miss a day on Wednesday

I decided to push our visit to the passport office until after we got back from dowsing for water.

I am not accustomed to coming up with excuses in order to dodge a day of work.

Because I try not to lie.

I have to come up with way too many explanations to try to explain away a phony illness

or a fake emergency.

I just cannot do it. I am very bad at lying.

At least I will have tomorrow to come up with a convincing story that I can tell Randall, so, that "I can stay home" on Wednesday.

And go out to Somervell County.

Then it's over to the Wigglers' property with Rhonda and her Uncle.

I cannot wait to watch the old guy walk around in a field and find water.

The Hybrid Diaries
Chapter Nineteen
Welcome to Tuesday. September. 11th, 1990

Well I went by the carbon fiber place and I spoke with a couple of designers.

I spoke with Christopher, and Dylan and they said that they were almost finished with the parts they were making for the F-16 and should have the fabrication room cleaned and prepped by next Wednesday and that they would be ready for the Slingshot by the following Thursday September 20th or by Friday the 21st at the very latest.

I told them thank you and that the timeframe would be fine with me.

But, I asked if they would be ready to go on Thursday because my anniversary was on Friday the 21st and I didn't want to have to move the project on my anniversary.

I had spent a lot of time planning a anniversary for me and Rhonda.

And I didn't want anything to mess it up.

I'll tell you more about it when we get closer to the 21st.

As I was getting ready to leave I was walking outside and I could smell the distinct smell of fresh paint.

And I saw a spray cloud coming from a building across the street.

The place was a paint and body shop called "Thunder Chicken Paint and Body"

And there was this guy applying a coat of clear coat over a candy apple red paint job on a 1952 Mercury street cruiser.

I asked Christopher and Dylan if you can paint on a carbon fiber body, and they both said "of course you can."

I asked them if they knew the man who owns the paint and body shop across the street.

They said that they didn't .

So I told them good afternoon and I walked across the street over to the paint and body shop.

I walked in and asked to speak to the manager.

The guy I approached said to hold on,

he got up and went to go and find the manager.

I looked around the shop and there was a showroom attached to the shop where you can see some of the projects that they completed.

There was something that caught my eye right off.

It was an old 1936 Ford Stake bed Pickup truck painted in Aqua colored paint with many layers of clear coating over the Beautiful aqua finish.

I could see my face and reflection in the finish.

And it looked brand new.

I could see inside the engine compartment and there was a big shiny V-8 engine of what looked like GM origins.

In place of the original straight six engine that originally came with the truck.

And I could tell that it had some more modern upgrades on it.

I was getting ready to take a peek in the cabin and see how that looked when a voice from behind me made me whip around.

"She's a beauty isn't she?"

I turned around and I stared into the chest of this very tall guy.

I could clearly see his name embroidered in red thread on his name tag which was

sewn on to his work shirt.

But, I had to crane my neck to see his face.

"May I help you with something?"

" Would you like to see the truck?" the guy asked me.

His name was Ed bell and he was the owner of Thunder Chicken Paint and body.

I introduced myself and I told him that

"Yes I was interested in the truck and wanted to test drive it and hear it running."

Ed went to his office and he brought out the keys to the truck and he started

his pitch by telling me that the truck had been completely restored and that it

has a 350 Chevy motor

in it and it pumps out around 400 horse power.

The engine is less than two years old and The truck has an GM 400 turbo-hydro-matic

transmission and a posi-trac rear end.

It's sitting on 17 inch rims with BF Goodrich tires all around.

Dual exhaust, a working A/C and a CD player.

It had seat belts and all new dash board lights and gauges.

I really fell in love with the truck and I couldn't wait to drive it.

We got in and Ed fired it up and I could hear the menacing rumble of the big block V-8.

Dugga dugga dugga dugga....Vroooom, vrooooooom.... Dugga dugga dugga dugga

Ed carefully drove the truck out of the garage and he pulled over so I can drive it down the street and around the block.

There were two niceties that Ed forgot to mentioned,

power steering and power disc brakes all around.

Also the bed of the truck was hydraulically operated and the bed could tilt just like a dump truck.

And it had ramps underneath the bed that slide out.

That would be useful if I was trying to haul something large.

And the fuel gauge worked too.

There was also a tow hitch and wiring set up to haul a trailer.

It was inspected and it had a current registration sticker.

I liked driving the truck.

It had bucket seats and an updated dash with a working A/C that blew out ice cold air.

And the heater put out heat too.

The wipers worked and it's motor sounded really powerful.

And it could really snap your neck back if you got on the throttle too hard or too fast.

When I got back I asked Ed, if he had the title, he said that he did,

So I asked him "how much he wanted for the truck".

Ed told me and I thought his offer was too high so I counter offered.

And we went back and forth until we both agreed on what we both considered would be a fair price.

So, I made a decision, and I let Ed know that I was interested in the truck and I asked him

if he could paint a sign on the doors and hood of the truck for me?"

Ed said that he'd have to see a picture of what I wanted on the doors and the hood.

Before he could give me an estimate.

I also asked him if he could paint a sign to place on the sides of the stake bed. Because this was going to be a company truck.

He said that he could do it and that he would throw them in as a bonus

If I decided to buy the truck and I said OK. I wanted this truck.

So, I went out to my car and I got a out a picture of the RepTex label and I also gave him a six pack just for him to try.

I also told Ed that I wanted a longhorn skull painted on the hood with the Texas Flag blowing in the breeze behind the skull.

I asked Ed for his shops phone number and I told Ed that I would be by later on to pick up the truck when all the signage was done.

Ed said that he's have his guys clean up and prep the truck for when I came by.

They would top off all the fluids and change the oil and make sure that there was gas in the tank.

So, I went home to pick up my check book and Dick was there at the house and he was very excited.

He was holding the newspaper and he had a big ear to ear smile on his face. There was a story about the Challenge in the newspaper and that was how I found out that a New England based University whose abbreviation has three letters in it was going to be entering the race as well

And so was the GM sponsored Solar Raycer.

And that was when I hit the roof.

"I can't believe it",

" THEY were going to be there"

" I had no idea that they would be there,"

" I had hoped and dreamed that they showed up and I competed against them.

220

But, I didn't really think that they would send in an entry

I kept saying their letters over and over again.

It sounded like I was singing the theme

song for the Mickey Mouse Club.

I can't believe it M.I.C..M.I.C.

but, I wasn't saying M.I.C. I was saying the initials for the best technical

University in Massachusetts.

I was repeating the name of the grand Wahzoo of all the North Eastern

technical Universities.

"Well I guess we just found out that corporate sponsorships really are

accepted." Dick said,

And we both let out a loud Cheer.

Rhonda and Candace were outside when they heard us holler and they came in

to find out what was the matter,

So, I showed them the article and Rhonda smiled at me and said

"Awww, M.I.C. you always wanted to go there."

"Hey, didn't you want to compete against them and kick their asses?"

"Yes, they are the ones. and those are the asses I was wanting to kick"

I answered

"Well it looks like you are going to get your chance." Candace said

"Yes, I think so."

We decided that this was a good occasion to go out for dinner and celebrate

the news.

(I hadn't mentioned the truck to Dick, because it was going to be a surprise.)

I was thinking that first off we needed a truck of our own to move stuff back

and forth

Without having to borrow my father in laws truck or using the truck at work to move things around.

Secondly, I was giving the truck to Dick and Candace and the brewery.

It was my first contribution to the company and my first duty as the director of promotions.

The truck would have the company logos and labels painted on the doors and on the side of the stake bed.

And it would tow the project around from Dallas down to Galveston and back.

And it can even haul beer kegs and cases).

But, I forgot to get my check book and drive back to Thunder Chicken Paint and Body.

So I called Ed, and let him know that I would be by there either Thursday or Friday to pick up the truck,

He said that would be fine.

So, I hung up the phone and I started cheering again.

The excitement was incredible.

All of the things that I had been hoping for were starting to come true.

All the pieces were starting to fall into place,

And I was in the mood for celebrating.

Dick was also inclined to do a little celebrating as well.

And we were all excited about going out to the property in the morning

(after we go over to my in–laws house to pick up Uncle Ronald of course)

We decided that steak would be in order,

So we decided to go to a place that claims to be from Australia.

And we ordered a couple of steaks for the men folk and chicken breast for the ladies.

We all had baked potatoes and veggies.

And we washed it down with a few Fosters (beer)

We decided that Australia didn't know how to make a decent steak,
And that they could learn a lot from Texas and Mexico when it comes to buying steaks.
In Mexico you can buy a sixteen ounce steak served anyway you want it,
 (rare, medium or well done) for around five bucks.
And in Texas we have a restaurant in Amarillo that serves a 64 ounce steak that if you can eat it all in a certain amount of time, the steak is free.
The steaks we bought were little meat rolls that barely resembled the cuts of meat they were supposed to be selling.
But, they were tasty. If a little on the smallish side.
We ate and then we stopped by the video rental place and rented a couple of movies
And we went home and watched them before we turned in for the night.
Well Tuesday came and went without any complications.
And It looked like Wednesday would be coming to our house in a few hours.
I laid in bed in the middle of the night and I couldn't sleep.
Thoughts danced around in my brain that kept me from closing my eyes.
I was trying to imagine what I was going to say to my old boss.
Am I going to quit today?
What am I going to do about not coming it today?
Will Randall fire me today or what?
I was thinking that maybe it would be easier if he did fire me.
That way I would be free and if he did lay me off and everything fell apart then I could still file for unemployment.
So many things on my mind.
Will Ronald be ready for us to come and pick him up this morning?
Will We be able to get our passports in order?

Will Ronald be able to find water?

Or will I embarrass myself in front of my sponsors?

I finally managed to drift off to sleep and I knew that the alarm would be going off very soon.

The Hybrid Diaries
Chapter Twenty
Wednesday September 12th, 1990

Buzz, Buzz, Buzz, Buzz… I slammed my hand down on the snooze bar and rolled over.

It seemed like I had just fallen asleep and the alarm started going off.

I looked at the clock through bleary eyes and I could see that it was time to wake up

And get ready for the adventure of a lifetime.

I had never seen a dowser at work (other than in the movies or on television.)

So, I was excited about that. But, I still had to duck out of work,

believe it or not I found that to be the hardest part to get my mind around.

I had decided on a daring plot.

I would go to work and open the store and make sure that everything was all set up

And that everyone was there, before I would fake a stomach illness and go home.

Then, We would leave for the drive over to my in-laws house to meet Uncle Ronald,

pick him up and then we would all drive over to the Wigglers property.

Perfect.

I rolled over to Rhonda and told her to wake up and start getting ready.

Then I got dressed and went into the kitchen.

Dick was already awake and he had already made coffee and he was pouring

over the hydrological and topography maps spread out over the kitchen table.
I told Dick that I would be back in about an hour,
I advised him to roll up the maps and not to forget to bring the maps just in case.
He said that he was on top of it, and he wouldn't forget.
So, I went into the fridge and took out the tea pitcher and I poured myself a glass of tea to go.
Then I went out and jumped into my crap mobile and headed off to the store.
I got to the thrift store and I opened up the store and let in my employees.
Then I made sure they had money in the cash drawers and that they had access to the change in the safe.
And I made sure that everything was cleaned and straightened out on the shelves
and that all the clothes on the racks were re hung back on their hangers and I had Diane start marking down the older items.
When Diane was finished I asked her to keep an eye on the sorting and make sure that the sorters had everything that they needed
Then a "wave of nausea overcame me" and I just had to go home and lie down.

I let Diane know That I was placing her in charge and I called my boss and I let him know that I had gotten ill at the store and I was going either to the doctor or I was going home to lie down
 intentionally left it vague so I would not necessarily need to have a doctor's note for Thursday. (I was lucky to get his voice mail)
So, I left the store and drove back home.
Fortunately Rhonda had gotten up and she had her coffee and she was dressed.
And blow drying her hair in the bathroom.

Candace was also awake and dressed.

She was sitting on the sofa watching the morning news on TV and was also drinking coffee.

I came in and asked Rhonda if she had called her parents yet.

She said "No, not yet,"

So I asked her to call them and find out if Uncle Ronald had shown up yet.

She said she would.

I asked Dick if we were all riding together or if we were taking separate cars.

It just made more sense to me for all of us to ride together in the Wiggler's Suburban.

But, We would have to get the kegs out of the back before we would have enough room for all of us to be comfortable.

So, Dick and I went out to the Suburban and opened the back door and unloaded the kegs and we moved them into the living room temporarily.

Then Dick loaded the maps and assorted paperwork and impact studies into the Suburban

And in a little while the girls came out and loaded their purses and assorted girl gear.

we got ready to leave.

I went back into the house and let the dog out for one more run before we left.

Rhonda had called and spoke to her Mom,

She told Rhonda that Uncle Ronald was there and he was having a cup of coffee.

And that we could get there any time.

And Ronald would be there waiting for us.

So, after Thor did his business and I let him back in the house we were ready to hit the road.

That being done, we loaded ourselves into the Suburban and headed out to the in-laws house.

On the way there we stopped at a fast food joint and we bought sausage and egg and cheese biscuits, hash browns and I got a Dr. Pepper.

We made sure that we brought enough for Uncle Ronald and Rhonda's parents.

We pulled up in front of Rhonda's parents house and we got out and walked up to the front door and rang the bell.

Rhonda's mother opened the door and let us in.

Ronald and Rhonda's father James were sitting on the sofa watching TV,

We all walked in and I did the introductions and Ronald stood up and he yelled at Rhonda

"Come here you! My, what a pretty lady you turned out to be."

"Come on over here and give your Uncle Ronald a big hug."

"I haven't seen you since you were knee high to a grasshopper"

"You just get more and more beautiful every time I see you"

Rhonda went over to Ronald and she threw her arms around his neck and she said that it has been a long time.

She said that Ronald was looking good for a man his age.

(Ronald is about 70 years old.)

Ronald smiled at me shook my hand and said that it was good to see me again (the last time I saw Ronald was about five years ago shortly after Rhonda and I were married.)

I introduced Ronald to Dick and Candace.

Ronald shook Dick and Candace hands. And he said

"Pleased to meet you" Then he asked if we had any maps of the property.

Dick said yes, and that he brought the maps with us.

"Great, then we can get started then." Ronald said.

Dick and Candace asked if they could unroll the maps on the table,

So, James and Ronald cleared away everything off the table and Dick unrolled the topographic map of the area

Ronald gazed down at the map and stared at all the features and then he looked up and smiled at Dick and Candace and asked what they were looking for.

"Water and the best place to drill for it." Said Candace and Dick nodded his head approvingly.

"Alright then," Ronald reached into his pocket and pulled out a crystal pendulum and he held it over the map with his elbows resting on the table.

He watched the pendulum as it hung motionless over the maps.

Shortly, it began to move on it's own.

First in a circular direction and then in a more

linear direction.

Then Ronald asked if he could mark on the map.

Dick said "Sure"

So, Ronald drew a border around a couple of areas that Ronald thought would be promising.

Then he got up and asked Rhonda's mother for a glass jar with a lid.

So, she went to the cabinet and brought out an empty jelly jar and handed it to Ronald who got up and went to the sink

and he filled the jar with water and closed the lid down tight.

And then he handed the jar of water over to Rhonda and he asked her to hold on to the jar, so, Rhonda stuck it in her purse.

Ronald made a few more marks on the map and he rolled the maps back up and he sat back down and handed them back to Dick and then he asked for his breakfast biscuit.

And he ate the biscuit and his hash brown and when he was finished eating, he washed it down with a cup of coffee.

He got up and walked over to the kitchen hutch.

Then he picked up a box that was sitting on a shelf inside of the hutch.

And after placing it under his arm, he turned around and then he asked us if we were ready to go.

We all said that we were and we told Rhonda's parents good bye and they wished us "good luck and happy hunting."

Then we piled into the suburban and we headed off to the Glen Rose area towards the Wiggler's property.

In about an hour and a half later we were pulling up to the gate that marked the entrance to the Wiggler's property.

Dick got out of the drivers side door and he walked over to the gate and unlocked the padlock that held the chain wrapped securely around the gate and the fence post.

He swung the gates open wide enough for us to get the Suburban through the gate

and then he came back to the truck and climbed back into the drivers seat and drove the Suburban onto their property.

Then he got back out and walked back to the gate and locked it behind us.

Then we drove to the area where Ronald had marked a box on the map.

"Wait, Stop here." Shouted Ronald.

So, Dick stopped the car and we all got out.

"Look over there" Ronald said

We all looked at a willow tree growing majestically out of a piece of land next to a rock outcropping.

And there were other willow trees in the area.

Now, Willow trees require a whole lot of water so they can live and there seems to be a lot of willow trees around here in this area.

He took his hand and made a horizontal circle in the air to indicate that the area where we are standing.

"So, do you think there is water around here?" asked Candace

"Well, the trees don't lie." Ronald said.

" And there are several willows around in this area, so that tells me that there is a lot of water on this property."

"The rocks and the outcroppings are another sign to help determine the presence of water.

See how the rocks sticking out of the ground appear like they were broken with something like a sledge hammer or a pick axe?

That is and indication of fractures in the rock,

That lets rain water permeate the strata and it drains down into the Edwards aquifer.

.
(Candace already knew all this being a geology major in college, their professor would take them out to do field research on University property.

And the students would go out to scout for signs of water.

So, she also knew what signs to look for when scouting out land to find water.

But, she was impressed by Ronald's findings.)

Ronald walked back to the Suburban and he pulled out the box that he had brought with him and he set the box on top of the hood.

He opened the box to reveal his dowsing rods.

Ronald had made his own dowsing rods out of coat hanger wire and cork wood grips.

Ronald asked Rhonda for the jar of water.

Rhonda went and got her purse out of the Suburban and she handed the jar over to Uncle Ronald.

Ronald took the jar and set it on the hood next to the box and he dipped the tips of the dowsing rods into the water.

He took the jar and he walked to what would appear to be the center of the box he drew on the map.

And then he poured some water around him on the ground in a circle.

He stood in the circle and gripped the dowsing rods in both hands

and started walking around mumbling something to himself.

I could not hear what he was saying,

"It sounded like he was saying a prayer or something like that".

He would walk around and say "fifty feet" and then stop and he would look around and then he would start walking again and then stop and he would say to himself

"one hundred and twenty feet" and then he would move on and stop and say something else I could not make out.

And then he walked around some more.

Then he stopped and he asked Dick.

"How deep do you want to drill?"

Dick said "Well shallow depth is good"

"No" Candace said,

"We need to be able to hit the water table smack dab in the middle to insure the best amount of water flow."

"Alright, so about a hundred feet to two hundred feet would be alright then?" asked Ronald

"Sure" Dick said

"OK then." So, Ronald went back to the box and he pulled out a can of bright green surveyors spray paint and he sprayed a circle on the ground.

and then he made an notation next to the circle. 75 dash 100 dash 120.

And then he said

"You can drill here."

"The water is down about seventy five to one hundred and twenty feet down."

Then Ronald walked back to the Suburban and put the can back into his box and he took the map out of the car and rolled it open on the hood and he pointed to another box he drew on the map and he asked Dick

"Now can you take me here?" Ronald pointed to the second box on the map.

It was a little farther out on their property.

So, we all piled back into the car and we drove over to the approximate area of the second box and Ronald repeated the procedure.

The area was also heavy with water and Ronald estimated that the water was a little over one hundred feet down and that it was well within the area of the aquifer.

"This area looks real good too." Ronald said.

As he sprayed another circle on the ground

and he wrote a notation 100 dash 125. and then he asked us to take him to the next spot.

And he repeated the process.

Ronald said that he was able to find water at all of the spots

his pendulum had indicated to him back at the in laws house.

And that the first site was the best site as far as the largest concentration of water at the most shallow depth.

somewhere between seventy five feet and one hundred and twenty feet down.

We had been at the property for about four hours watching Ronald walk around with dowsing rods held loosely in his hands.

We would get excited when we saw the rods move closer together as he got closer to the water sources.

And then when they would pull his arms down,

That is how we knew he had found Something, hopefully water.

We spent about another hour just watching Ronald walk around and we would watch the rods pull his arms down over and over again.

And then he would spray another circle on the ground and make his notations.

And then move on to someplace else.

It was starting to get late and we were getting hungry,

so, Ronald finished up his dowsing .

and we loaded up everything back into the Suburban and got ready to leave,

Candace sat in the back seat with Rhonda and me and the girls decided that we would pull over and find some place to eat when we were back in Dallas.

We all reluctantly agreed and we headed back to Dallas.

On the way back Ronald regaled us with stories of when him and James were young boys

and what they did and how much trouble they got into. and so on.

I am sure that James would have been humiliated if he knew what Ronald was telling us.

James was a very religious man who was into fundamentalist Christian beliefs.

Kind of on the stodgy side.

While Ronald kept more of their father's spirit alive.

It was their father who showed them both how to witch for water,

but, because of his religious beliefs James did not practice dowsing anymore.

believing it to be the "work of the Devil"

Ronald on the other hand had made a nice little living out of dowsing.

And he had been all across the southwest helping people find water.

And drill for wells. He was quite famous among the people who use his services.

And almost all of his fame was spread by word of mouth.

Ronald believed that Dowsing was a gift that was passed on to him by their father.

And that this gift should be used to help people.

That was more of a payback to him than the money he earned as a dowser.

He loved to help people and he really enjoyed dowsing.

And he was happy that something he did would benefit others.

And help them bring forth life and crops from a dry arid looking piece of land.

That was what made Ronald the happiest.

We got back into Dallas and we started drilling the girls for input.

Where would they like to eat and what type of food would they like to have and so on.

We decided on a place called "Norman's old style Café" basic comfort food type food.

Hamburgers, Chicken fried Steak and mashed potatoes and meatloaf.

Nothing too fancy, just the basics.

I told Dick that I consider the "Chicken fried steak to be the yardstick on which the entire menu should be based on.

Because it is really difficult to screw up a chicken fried streak.

And if they serve you one that tastes like crap or else, the meat is nothing but breading.

or the mashed potatoes are bad, then Do NOT eat at that restaurant again.

We were fortunate that the Chicken fried Steak and mashed potatoes and gravy were edible and actually quite tasty.

Everyone was enjoying their meals and we all had a good time.

It was starting to get late so we left the restaurant and took Ronald back to the in laws house.

We thanked Ronald for helping us out with the dowsing and we paid Ronald for his services,

Ronald told us " I just wanted to thank you all for everything.

I hardly ever get the chance to see my brother and Gosh,

I can hardly remember the last time I saw my niece.

I am usually out on the road dowsing for wells a lot and you guys were very lucky that I was home at the time of your call."

"I think that you guys are some of the nicest people I have ever had the opportunity to work for.

And I would like to wish all of you much success in all of your future endeavors."

Rhonda I think you have grown into such a beautiful woman I am so proud of you".

Andy, It was good to see you, you take good care of my niece now."

Thank you all for breakfast and dinner, it was delicious".

We all told Ronald that he was more than welcome, and that we really enjoyed watching him work.

And we thought that what he did was fascinating and we really enjoyed his company.

So, we shook Ronald's hand and Rhonda gave Ronald a big hug and a kiss on his cheek.

I told Uncle Ronald that "I promise I will take good care of Rhonda."

We told Ronald Goodbye and we said goodbye to Rhonda's parents.

And then we drove back home to my house and let Thor out to do his business.

Then we moved the kegs out of the living room and back out to the Suburban. We then changed our dirty sweaty clothes, got into some comfortable clothes I checked my phone for messages and I checked in with the store to make sure everything was alright.

Everything was fine. And I hung up the phone and I went into the kitchen and Dick and Candace were pouring over the maps and they were shouting "No way" or "I don't believe it" and "Wow, I am amazed"

The Wigglers had brought another copy of the map with them and they kept it separate from the other maps.

This map was a hydrological map and it was different because it had been surveyed by the Professor of Geology at the University of Texas.

And he had highlighted the spots on the Wiggler's property that had the highest probability, and the greatest potential for finding water on that site. Candace had taken the map (the one that we took, that Ronald had marked on) and laid it on top of the map with all the high probability targets circled with a high lighter marker.

And the results were astounding.

Ronald had hit every place that was indicated on the maps and he also gave a depth that was almost dead on target.

He had also found another site that the UT Professor did not identify.

Candace was truly amazed by Ronald's accuracy and the fact that he had a detailed knowledge of the terrain and its features.

And how water was stored within rock strata's.

It had taken Candace years of hard studying at UT before she could understand how all of the rock structures and the soil types work together and the way water flows through fractured rock outcroppings.

236

And the way the vegetation grows around sources of water.

it is not common knowledge and it takes a trained eye to be successful at seeking out sources of water.

Now, with the results in hand,

The Wigglers were now confident that they could locate water based on their maps and the begin construction of the new brewery as soon as the wells are dug and the hook up from the municipal water company was run out to the property.

Now it will be up to the architect to do the design and the engineers to do the preliminary work ups.

Then they'd need to hire a drilling company to drill the wells and get the city to come out and run water pipes, sewer lines, and electricity, phone and gas lines.

Dick and Candace Thanked us both for all of our help and taking the time out of schedules to go with them they said

"That we had taken a load off of their minds."

At last, I feel like I had made a contribution to the cause.

And now, I felt like I was doing something productive for my benefactors.

It was what I needed so I would no longer feel like I was a drain on their resources,

and I was really trying to do something that would benefit Dick and Candace.

I felt really good,

I didn't feel like I was like some albatross hovering over the Wiggler's heads anymore.

Please, don't get me wrong,

Dick and Candace never did anything that made me feel inferior or inadequate in any way.

In fact the only thing they ever offered was encouragement and financial support.

All these feelings were all in my head and I desperately wanted to do something that would banish those negative thoughts from my mind once and for all.

And I did it.

We sat around the house and relaxed and watched Star Trek the Next Generation and played Uno, until we were ready to go to bed.

All that walking around and all the driving took a lot out of us.

And besides the Wigglers were heading back to Austin on Friday..

I wanted them to stay at least until I picked up the truck and showed it off before I presented them with the keys.

But, Dick said that they had some things to take care of back in Austin.

I know that Candace had been talking to Rhonda about us moving closer to Austin so that I would be closer to my job when the project was completed.

I did not have any objections about moving to Austin.

In fact I always said to Rhonda that if the opportunity ever presented itself then I would move to Austin in a heartbeat.

But, up until now, that option never presented itself.

So, I was receptive about moving. But, not until the project was completed and the race was over,

because my sponsors are here in Dallas, and I still need their help to finish the vehicle.

The Hybrid Diaries
Chapter Twenty One

Thursday morning I got a call from Toby.

He had been working on the wheels and the flywheel generator and the rear drive wheel.

And he said that he was almost finished with the preliminary testing and he would be ready to help me with the final assembly and the testing.

I let him know how grateful I was for all his hard work and I asked him if he would be interested in coming to Australia with us because we could really use him if he should decide to come along.

I told him that the Company would pay for him and his wife and I needed someone I could trust to ride shotgun with the vehicle on the container ship.

(After meeting Toby's wife, I knew that she was the type of person who would not let him go to Australia without her,

So, I had already suggested that we invite her along as well to insure that Toby would be along for the ride with her.)

I let him know that if he came I would make him the "Official Crew Chief and Lead Engineer"

He said that he would have to think about it and he would let me know something soon.

That was strange I thought as I hung up the phone.

I thought Toby would have jumped for the chance to go to Australia and see our vision racing in the sun.

But, he kind of sounded distant and preoccupied.

So, I just said goodbye to Toby and I hung up the phone

I got dressed and got ready to go to work.

I planned to call Mr. Bell at Thunder Chicken Paint and Body from the office after I got everything all set up and running smoothly at the store.

So, I went to work and did the payroll for my employees and cut checks for my phone solicitors and I worked around the store.

Then I got on the phone and called Ed at the shop and I asked if the truck was ready.

Ed said that everything looks good and the truck has been painted and washed and the signs have been placed on the outside of the stake bed rails.

All I needed to do was come by and pay him and pick the truck up.

I told Ed that I would be there around lunch time.

And he said that would be fine, and that they will be waiting to see me when I got there.

So, I said ok, and I hung up the phone.

Then I called Rhonda at home and I asked her to see if Dick would be up to running an errand with me at lunch time.

So, she called out to Dick and Candace and confirmed that Dick would indeed be willing to run an errand with me at lunch time.

"Good" I said that I would pick him up soon and I hung up the phone and I went out on the sales floor and swept and mopped then I went back into my office

I closed the door and did some work until lunch time came around.

And I got into my car and I drove over to the house to pick up Dick

Then we headed out to Grand Prairie to the paint and body shop where the truck was supposed to be ready and waiting for us to get there.

Well, Dick and I pulled up to the shop and we got out of the car.

I asked Dick to wait outside while I ran inside and spoke with Ed.

I told him that the guy I was buying the truck for was right outside and I haven't told him anything about it. So, Ed, had one of his guys throw a big tarp over the truck and I went back outside to get Dick.

Together, we walked back into the shop and we walked back to the work floor and we stood in front of a big tarp which was thrown over something that we could not see.

Then Ed came over and introduced himself to me and Dick and he asked me.

"OK, are you ready to see it?"

"OK, let her rip" I said.

"OK guys, here we go, One, two, three…UP!" shouted Ed.

Voila! And off the cover came.

We both just stood there. Staring at this beautiful Aqua colored truck with brand new "New Paluxy Brewery" signs on the back of the truck and Rep-Tex Beer labels on the doors .

And then we focused our gaze at the longhorn skull laying over the Texas flag waving in the imaginary breeze.

"Oh my God" Dick said "Did you do this?"

"Yes, I designed the artwork but, No, but, I had it done." I hope you like it"

I handed the keys over to Dick and told him that

"Maybe he should give it a test drive."

He said "alright" his hands were shaking and he was visibly excited.

So, Dick opened the door and looked at the inside of the cab. then he climbed in behind the wheel.

And started the truck up. Vrrooooommmm, the engine roared to life and he gently eased the shifter into drive and he drove it outside and out onto the street and took it out onto the street and down the block on his first drive.

Meanwhile,

I had whipped out my plastic and I put the purchase on my expense account and Ed signed the title over to me.

In a few minutes Dick drove back up to the store and he had a big grin on his face.

He pulled the truck up to the front and parked, and shut off the motor.

And he got out.

He did a walk around the truck and he kept admiring the signs and the logos and the flaming skull on the hood was what he liked the most.

"Wow, this is an awesome looking truck ."

"What year is it?"

" What color is this?"

And I've never seen anything this cool looking before."

I told Dick that the truck was a 1936 model and it was spray painted in an aqua color and the entire truck had been completely redone and restored to almost new condition.

And I had the signage and artwork done special for the Company.

I also told Dick that the truck was ideal for our purposes and it would also be a great way to "Show the flag" so to speak.

And that the truck would and should accompany us to any events that the brewery was sponsoring.

Dick was thrilled by the idea, and he could not wait to drive the truck back to my house.and show it off to Candace and Rhonda.

So, we both got into our vehicles and we drove back to the house.

(Dick followed me home) and we got out and Dick and I went up to the door and knocked loudly.

The dog went off immediately, and I heard Rhonda tell Thor to stop barking,

She opened the door and saw us both grinning at her.

I asked Rhonda to go and get Candace.

So, she went and brought Candace up to the front door.

Then when Candace came to the door ,

we both stepped aside and let the ladies get a good look at the truck.

Candace and Rhonda walked out there and stared at the truck for a long time.

Just taking it all in.

And finally, when we asked if Candace liked the truck, she screamed out

"Hook em Horns!" and pointed to the skull on the hood.

She noticed that the skull had an dark orange background around the skull to give it some depth and contrast against the waving Texas flag in the background.

"I LOVE it"

" It looks so cool. I never knew trucks actually looked like that".

" That is so awesome."

Candace was very excited.

Dick was almost beside himself with joy.

"I just wanted to thank you for everything you've done for us", Dick said

I know that anyone can go out a buy a new truck from any dealership,

but, to do what you did, takes an incredible amount of thought and creativity.

And Candace and I really appreciate everything you guys have done for us.

putting us up for the week and taking care of us and going out to our property and locating water, and now this. I just don't know what to say.

I am so honored by your gestures"

"Please, don't take this the wrong way."

" We LOVE the truck,"

"And, yes, we should use the truck as a promotional vehicle."

And It looks so good The way it is. "

"But, I know that you need to have a truck up here while we work on the

Project"

" So, why, don't you hang on to the truck for right now and use it to haul the project."

" I know you are probably tired of having to schedule everything around when you can get access to the thrift store truck."

" Or having to rent a truck just to get the project to the other side of town."

" This should solve a lot of problems in transport."

" And maybe you can put some flashing lights on the roof and on the back of the bed and in front of the grill and we could use the truck as a pace car / sag wagon.

In case of a break down."

"That sounded like a great idea to me." I said,

" But, I bought the truck for you. Dick I didn't buy it for myself." I said.

"That's OK, Andy. I know what you were intending for the truck and It is a tremendous gesture. "

I was going to do the same thing for you too."

" I know that you will lose access to the thrift store truck as soon as you leave the company."

 And you will still need to be able to move the project from place to place.

So, I understand the necessity of needing a truck."

"But, Dick, I bought the truck for you. And Candace and the brewery."

" I did not buy it for myself. I was going to go out and spend 1500.00 on some old used pickup. That was all I was needing."

"The truck I bought for you is way too nice to be hauling around my project in the back;". I responded.

We both sat there staring at each other, both of us were now caught in a dilemma,

He really liked the truck and I wanted him and Candace to like it and take it home.

But, Dick was trying to help me out.

so, he wanted to leave the truck with me and let me use is for the project.

BUT, he didn't want to offend me by refusing the gift.

So, there we were desperately trying to figure out a compromise.

When the girls came into the room.

"What's up guys?" Rhonda asked.

So, I explained the situation to the ladies.

"Well, Dick, why don't you let Andy use the truck and Andy, why don't you borrow the truck until after the project? Implied Candace.

"Yeah, that makes the most sense". said Rhonda.

"Look Andy, We all know that you bought the truck for the brewery and for promotions.

"So, We all know that the truck belongs to the Wigglers"

"And If they agree to let you borrow the truck then that should be reason enough for them to let you use the truck for the project."

"Besides Andy, you already signed the title over to Dick right?" Rhonda asked.

"Yes, I already gave Dick the title." I said.

"Good, so it's not your truck anyway." snapped Rhonda

"And if they agree to let you borrow it, then what is the problem?"

"I bought the truck for them, not for me to use."

"that's like me buying you a box of chocolates and me eating it before I got home and gave it to you." I said.

"Rhonda said, "Look, we all know you didn't buy that truck for yourself."

" We all accept that the truck belongs to the Wigglers

Therefore, they can do whatever the hell they want with the truck.'

" So, If they say to you to, that they want you to use the truck with their permission"

"then you better do what they say and stop trying to argue with them."

"End of story." Rhonda pronounced.

"No, It isn't, I said.

"Why the Hell not?" they all asked

"Because, I still need a trailer to haul all my stuff in." I responded.

All three of them let out a huge sigh… and Rhonda sat down at the table, and placed her elbow on the table top and then she hung her head down and placed her right hand palm against her forehead and shook her head from side to side in disbelief.

"It's always going to be something with you, isn't it?" Rhonda sighed.

Well, I guess, I'll go and buy a trailer. I said.

Well, we all got a laugh out of it all and I looked at the clock and I was late.

So, I had to get back to the store.

So, I asked everyone to be thinking about what they wanted for dinner.

And I headed back to the store.

When I got there, there was a problem.

David had just broken down on the side of the road.

So, I got his address, and I had to call a wrecker to pick up the truck.

I had a padlock that I use to secure the back of the cargo box.

When we have to leave the truck unattended.

So, I had to take the padlock and drive out to where the truck died and rescue David.

And bring the truck back to the store and find out what is wrong with it.

then have it towed to the repair garage and then go to the truck rental place and rent another truck while the company truck was in the shop getting repaired.

That took the rest of the day and the rental truck would not be ready until in the morning.

So, we had the truck brought to the store where we hastily unloaded the cargo box, and then we had the truck hauled over to the auto repair place down the street,

I asked them to call me when they found out what the cause of the problem was.

It didn't take them too long to figure out that the timing chain had broken and it would need to be replaced.

I could do it, but, I was not going to waste my time doing auto repair work on the Company truck when I can pay a mechanic out of my store's expense account.

So, I asked the mechanics what the damage would cost me, and they said, about 200.00 dollars so, I told the mechanics to go ahead and repair the truck

It was time for me to go home at last.

I did my last walk around before I locked up my office and headed out to the parking lot.

I opened my car door and I got in and fastened my seatbelt.

And started up the car. I backed up. and drove out of the parking lot and onto the roadway.

Unfortunately I have to pass the auto repair place on my way to my house,

And I saw the company truck sitting forlornly in their parking lot.

"Stupid timing chain" I kept thinking to myself I wish I could just get rid of that stupid chain.

All they do is break.

So, I started thinking about my vehicle.

It's got a big ass bicycle chain that drives the flywheel.

Now what would happen if that big stupid chain broke?

I'd be stranded.

Left high and dry, completely at the mercy of the elements.

"Why do I even need a bicycle chain, and why am I hanging on to the original design

When I can use new concepts and ideas to make everything better.

Truth be known, I only used bicycle parts and tubing because it was plentiful and also available and very cheap.

And since I was working on the concept all by myself, and I didn't have much money for parts, I decided that bike parts would be fine for making the original prototype,

But, they would be unacceptable for our actual solar racer purposes

because it would not be strong enough to handle the strains and forces of long distance travel and it would not be sturdy enough to handle all the forces that the frame would be subjected to..

So, I started reasoning why I would still need that long bicycle chain to drive the flywheel.

Why not use a electric motor to spin the flywheel and I could use my legs to spin the crank on a hand crank generator and have that charge a battery that runs the motor that will spin the flywheel at much faster RPMs than my legs could provide.

Even with a whole lot of up and down gearing.

And, it would have the added benefit of shortening the nose of the vehicle and allow for more room in the cockpit for possibly two people.

And it would be much safer and offer some measure of protection for the driver / rider.

And hopefully save weight.

This was all very exciting to me.

All of this designing and building and dreaming up new ideas to improve the over all look and performance of the machine.

I decided that I was going to have to call a meeting and get everyone together for a brainstorming session just as soon as the carbon fiber people are ready for me to bring the vehicle over to their shop for the all important body construction.

While I was thinking about all I had to do, I let my mind drift for a little bit.

I always wondered to myself,

"Why was I selected to be the one with this idea stuck in my head?"

"Why was I chosen to design and build this project?, Why me?"

Honestly, the whole project has invigorated me.

I have never felt so alive and so terrified at the same time.

Rhonda always said that I was a big mass of insecurity, worry and self doubt.

And to a greater or lesser degree I always felt that she was right.

I am a big mass of self doubt.

Only because I have never done anything like this before.

And yes, it can be very intimidating to someone who has no experience undertaking a project like this.

This is "On the job training" of the highest caliber.

And with all the possibilities of things going terribly wrong it is quite scary.

I am glad I have Rhonda,

She helps keep me rooted and focused.

And I do not have to fight the urges to throw my hands up in the air.

Or pull my hair out by the roots and run screaming naked down the street.

All I do know is that this idea has been in my mind for years and now I have gotten this golden opportunity to build a once in a lifetime dream and make it a reality.

And show the world that I was not crazy.

And that my concept will work and that I knew what I was talking about.

The crap mobile pulled gracefully into the driveway and I parked the car and got out and walked up to the front door.

I went inside and everyone was in the living room talking and having a good time.

So, I changed into some clean clothes and we batted a few restaurant ideas around before we decided on sea food.

So, we piled into the Wiggler's Suburban and we headed out to

"On The Shore Restaurant and Bar" I called before we headed out to make sure that Sonny was on duty tonight.

And since he answered the phone, I was sure he would be still there when we arrived.

So, I told Sonny to stay there and We would be dropping by for dinner.

Sonny said

" he would hang around until we showed up."

So, we drove over to the restaurant and as the girls and Dick were getting out of the Suburban I noticed that something was not right.

I could see inside through the big plate glass windows and I could see Sonny with his hands up behind the counter and some guy in a hoodie and a baseball cap pointing something that looks like a gun at Sonny's face.

I didn't notice the robber looking up or out of the window

so, I assumed that he didn't see us pull up in front of the restaurant.

Now, I acted like I dropped something and I bent down behind the door as I was stooping down to pick up the imaginary object and a whispered to Dick,

"Hey Dick, there's someone in the restaurant and they are robbing Sonny!"

"I can see them pointing a gun at Sonny's face."

I told the girls to crawl over the kegs (which were still in the back of the truck) and get down behind the truck and crouch down and stay away from the windows.

And while I was getting the girls situated Dick had reached under the seat and he pulled out a mobile phone.

And he called the police.

The cops said that they were already in the process of responding to the silent alarm at the store.

I guess somehow Sonny must have tripped the alarm.

So, Dick told them that we would keep an eye on the restaurant and the suspect and he gave them a description of the perpetrator.

I borrowed Rhonda's camera out of her purse and I crept up to the side of the window without being noticed and I hid behind the wall.

I watched the robber .

His attention was clearly on keeping the gun in Sonny's face with one hand and grabbing the money that Sonny was setting on the counter and stuffing it into a pillow case.

He had just stuffed the gun into his waist band of his pants so he could grab hold of the pillow case with one hand and stuff the loot in with the other hand.

And right at that very moment I rolled across from behind the wall.

I stood there out in the open and in plain view.

In front of the plate window and I snapped a couple of pictures of the robber.

He must have saw the camera flash in the corner of his eye because he looked up.

As he was spinning around to leave.

Just at that very moment I snapped a quick picture of his face.

That was when he pulled out the gun again and fired a couple of wild pot shot rounds at me in my general direction.

I ducked around to the other side of the restaurant building.

And darted between two big metal trash dumpsters.

And then I ran around to the other side of the building.

Just as the cops pulled into the parking lot.

They surrounded the entire area.

When they clamored out of their squad cars they shouted

"POLICE!!!"

"Hands in the air!", and,

"Get on the ground!" and

"drop the gun!!!" and my all time favorite classic cop phrase

" freeze!"

I immediately dove to the ground.

And the cops were quick to pounce on top of me and push my face into the asphalt

Then they did the same thing to the robber.

Once the police were able to sort out all of the chaos.

They released me and removed their handcuffs and apologized to me.

I brushed the gravel off my clothes and wiped the sand and grit from my face.

And I watched the police pick up the bad guy by his arms and place him in the back of a squad car.

Dick and the girls were in a much better position to see the cops apprehend the crook

And they alternated between telling me that I was brave.

And being extremely stupid for risking my life like that.

Their insults and adulations made me feel uncomfortable.

In fact it wasn't until they started retelling the story to Sonny,

That I started feeling queasy and inside I started shaking.

Because it was just now soaking in, what had happened and what I had done.

At that moment I was just happy that the cops apprehended the bad guy without shooting him or me.

After the officers let me back up and frisked me for weapons

I told the officers that I have some photos of the bad guy in action robbing the place and pointing a loaded hand gun into Sonny's (the restaurant Manager's) face.

The police said that they were grateful to us for having the camera and they were nice enough to confiscate Rhonda's camera as evidence and they said we could have it back in a couple of days after they developed the pictures of the perpetrator and the robbery in progress.

Meanwhile I could hear other police officers reading off the Miranda rights to the crook and padding him down in a search for more weapons or evidence.

The cops asked me if I wanted to file charges against the bad guy for assault with a deadly weapon and attempted murder..

I said no, because there was no harm done, and I really didn't think he was trying to kill me or Sonny.

I think he was shooting wildly at whatever he perceived to be a threat..

The cops said that they were going to prosecute him anyway and that they really didn't need my charges or testimony to put him behind bars.

So, I said "OK, do what ever you guys are going to do" and I left it like that.

So, after being interrogated for the next couple of hours, the police took down their crime

scene tape from around the building and their red thread that they used to highlight the flight path of the two discharged bullets.

they left with the perpetrator and all of the information that they needed, Sonny closed and locked the front door and we all sat down for several rounds of strong drinks to steady our nerves and all the free seafood we could eat.

Everything was "comped" which basically means "On the house"

That was just Sonny's way of thanking us for saving his life and foiling the plans of a dangerous crook and stopping an armed robbery in progress.

Except for being shot at Twice.

And costing Sonny a new plate glass window,

we all got off very lightly, and no one was harmed.

Of the two bullets that the bad guy fired me the first one went into the top of the window and out into the street.

And the second one slammed directly into the wall next to the window frame.

Fortunately the first wild round that went through the window missed the girls and the truck.

It then sailed across the street and crashed into the window of a second floor office which was luckily empty at the time.

Sonny kept pouring us round after round of "the good stuff" not the regular tequila that he sells his patrons but, Patron Silver and Gold, and Herradura.and Grand Marinier an

orange flavored Liquer,

Which is used for added kick in top shelf margaritas.

Speaking of margaritas, the girls had several Top shelf margaritas .

"More margaritas Ladies?" Sonny kept asking

"Yes, Please!" they would both say. Over and over again.

We finally had to stop him because we were all getting pretty drunk and we

254

were so stuffed with shrimp and crab legs and crab cakes and lobster tails and claws and sour dough rolls and baked potatoes and steamed veggies that couldn't possibly eat another bite.

We didn't even have any room left for dessert. So, Sonny stuffed a whole Cherry Cheesecake in a cake box and he let us take it home.

Plus, he kept calling me and Dick heroes for saving his life and his livelihood.

We kept telling him that I was not a hero and that Dick was not a hero.

We only did what Sonny would have done for us if the roles were reversed.

"You know what?" Sonny said,

"No one in back saw what was going on up here"

" I was up here alone closing the register and working on the daily books and there was no way I could have called the cops because I couldn't make it to the phone before the guy pulled the gun out on me.

"They were all cowering in the back,"

"They would have just let him shoot me."

" I barely had time to push the silent alarm button.".

Dick let him know at someone in back had indeed seen something because when we called the cops they were already in route to the restaurant because someone there had called the cops and let them know that there was a robbery in progress.

 The police did not mention responding to a silent alarm,

they said they were responding to a call.

But, he still insisted on calling me a hero and calling Dick and Candace and Rhonda heroes for saving his life.

He was starting to embarrass me.

We ate and drank and we celebrated life and had a great time at the restaurant.

And we eventually figured out who called the police in the first place.

There was a very pretty young girl about twenty two years old named Monica.

Who was working at the restaurant part time while she was going to the university down the road.

It was SHE, who saw the bad guy walk into the restaurant from the store across the street.

And it was She who called the cops from the store's phone.

She really liked Sonny and must have a huge crush on him.

But, Sonny never noticed her before.

He saw her as a waitress and an employee and nothing more.

Until he found out that she was watching out for him.

And watching his back.

Meanwhile our girls were switching from top shelf margaritas over to coffee so they could try to stay awake.

Nothing funnier than two drunk girls who are amped up on coffee and unable to pass out.

They were hooting and hollering and laughing and cutting up and just having a great time.

The alcohol must have helped them forget the experiences earlier in the evening.

Really funny stuff.

We took the opportunity to leave a nice large tip on the table and we still hid about fifty

dollars under our plates.

So, we still paid Sonny for our food. I didn't think that Sonny

was in a position to remember all the little details such as the fact that he had just been robbed, and that he had just lost the restaurant's entire nights till and credit card receipts,

we didn't bother to ask him if the robber had forced him to open his safe or not. and there was nothing in the tip jar to tip the wait staff.

At the very least (thanks to us) Sonny would have some tip money to divide up amongst his workers for the night and a little cash to make a deposit in the bank for the next business morning or else he could use the money for change in his cash register.

I think Sonny was still in a state of shock.

We walked out to the Suburban and we started to climb into the truck when Sonny ran out and he gave me and Dick a bottle of his best tequila and he told Dick that he would be buying his draft beer from Dick in the future and from here on end.

So, not only was it an exciting night it was also profitable for Dick as well. He had just made his first foot hold into the vicious and cut throat Dallas alcoholic beverage markets.

Dick gave Sonny his business card and told Sonny to call him next week and he would be working on processing his order.

So, we went back to my house and let the dog out and then let him back in when he was finished.

Then we watched some TV while I worked on some new designs for the project.

We got into comfortable jammy type clothing and got ready to turn in for the night

because Dick and Candace were going back to Austin some time tomorrow.

Rhonda and Candace were discussing the Chili Cook off in October and Candace was offering to sponsor the Cook off and supply the beer and

Rhonda said that she would contact the Regional Director for the Leukemia Society on Monday and see if he was receptive to her idea.

Knowing him like we do, that shouldn't be a problem.

So, that evening, we decided that we would be competing again this year and that Dick

and Candace have agreed to be on our cook off team in Mid October.

They both said that they would be looking forward to being on my team.

And that it sounded like a whole lot of fun.

We both told them that it was a lot of work but, it was for a worthy cause.

And yes, it is a whole lot of fun.

We showed Dick and Candace some pictures that we took at the last cook off and I took Dick into the den and I showed him some of the posters from the last four cook off years that We had competed in.

I had taken each poster and framed them and hung them on my wall.

And I was saving some room on the wall for this year's poster.

These posters are full of Texas references and are very State specific.

Texans are very proud about their chili and Bar-b-que.

Seeing the posters hanging on my wall made them more determined to join our team and participate in the cook off with us.

I was excited because, after doing the cook off for four years, most of my friends have already participated and they try to make themselves scarce during the month of October

So they will duck. and hide whenever I come around.

It really is a whole lot of work setting up campsites and setting up the smoker.

And you have to truck in your own wood because there is nothing there at the cook off site to burn.

Everyone liked my retelling of "What happened last year at the last cook off".

And we all had a good laugh at my pain and expense and we joked a little more about some of the funnier things that can and do happen sometimes at these kind of events.

Well, after all the excitement of the week still fresh in our minds such as buying a new

(new to us that is) truck that looks really cool and hiring a dowser and then finding water on their land and places to drill for wells.

And then on top of all that. Going out to eat and thwarting an attempted armed robbery and saving the life of a good friend.

Being shot at not once but, twice, and then being arrested and interrogated by the police,

and watching the crime scene police take pictures and testimonies and statements and then finally we loaded ourselves down with all the eating and drinking and all the celebrating that we could .

After an evening like that we were all pretty worn out. And since it was almost 3:00AM when we got back home and after changing and getting comfortable it was now much closer to four AM than three AM.

We all had decided that it would be a good time to turn in.

We will have a busy day tomorrow and the morning was going to be here in less than three hours.

We needed to turn in and get some sleep.

Rhonda and I said good night to our guests made sure that they had everything they needed and then we headed back to our room.

and to our big, soft, warm, fluffy and comfortable bed and off to dream land with a rude awakening waiting for me in around three hours.

A blaring alarm clock broadcasting the arrival of morning.

"Good night Honey, I whispered into Rhonda's ear, and I kissed her forehead, because as soon as her head hit the pillow, it was lights out for her.

I wouldn't be far behind her.

The Hybrid Diaries
Chapter Twenty Two

Morning has broken. Today it is Friday, September 14th, 1990

Today is the day Dick and Candace have to go back to their home in Austin.

So, I woke up extra early and set up the coffee maker and I made some iced tea and put it back into the fridge so, it'd be cold when they all woke back up.

Then I pulled on some clothes and drove over to the doughnut house and picked up a dozen glazed and assorted and brought them back home for breakfast.

Just as I walked back inside the house the phone was ringing and I answered the phone and it was Bill the guy at Global Composites of Grand Prairie carbon fiber place and he said that they had finished up the tail section of their F-16 project for Lockheed

and they would be ready for me to bring my project over next week probably around Wednesday or Thursday at the latest.

I said "That would work out perfectly for me because we were going on a little vacation On Friday and we would need to have it at their shop as soon as possible to avert and potential conflicts in scheduling".

I also told Bill that "I would like to have a brain storming session on either Monday or Tuesday which ever would be more convenient to them."

And Bill said that he would call me later on today to "firm up a time to meet with me."

I also let them know that I would be bringing my engineer along with me to the meeting so we can discuss "The integration of our technology and their carbon fiber expertise."

So, I hung up with Bill and I called Toby to let him know that we were meeting with the carbon fiber body people on either Monday or Tuesday, and I asked if he could be there with me.

He said he had some comp time coming up and that "he would be happy to come with me when we decided on a time to meet with them."

I also mentioned that I would need a hand loading the project from the mini warehouse onto the back of the truck.

Toby said "Just let me know when and I will be there."

"Good," I said. "I'll call you when I am ready."

And I hung up the phone and I also called Mr. Ed Bell at Thunder Chicken Paint and Body. (The same body shop where I bough the truck for Dick and Candace)

"Hi Ed". I said into the mouthpiece. Of my phone receiver.

I identified myself and I asked him if he could do a custom paint job on the project for me.

He said that I would have to speak with "Max" and see how his schedule looked.

"Who is Max?" I asked.

"Max Powers" He's my graphic artist who did the actual work on the delivery truck.

"All the signage and the murals, that was Max's work, Based on your designs of course."

I told Ed, that "I would be across the street at the Global Composites of Grand Prairie

home offices on either Monday or Tuesday, and I could meet with Max then, once I am able to firm up the actual time of my meetings."

Ed said that would be fine with him, and he said to "call him when I knew something."

I told him I would as soon as I find out. And I hung up the phone.

Dick was standing behind me and I didn't hear him come in the room.

"Man, this is starting to get good!" Dick said

"Starting to???" I asked.

"I'm on the edge of my seat." I replied,

"Plus, I have to work on my anniversary next Friday, and I need to get everything ready for our trip."

"Anniversary?, wow, So, How long have you guys been married?" Dick asked

"This will be our fifth anniversary. And I want it to be special."

"So, I called Amtrak and made my reservations for a drawing room on "The Texas Eagle."

"We are going down to San Antonio, by train. Just like they used to do way back in the nineteen thirties. I've been shopping at the thrift store and I purchased a lot of old antique dresses and a couple of old suits, shirts, ties and a couple of hats for me."

" That way we can dress in period clothing and pretend we are going back in time to the late twenties or early thirties."

" I have been planning this trip since May and I had bought Rhonda a string of genuine pearls for her neck."

"And some cultured pearl earrings to give her on the day of our anniversary."

"I had contacted the Railroad Concierge and had them scheduled to bring a cake and a bottle of wine to our stateroom after Dinner, so we could celebrate in private."

Dick said "That sounded like a great idea to him."

And I was almost certain that he would invite himself and Candace,

but, he didn't make that suggestion (much to my relief)

I guess he also understands love and romance and the need for privacy when you are on a trip that has a certain amount of romantic intent to it. …

"I also booked our old room at the Emily Morgan Hotel (formally the old Medical Arts building)" in Downtown San Antonio..

"That is where we had our honeymoon way back in 1985."

"And I called a carriage company and I made a reservation for them to tell the concierge that they are supposed to pick us up in front of the Hotel for a romantic horse drawn carriage ride around the City after dark."

"Did I mention that the Emily Morgan is directly across the street from the Alamo?

"We can see down into the courtyard from our Hotel room window."

" Also the River walk is on the other side of the street and the Hyatt is down the block.

"And all the shops and restaurants are well within walking distance to where we will be staying."

"Plus, I have secured a rental car for when we get there so we won't be limited to the down town area."

"I made an appointment with the rectory at the catholic church and I planned to have the Priest at the Catholic Church downtown, renew our wedding vows and bless our wedding bands."

"Well, It sounds like you have everything well planned out." Dick said.

"You know what? I am going to be so stoked when you finally quit that thrift store job of yours. " Dick blurted out.

"Why?" I asked "Why are you going to be stoked?"

"Because, You are one hell of a loyal employee."

"And I can hardly wait for you to come on board at the brewery."

That guy you've been working for can't possibly pay you what you are worth."

"Plus, he doesn't respect you or your ideas."

" And he sure as heck doesn't appreciate you, or all the sacrifices that you had to make to keep his store up and running."

"I've been watching you for quite a while now."

"And you have been nothing but honest and hard working,"

"You are a very good boss and you are very professional with your employees."

" I can tell that they really like you and would probably do anything you ask them to."

" You can't buy that kind of loyalty with intimidation."

"Besides, when the restaurant was being robbed you stepped up and stopped that robber from shooting the restaurant manager."

" And you took his picture even when he started shooting at you."

"Not once but twice! I could not believe that!"

"If I didn't, then that guy might have shot Sonny." I said.

" I couldn't just stand there and let that happen." I sighed.

"I wasn't trying to be heroic I was trying to distract the thief from shooting Sonny."

" And I figured that he would be less inclined to shoot the Manager if there were Witnesses right outside the door watching everything transpire."

"You didn't think anything about your own personal safety,"

"You made sure the wives were out of harm's way and then you acted."

That shows a considerable amount of situational awareness."

" You can't teach that to people."

" You have to be born with instincts like that. "

"I can really respect anyone who puts the safety of others above them own."

"That is the kind of person I want working for me." Dick said

I agreed with Dick, and turned my head away to try and hide my embarrassment.

"By the way, "

Have you given any thought to getting a real estate agent to find you a house in Austin?"

" From here to Austin is a pretty long commute." Dick added.

I told him "I would look into it as soon as we got the race out of the way."

Dick said that would be fine and then he asked me "When I would be turning in my two week notice at the thrift store?"

I told Dick that "I was originally planning on doing just that around Halloween but, time was getting shorter and I needed to start focusing on the project and being available for the body people. "

"And then, I would have to be ready to assemble the Slingshot and begin shake down testing to work all the bugs out of the system before we load it into a container and have the container loaded onto the ship."

"But, my first priority will be calling my boss to see if, he remembered that I had been asking for this time off since May," and let him know that "

"I will be off for the three days and two nights that we will be spending time in San Antonio. "

So, I called Mr. Jerkoff, and left a voice message on his machine.

I know, he'll want to call me at the store to make sure that I opened on time and that I am there, and I am hard at work.

So, he can yell at me over the phone.

I prefer the phone as opposed to him actually being there yelling at me in person.

It is belittling enough having to listen to him prattering on and on about how I was "robbing him, picking his pocket and stealing him blind."

It's really hard to swallow, and a huge blow to my ego and self esteem.

It's character assassination and he's besmirching me on a personal and professional level

But, it's even worse when he's standing right in front of you and spit flies from his mouth when he's yelling and it lands on your face.

It undermines my authority at the store when he tries to redress me in front of all my employees.

Mr Jerkoff, likes to do crap like that because it makes him feel big and he shows my employees who is really running the store.

They all hate him. I don't blame them at all in the least.

He's a detestable old fart.

If he was around all the time, then I would have quit months ago.

That's why I would prefer him yelling at me over the phone, at least I stay a little drier.

And he can't redress me in from of my workers and I save a little dignity.

In all fairness, I had told Mr. Jerkoff that I wanted this time off back in May when I first started planning this.

And I had periodically reminded him that my anniversary was coming up and he would say, "yeah, right, and OK."

I had been busy planning this trip and I had already done most of the work and now,

I will have to get all of my other ducks in a row,"

"And then I'll have to drop off the dog at the In-Laws house next Thursday evening

Before We can head to Union Station in downtown Dallas and catch our train early on Friday morning.

Then we can board our "magic carpet" and be whisked away from all of the crap at home

And we will try to relax and sit back and enjoy our ride as we head out to our anniversary trip to San Antonio.

We get all of our meals served in the Dining car.

That's breakfast, lunch, and dinner.

(usually prime rib.) Plus we can drink in the Bar car and look at the great outdoors from the comfort of the sky car.

There is even a smoking car if you are so inclined.

(I know I am sharing more personal stuff that I did not want to divulge, but, after much thought I decided that it was a necessary part of the story,

So, I'll try to sum it all up here in this chapter and then we can get back on target with the project. Unless you like this sort of stuff.

If so, then please feel free to continue on.)

I personally enjoy all the thinking and planning and handling all the minute details and getting everything together for an adventure like this.

I have always been pretty good with logistics and now,

I have more on my plate than I ever had before.

I think I will have to let something go.

I do not understand why I feel so bittersweet about leaving my job at the thrift store.

I guess it is because, I liked being in charge of the store and running the business.

Plus, my employees like me and they respect my judgment and they listen to what I have to say to them.

I know I will miss each and every one of them.

OK, I got to the store just in time to answer the phone and listen to my boss scream about my taking too much time off.

And If I wanted to go somewhere then I was supposed to give him at least two weeks notice before I can ask for any time off.

I reminded him that I had already told him all about this way back in May, but, he conveniently did not remember anything about that.

I was really getting tired of all of his BS.

And, now it was finally starting to wear on me.

After all, I do my job, I work six days out of seven, and even when I am off duty,

I am always on call until the store closes for the night.

That was a duty that was drummed into my head one night.

We needed groceries, so Rhonda and I went to the store after I got off work, and when we got back home there were several messages from the store and a couple from the boss.

One of my employees had locked the keys in the safe, and they could not close the store without them.

The employee called the other store in Garland, and Dufus, (the store manager),

decided that this would be a great opportunity to show off to my boss and make me look bad at the same time.

So, he thundered over to my store from his house down the road

And he kicked my office door in to try and find the extra set of keys.

When I got back to my store with my keys Dufus started trying to bitchtate all the errors of my ways.

And he tried to make it out like I was the one at fault.

And I was the one who locked the keys in the safe.

And that I was inconveniencing everyone.

By not being where anyone could reach me in a moment of crisis.

Then he tried to cover his cop like action by saying that

" He was looking for my extra set of keys"

And then he had the nerve to say.

"Now fix that door"

I told him to go "frack himself and die" and

"Get the hell out of my store"

I never liked Dufus, he was really untrustworthy.

Very weasel like, he even has a weasel like face.

In fact, he gives weasels a bad name.

One payday after I had just started working for Mr. Butthead.

I had just cashed my check and I had stashed the money in my briefcase that I carry.

And I locked it in Dufus's office.

Then when I went back into the office I saw Dufus leave to "Go and run an errand."

So, I went into the office to make a phone call and check on my stuff.

My briefcase was opened and my money was gone nothing else was taken.

So, when I told Dufus about it all he said was,

"well, maybe you shouldn't leave your stuff in my office.".

I knew the son of a Bitch took it because no one other than him knew where my money was or where I had stashed my briefcase.

He was a classic asshole, and The more I think about all the crap and stuff I had to put up with them just to bring home a measley paycheck,

The more I thought about ways that I could leave, and burn them for their abuse.

But, I guess I didn't have the balls to just up and quit on them.

That was then.

But, Like Good, Ol Bob Dylan used to say, "The times they are a changing"

I was starting to re-evaluate my position and my worth to this organization.

And I was coming up short every time.

My paychecks always leave me at least two hundred dollars short every week.

And I was just barely holding my own and still falling short with no end in sight.

It wasn't just the lack of money it was the way that they were treating me.

No job is worth that kind of abuse.

But, I was afraid of letting Rhonda down.

So, I swallowed my pride and dealt with it.

But, that was before I accepted Dick's position at the brewery.

And since I became his VP in charge of Promotions…

Money has been the last thing I think about.

In fact, since I have been working with Dick,

I have had zero financial worries.

And I was able to bank my thrift store paycheck and most of my salary from Dick.

I have used my expense account sparingly and only for the project related expenses and I have been reimbursed for the items that I had paid out of my pocket.

I was now in a position where I actually had some savings that can hold me and Rhonda for almost six months.

In the event that everything fell through.

I have never been in that position before.

And believe me, I was beginning to enjoy it.

I like having paid all of my bills in full and on their due date (for a change).

And I especially loved being able to take a little better care of Rhonda.

She's my life, and she's always been there for me.

And she has pulled the oxcart when I was down and out and out of funds.,

That is why I wanted to do something special for her and for our anniversary.

So, I wanted to take her back to the scene of the crime so to speak, and renew our vows and just enjoy ourselves.

I had already paid for the trip in advance out of my own pocket (this was prior to me excepting my position with Dick)

I believe that I had accounted for everything we would need for the trip.

I tried to plan it out down to the smallest of details.

This would not be the first time that Rhonda had been on a train.

Once when she was a little girl.

Her grandmother took her on a trip where they rode the train.

She recounted the story to me.

The thing she remembered most was she was little, and she had gotten hungry on the train so, the porter brought her a ice cold bottle of Coca Cola and a Hershey's chocolate Bar that the train kept in the refrigerator.

She remembered how cold the Hershey bar was and how cold and refreshing the coke tasted to her. (I personally prefer Dr. Pepper,)

So, yeah, for our trip,

I had made arrangements for the porter to bring her a iced cold bottle of coke and a cold Hershey bar when we get on the train.

(If you are interested in doing something like this check with Amtrak and their concierge Service, they may be able to hook you up.)

But, just in case, We were bringing an ice chest full of beer and cokes in bottles and other types of drinks (Like Big Red) and I had also carefully placed some extra Hershey bars in a zip lock bag so they would not get wet in the ice chest.

We were all ready for our trip to begin.

So, as far as I know, everything is on track for our trip the only X factor will be Mr.Butthead.

Well, I got to the store on time and We opened up on time (as usual) and of course the phone rang, and it was my boss. Mr. Butthead.

He didn't say "Hello" or "Hi" he went straight into bitch mode.

"What do you mean you want the weekend off?"

"No, I said , I want Friday Saturday and Sunday and Monday off."

"I told you all about this in May."

" You said OK, so, I do not see what the problem is".

"Well, you didn't tell me in June." He snorted.

"Yes, I did. I told you in June on the 17th,

And I also mentioned it in July and again in August..

I marked it on my calendar every time I tried to remind you of it."

" And you would just say "OK" so I thought you were listening." I replied,

I was trying my best to sound professional and not lose my temper with him.

But, he was doing his best to try and force an issue.

So, I decided to be calm and cool and if he was going to be a jerk,

Then I was more than ready to give him my two week notice

and head out the door.

"Well, I don't remember you telling me anything about it"

" And since I did not know about this then I cannot approve your request for time off. Sorry."

Well, that was what I had thought he would try to pull off.

So, I was going to ignore him.

"Well, Like I was saying, I am going on my vacation this Friday and I will not be back to work until Tuesday of next week."

"I'm just giving you a heads up."

" Like I had done for the last few months."

" You knew this was coming and now is the time for my vacation." I said.

(Now, much to my surprise, I kept an cool head.

And I acted professionally, and I was not disrespectful). But, Mr.

Butthead kept yelling into the phone.

"Well, if you don't show up for work on Friday then consider yourself fired."

He shouted loudly into the phone even more forcefully than before..

"Alright then, I said "I wasn't going to do this."

"But I am giving you my two weeks' notice." I said.

"No, No, NO, You are not quitting on me."

" You can take your two weeks and cram them". He screamed.

"OK, then, Your choice. I'll leave my keys in the store for you to pick up when you get here."

"Fine then, you can just say that Friday will be your last day,"

"And don't worry about the store, you will be easily replaced."

Then I hung up the phone.

I was feeling so relieved. But, I was also pissed off at the same time.

but, I could not have expected to be treated any better than I was, because, that was what I had become accustomed to.

I started asking myself, why did I put up with all this crap for so long?

Probably, because I was afraid.

Afraid to be out of work, and losing our rent house,

that we didn't even own.

Afraid to take a chance, and maybe move somewhere else with my wife and \ start life anew.

I sold my soul to this asshole for twenty thousand a year.

It sounded like a lot at the time until you try to live on that small of a salary.

Then you can see just how far short the money will go.

Trust me, not very far.

Now, suddenly, I was starting to lose the weight that I have been carrying for over a year,

and at the same time I was starting to feel sad for the future of my employees.

So, I called a meeting of my staff and told them the news.

There was a gasp followed by stunned silence after I made the announcement.

And when I dismissed them after the meeting.

They each came by my office to tell me how much they enjoyed working with me.

And the store and how much they each will miss me.

because, they all said that they would quit if Mr. Butthead came up to run the store instead of me.

They all knew Mr. Butthead, and no one would like to have him breathing down their necks waiting for one of them to make a mistake so he could jump all over them and make himself feel better at their expense.

That was MY job. And I had enough of that position.

I thanked them all for all of their hard work and for everything they have done for me.

And I let them know that I would be alright, and that I had something lined up "just in case"

They smiled when I asked them if they would be OK.

They each said that someone was opening a new thrift store down the street.

And that they would try to get a job with them.

Now, I knew about the new thrift store because I drive by the building every day on my way to the bank to make my deposits.

And I have been watching their progress with a knot in my stomach.

The area where I had my store was already heavy with competition and I had tried to have some pretty decent sales to try and keep my numbers up.

But, I knew that would be much harder to do when the new store opens up..

So, I guess it's all a matter of timing.

And I have finally accepted that now is my time to go.

I have done everything I could to keep the business going and to provide for my employees and my boss.

And to try to leave things better than they were when I first arrived.

I had a suspicion that this would be how it ends.

And I was right. I was seriously wanting to have a amicable parting of the ways,

"Just in case" but, I really didn't expect that to be the way it goes.

All I care about now is finishing up my day and going back home and letting Rhonda know,

Then I can call Dick and let him know (what Dick would call " the Good News")

that I have left the thrift store and I was now a full time employee of the New Paluxy Brewery.

Well, the day passed by quietly, except for a call from Toby.

letting me know that he had cleared a couple of days off so we can meet with the carbon fiber body people.

I let him know that I would be by to pick him up next Tuesday (I thought about doing it

on Monday but, I wanted to have it on Tuesday to allow time for everyone to clear their calendars and get ready for the meeting).

And together we can go by the warehouse and his garage on Monday and pick up all the parts and the frame and take it out to Grand Prairie.

I also called the solar panel people and spoke with Rich again.

He said that they would be happy to meet with me.

I had decided that we would have the meeting at the carbon fiber place on Tuesday.

So, I had Toby (electrical parts and motors) and Rich (the solar panel guy)

Toby said that he had already talked to the battery people and they will also be in attendance.

And of Course Bill and the carbon fiber body people and last but not least I had Ed Bell, and Max Powers, the paint and body guys.

All were clearing their calendars for our meeting on Tuesday.

With that out of the way, I finished up my day early, and I left Diane in charge for the evening, and I headed home.

I pulled into the driveway and I parked behind the brewery truck and I saw that Dick and Candace were still at my home.

Because the Suburban was parking in the street in front of my house.

So, I walked up to the door and opened the lock and I let myself in.

"Welcome Home!, Rhonda shouted as I walk in through the door.

"Wow, you're home early" Rhonda smiled at me as she walked up to me and threw her arms around my neck.

I hugged her back and I gave her a kiss, and then I saw Dick and Candace.

"Hi Guys!" I shouted to the Wigglers, they smiled and welcomed me back.

So, I told them that I was hungry and I wanted to get something to eat.

Plus, I had some news that I wanted to share with everyone.

"No Problem" said Dick, "Candace and I could go for something good.

What are we celebrating?"

"My new position as the VP in charge of Promotions" I said.

"What? do you mean you turned in your two weeks notice? Asked Candace

"Well, I tried to, but, Mr. Butthead to me to cram it,

So, I guess, I am no longer an employee of the Vietnam Veterans Foundation anymore."

I am an employee of the New Paluxy Brewery if you guys will still have me."

I sighed

Dick tried to contain his joy but, he's lousy at poker too.

"So, where do you want to go for Dinner?" Rhonda and Candace inquired.

"I would like to have Mexican food." Was my answer.

So we piled into the Suburban and I got shotgun and Dick drove and I gave him directions to the Mexican Cafeteria.

I love this place, they have their food served Cafeteria style and it's all you can eat.

So, I usually load up a plate or two of tacos rice and beans,

with a beef burrito no sauce no onions, beef enchiladas and tortillas with

butter and lots of sopapias and honey.

Rhonda likes to order the flautas and taquitos, with rice and beans

The Wigglers also loaded up their plates.

We took our trays loaded with food and we went to go and sit down on some

red velvet wooden high back chairs around a wrought iron and glass top table.

We sat , and the waitress asked us what we wanted to drink, so Rhonda and

Candace ordered unsweetened tea.

Dick and I had already sweetened tea.

I cannot remember how many times we refilled our trays but, it was at least

several times.

And I got too stuffed to jump.

While we were eating I told everyone about the way I was let go.

And when I was finished with my tale Dick said

"Good riddance, and Welcome aboard!"

Then Dick said, I want to give you a toast but, I guess we'll have to do it with

tea.

"May fortune favor us with a strong wind at our backs."

"Here, Here" I responded.

"Salute!" the girls all cheered.

Well, since I am now going to leave the "safety and security" of a crappy

position for the uncertain future of a new career.

I was filled with excitement and trepidation.

Mostly fear. I know, that I will be alright, and that somehow we will continue

to survive.

But, Other than that, I had no guarantees that anything will work out.

Now, First and foremost, let me say that life never has any guarantees of anything ever working out for anybody.

And anytime we believe that we have "everything under control" that is just imaginary.

We really have absolutely no control over our lives and circumstances.

We just react to our conditions and try to do our best to make things better for ourselves.

But, in reality, we have no actual control over our lives.

So, there are really no guarantees for anything.

I have always tried to make things seem to go as smoothly as possible and try to take the bumps in stride.

I really didn't know why I was worried about leaving the thrift store.

It's just that I had been there for almost two years now, and I had gotten settled into my position.

And I didn't want to change. I had gotten so wrapped up in the position that I had forgotten that other people depend upon me,

and that the job was just that a job.

It would never turn out to be a career.

I guess I had gotten lazy, and fearful about my life.

It was like I was living with Blinders on my head that were keeping me from seeing all of the opportunities that were still out there.

Opportunities that I would have never had a chance to experience had I not spoken up.

So, I feel that now is my chance to make a difference in my life and in Rhonda's.

And to finally get the opportunity to show the world that I am a creative person with ideas and inventions that I can now share with the rest of civilization.

This will be my big chance to make a difference.

So, I will look back at my experience in retail management with bittersweet memories of how I was in charge, and how I was as a manager.

And how people can treat you like crap no matter what you have done for them or even how recently.

Well. Because of me coming home and making my big career announcement.

Dick and Candace did not leave as originally planned.

So, they will probably leave tomorrow.

It doesn't matter, We really enjoy their company and They are really nice people.

We went to bed late and I still had to get up early tomorrow.

So, as much as I tried, I couldn't sleep very well.

I kept thinking that this was not the way I wanted to leave the Vietnam Veteran's Foundation.

I wanted to leave on a more positive note.

After a long while of trying to relax,

I drifted off to a fitful sleep.

I woke up at my usual time that I would normally wake up on if I had to go to work on a Saturday.

So, I got dressed and I drove up to the thrift store.

I had opened the door and locked it behind me (the employees had not shown up yet because it was still too early.)

I went into my office and finished up my books for the week

when I heard the back door being opened with a key.

My boss Mr. Butthead had shown up.

So, He walked in and I was sure that he was surprised to see me there.

The first words out of his mouth were "Why are you here?"

I wanted to finish up my books for the week and drop off my keys.

And, I wanted to make sure that the store was opened up on time .

"Why are you here?" I asked

"I told you that I wasn't leaving on my vacation until next Friday"

"Well, I thought you quit, and there wouldn't be anyone here to open the store."

" So, I had to come up to Dallas, and open the store". He growled at me.

I could tell he was more angry at being inconvenienced due to my untimely leaving.

He always was on the dramatic side.

"Look", I said," I was going to turn in my notice to you anyway,"

"but, your crappy attitude made me want to leave your employ that much more."

" So, Please, let me say that I am leaving the company,"

"And I have found another job with another company."

"Well, I went by your house on my way here, and I saw some Company truck in your driveway."

" So, I guess you are going to be a truck driver for some brewery?" he asked

"Yes, that is exactly what I was going to be doing."

I'm going to be a driver for a small local brewery.

They aren't paying me as much, but, it is a good Company to work for."

(now, I did not want to tell him the truth about my new position or my project.

It seemed to make him feel almost "Happy" about me driving a truck and working for less money than he was paying me.

It made him feel like I was taking a huge step backwards,

and that if not for him I probably would have starved to death long ago.

That is the exact opposite of the truth.

but, it made him happier than the truth would otherwise have.

So, It made me feel better as well,

because, I had my opportunity to meet with him one last time,

One on one and finish up my loose ends.

And turn the keys back over to him.

So, I spent the rest of the morning gathering up my personal stuff out of my office and Loading it up in the Crap mobile.

He really surprised me when he said that he was sorry for treating me the way he did,

He even said that if driving the truck didn't work out for me…

Then I could come back and work for him again.

I told him thank you, and I walked out of the store.

He followed me outside to the car and I had one last thing to share with him.

I said

"Hey, did you know that someone is opening up another thrift store right down the street?"

"No, I did not know that". He replied

"That was just an FYI." I said as I opened the car door and I climbed into my ratty car

seat and got ready to start the car.

He came right up to the car door and he shook my hand and wished me good luck.

And I drove out of the parking lot as the Manager for the last time.

I headed home, feeling bittersweet again.

I drove into the drive way and I got out and walked up to the front porch.

The mailman had just let and there was mail in the mail box.

Plus my paycheck from the brewery.

Suddenly I didn't feel quite so gloomy.

I walked into the house and I kissed Rhonda, and said

"Hi" to the Wigglers, and we went out for lunch.

This time we when to "Ribbies" and I had (yeah, you probably guessed it)

Barbeque ribs (Rhonda will not eat ribs because they look like what they are, and to her that's a little too much reality for her to stomach.

That was OK, though, because the girls ordered grilled chicken breasts for their dinner

and Dick and I ate like the carnivores that we are.

Tearing hunks of meat right off the bone.

Then we washed it down with a couple of beers and I bought us all a slice of cheese cake.

Soon, we were almost immovable. Stuffed to the gills with food and drinks.

I put the entire lunch on my expense account card, and we went out and walked around the mall looking at stuff. trying to walk some of that lunch off.

And I was thinking that I may need to buy some more luggage before we leave for San Antonio.

Dick and Candace told us that this time they really do have to go back home and start getting everything ready.

Candace said that she would handle all the travel arrangements

And Dick will see if there is something that he can do to help .

I told them "That would be great.."

So, we headed back to my house and we spent the rest of

the evening working on our plans and then watching Saturday Night Live on Channel 5

We drank and laughed, and then we turned in.

The Hybrid Diaries
Chapter Twenty Three

Wow, My first Monday morning without a job.

But, I have so much to do, I hardly missed not going to the thrift store.

I got up made coffee for Rhonda and I drank a big glass of tea and threw some work clothes on.

Then I took the keys for the stake bed truck off the key rack and I walked out the front door and I climbed into the cab and I slide myself behind the wheel.

And I turned the key and the beast fired right up.

I could feel the vibration of three hundred horses rumbling under the hood.

So, I gently eased the car out of park and I put it in reverse and backed up out of my driveway and out into the road.

I took the stake bed truck over to the mini warehouse and I started loading the slingshot into the bed of the truck and I decided that

"Maybe we should invest in a trailer as well to keep the stuff out of the open.

This way we could lock up the project out of the elements."

So, after I got it all loaded up, I went back home and I called the carbon fiber people.

They said that they cleared out the conference room and that they were ready for my meeting tomorrow at noon, at their Company.

I let them know that I was looking forward to it. And I hung up the phone.

Then I called Toby, and he said that the battery technician would also be with us,

so, he can meet the other sponsors and show them what he has to offer us in the way of battery technology and inverters and speed controls.

I let Toby know that I would be meeting him where ever he wanted to meet.

I could pick him up at the Shop or else I could meet him and the battery guy at his house.

Toby said that he was going to be "Sick" tomorrow and that I could come by his house and pick them up.

So, I said that sounded fine to me, and I told him I would be there around 11:30 in the morning to pick them up.

And Toby said he would be ready.

I hung up the phone and I called over to Rhonda.

"Honey? We need to get our passports in order."

Rhonda said, "Well, I am NOT going to get my picture made until I get my hair done and I look presentable.

I said, then, take the car and go get yourself all painted up and then we can go".

Rhonda said that she'd be ready after she takes a shower.

I said that This would be acceptable.

And she went off to get ready for her shower.

After about an hour and a half.

Rhonda was ready and we got in the car and drove to her beauty parlor.

About an hour and a half later,

Rhonda emerged looking very pretty.

Then we got back in the car and we drove downtown to the Passport offices and filled out our paperwork and brought in several forms of identification. We paid our fees and they told us where we needed to go to get our pictures made.

So, we took the info with us and headed out to the photo place for our pictures.

They gave us an envelope with a return address on it and we were supposed to take the pictures and place them along with our paperwork in the return addressed envelope.

So, we got our pictures made and we mailed them off with our paperwork and we were done.

We went back home and Rhonda cooked Dinner for me while I drew some more plans for the meeting tomorrow morning.

The Hybrid Diaries
Chapter Twenty Four
The First Big Meeting

OK, welcome to Tuesday the big day for our first team meeting.

OK, now, I was thinking that the meeting would be me showing the carbon fiber body

people what I had built and what I would like to try and do with streamlining the body of the vehicle and making it more aerodynamic.

And they would go "Wow, what an incredible design" and

" Oh, please, share with us your wisdom and genius."

And then the butt kissing would begin. HA!.

Reality has a way of kicking the crap out of you when you least expect it.

And fate loves to throw curve balls.

I dreamed that they would hang on my every word and that they would take orders from me.

And I would be the one that they need to discuss all of the plans and designs,

I would draw out the designs and they would manufacture the actual parts based upon my designs..

I would be the Designer and builder and they would follow my designs to the letter.

Boy, was I in for a shock.

I went over to Toby's house and picked him and Earl from All Nationwide Batteries up and we all climbed into the truck and headed over to Grand Prairie.

When we got out to Global Composites of Grand Prairie,

We parked up front and we walked into the reception area and we let the receptionist

know that we were there for a meeting with Bill, and his design team.

She said that "they were expecting you and to have a seat.

And Bill would be there to meet with you and a few minutes."

We said, "Thank you" and had a seat waiting for Bill to come

And take us back to the conference room.

Instead, one of the technicians came out.

It was Christopher.

He asked us to follow him back

So, we all followed Christopher back,

not to the conference room but, to the production floor.

There was a tarp in the middle of the room covering something.

And the rest of the design team was just standing behind the covered lump on the production floor.

They had set up several chairs on the opposite side of the lump,

The other members of my team had already arrived.

They were just waiting on Me and Toby and Earl.

So, we were all assembled for the first time together.

Bill said, "Everyone, Please, have a seat. So we all complied. And sat down.

Bill cleared his throat and began to speak.

"Hello, My name is Bill D. Cabumba and I am the owner of Global Composites of Grand Prairie, and we have been working on the "Slingshot project almost since the day when Mr. Barrientos came into our offices.

And proposed his racer project with us."

We recognized the technical issues a project like this would face and we did a design survey based on the plans and designs from Mr. Barrientos, and from Mr. Caliente.

And we took the dimensions from the plans

And we had handed the plans over to the design crew headed by Dylan Andrews.

Now, I would like everyone to stand up and introduce yourselves.

Starting with My chief Designer.

Dylan stood up and smiled and announced to all present

"Hello, My name is Dylan Andrews and I am the head designer and fabricator for Global Composites.

He sat down and the next person stood up. "Hello, I am Barney Nobles and I am the chief fabricator and autoclave operator for Global Composites of Grand Prairie.

Barney sat down and the next one stood up.

"Hello, my name is Christopher" and he sat down.

We smiled. As the next one stood up.

"Hi, I am Sam, and I work in Carbon fiber and Plexiglas.

And I also run the autoclave."

As Sam sat down, someone next to Sam stood up and said

"Hello, I am Rich Richardson and I work for Tri Solar Confed. Solar Power Panel Corporation"

"We design and manufacture commercial and residential solar panels for home and industry."

(How the heck did they know about Rich and "Tri Solar Confed. Solar Power Panel Corporation?"

I didn't remember giving them any info concerning the panel suppliers that I was looking into.

I wasn't going to contact Rich until I had something to show them.

because, I wanted them to think of me as a serious contender for the energy challenge,

not just some kook off the street talking about flying saucers or bigfoot.

I must have left their number on some of my plans.

How else would they know.?"

Either way, I was delighted to see him there.

Rich Sat down and Toby stood up.

Hello, my name is Toby Caliente, and I work for Unidynamic Motors of Dallas, Texas

Where I am the head motor designer.

Toby sat down and Earl stood up.

"Hi, my name is Earl and I work for All Nationwide Batteries Inc.

"Where I am the chief designer for product development and research."

Earl sat down and it was my turn.

"I am sure you all know me, My name is Andy Barrientos, and I am the designer and inventor of this project.

And I would personally like to thank Bill Cabumba and his staff for allowing us the opportunity to present and develop my ideas into a viable Human/solar powered hybrid electric vehicle.

And I would like to thank everyone who has shown an interest in the project and my sincere thanks go out to all of you.

I sat down and Bill spoke.

"Alright, now that being done, it's time for us to show each other what each of us have been working on. And assemble the pieces into a cohesive viable racing machine."

" I am going to let Dylan begin by presenting our first "proof of concept" body design.

Dylan?" We all gave Dylan a big round of applause then Dylan stood up and he began his debriefing.

"Gentlemen. He have been very busy around here."

" We finished up the tail section of the F-16 that we were modifying for the Air Force early, and we immediately started working on a prototype body based on the blueprints and designs that were submitted to us by Mr. Barrientos.

We used his original measurements and we did a substantial amount of redesign due to the substitution of carbon fiber as opposed to tube steel. Or aluminum.

Please, allow us to show you our first prototype…

(OK, now here is where reality comes crashing into the dream and it leaves you standing there with a stupid expression. Like a idiot.

While everyone around you is laughing at your expense.)

There I was standing next to a tarp, with my invention still out in the truck.

(And all the stuff that I had done was getting ready to fly out the window… Wait for it…)

"One… two… three… UP!" Shouted Dylan as Sam and Christopher pulled the tarp away from what it was hiding and protecting.

There it was. Their version of my idea based upon my designs.

And their knowhow.

It looked like the front half of an F-16 with a canard,

And the rest looked like a formula one race car.

It was a fantastic effort.

And one I could have never pulled off by myself in a hundred years.

We all let out this huge gasp.

At last I could see everything actually coming together.

The body was streamlined and I could not believe this but, there appeared to be four solar panels along the body mounted on four 3 X 8 carbon fiber panels that were hinged at the centerline of the body.

It looked a lot like my original plans but, It was more curved and much more streamlined.

The body was riding on bicycle wheels because they did not have access to Toby's electric wheels.

When I first designed the Slingshot, I intentionally used straight lines with very little curvatures

because I did not have the tools to design the body the way I wanted.

So, when I laid my eyes upon their prototype I almost teared up.

OK, it wasn't almost.

I just stared at the prototype and I had to dismiss myself.

 so I could wipe the tears from my cheeks and my eyes.

Then I went back and everyone was surrounding the prototype and taking pictures and

measurements and discussing the how to and why things are designed like they were.

And when I came back everyone got quiet.

I apologized for leaving when I did and I explained that I was completely overwhelmed by the prototype.

And I was amazed by everything the design team had done in the short amount of time that had transpired.

Dylan tore right into all the changes and the differences from the original designs and all

the "Why stuff" started pouring out of him like someone who just let all of the air out of a balloon by letting go of the nozzle.

And watching it fly around the room.

That is how Dylan tore into their presentation.

He cleared his throat and began.

"We started with changing the steering to a more conventional upper and lower A arm design with a control rod based steering gear as opposed to a rack and pinion system.

"The car is very light by car standards and it doesn't need a power assist to augment the linkage".

"It does have shocks but they are very lightweight and very efficient.."

Then we completely rounded off the body and made it very aerodynamic."

" With an extremely low drag coefficient.

" And then we saw the notes and the phone number for the solar panel company on your original plans and we assumed that you wanted us to call them so we called them instead,

and gave them the specifics about the design and shape. "

"The panels automatically fold out when the car is parked and they retract against the body when the car is in motion."

" Plus the panels have 180 degrees of motion across the spine of the car and they can also move forward forty five degrees."

"Yeah, You can use them like airbrakes! " Laughed Christopher.

"Yes," said Dylan who is obviously the more serious of the designers.

"Like I was saying,"

" We placed a canard on the front of the car to improve down force and to keep you on the road."

" We applied solar panels to the canards to help with energy production.

It's OK if they get broken off in an accident.

"They are designed to break away without damaging the body.

So, Don't worry if you accidently break one off,

"We have made plenty of replacement panels.

Also, we redesigned the cockpit to a more functional design with a heads up display that we had borrowed from another aircraft project.

We also enlarged the interior dimensions.

So, you can accommodate an additional rider in tandem or else the equivalent weight of an additional rider,

Which is in line with the copy of the rules that you have supplied to us."

"And then we constructed a carbon fiber safety cage that surrounds and protects the driver from impacts and accidents."

"We have an unconventional control system in the cockpit for steering and systems management.

With a dual joystick control system"

" The stick on the left controls the Steering and braking, while the stick on the right hand side controls steering and the energy management system and the accelerator.

The dash is split in the middle to allow for the pedals,

crank and to provide sufficient legroom for a rider. "

"The rider's pedals are for cranking only."

"There are no controls provided on the pedals."

"As for exterior lighting, "

"There are standard light housings in all the normal places for mounting Headlights taillights and a brake light for stopping."

" We have even planned to put in windshield wipers. Just in case."

"The cockpit is protected by a one half scale wind screen.

Which was modeled from an F-16C, (This is the trainer version of the F-16 not the standard fighter version."

It comes with an extended cockpit canopy to accommodate the instructor in the back seat.)

The canopy has a gold colored light reflective anti glare coating and two rear view

mirrors mounted on either side of the cockpit canopy" Dylan concluded.

THEN, They all turned to me, and that is when all the questions started coming in.

"How do you do this?" and "why is that where it is?".

While I was gazing at the proto type,

Dylan walked over and opened up the side door on the mock up to show us where everything is supposed to go.

And Earl was taking measurements and pictures to see how much space was left for battery accommodations

Then Dylan spoke up.

"If you have any questions about the design or need measurements,"

" Christopher has a stack of copies of the plans for your personal use."

" And we ask you to PLEASE, keep all of this information to yourselves."

Dylan added.

So, we started fleshing out the design and Toby excused himself and I followed him outside to the truck.

"Do you want me to bring in the wheels and get it ready to mount onto the prototype?"asked Toby.

I said, that "Those are probably the only thing from my prototype that they can actually use."

Now, let me be the first to admit that I NEVER EVER intended to race my original

prototype in the energy challenge.

I only made it as a proof of concept vehicle to try and obtain some kind of sponsorship.

I even mentioned this in the first chapters,

But, If I dared to take it off the truck and place it next to the magnificent looking carbon fiber mock up, it would look like a Conestoga wagon next to a jet fighter.

Everything I had done on my own looks like it was thrown together (which it was)

And I was feeling extremely self conscious.

Worrying if I even deserved to be in the same room with these creative individuals.

But, I sucked up all the self doubt and I decided,

"Heck" lets' just bring it all in."

" We have to unload the truck anyway." I said to Toby.

So, I swallowed hard, and I asked for volunteers to help off load all of my stuff from the back of the truck.

And The guys jumped at the chance and they followed us out to the truck and they started yelling, "Wow cool truck!" They walked all around it and checked out all the cool additions and the restoration.

Then they started taking the boxes and the frame and the other components that we had

been assembling and brought them in and set them next to the prototype.

And they began unpacking the parts and started laying them out on a big table in the back of the room.

Toby started to talk to the group.

"These are bi-polar dual induction reversible electric motors.

And they---- "

OK, As per my promise to Toby, I cannot divulge any thing about the wheels, sorry,

Suffice it to say "They work, and they work very well. And Yes, the other guys were extremely impressed with Toby's work,

Toby told them that all of his work was based on my original designs.

(That made me feel proud, and a little better in the self esteem department.)

But, I CAN tell you what they are designed to do.

They work off electricity.

They are reversible, so they can stop the vehicle by reversing the electric field.

And also they can go backwards. So, technically, it has a reverse gear.

(We did add disc brakes as a back up.

The brake control is on the right hand side Directional controller in the cockpit.)

The crank turns another generator which supplies power to another high output high RPM motor that spins the flywheel much faster than the original crank and chain design that I had come up with.

The flywheel which in turn produces electricity for the batteries.

And the rear wheel is also an electric motor which provides motive power for the project

And also serves as a generator when it is not being used to provide momentum for the front wheels to start generating electricity and vice versa."

It can be used to stop the vehicle in an emergency. Should one arise.

The solar cells are large enough and are able to charge all of the batteries in about an hour and a half.

I looked over at Toby and I put my hand in front of my mouth and I leaned over and whispered into his ear

"These are bi-polar dual induction reversible electric motors???" I said to Toby?

"What the heck are Bi-Polar electric motors? "

They were throwing tech terms around and I didn't want to seem ignorant, so, I just made it up."

Sounded good though right?" Toby whispered back

"Yeah Toby, We'll just let the name stick." I replied

Paper and pens and photos were flying all over the room and everyone was talking either to other designers or else talking to themselves and calculating measurement s on calculators and on paper, and coming up with ideas.

It was several hours later when we decided to order pizza, so, I took a few minutes break

to call my wife and let her know what we were doing and it looked like this could run late.

Rhonda said "No problem" and She was just happy to hear from me.

So, I hung up the phone and we got back to pulling and poking on the body and talking about how all of the mechanical designs have all been replaced by electric equivalents

It's just like what Toby was telling me.

"For every mechanical connection there is an electrical solution."

Gone were the bicycle components.

Gone were the chains and the bike parts, and the tube

steel and angled iron and aluminum tubes which I had originally used.

We were now working with carbon fiber, reinforced aluminum tubing and steel inserts for proper grounding.

And everything was custom made for just this vehicle and just for this application.

This was better than Christmas for me.

Better than revenge.

This was a dream come true.

And it was sitting in the middle of a room full of technicians and design geniuses.

All of them excited about the possibility of going to Australia and competing for the glory of victory.

I have longed for this sight in my mind for so very long, and now it was happening.

Everyone trying their best to all wind up on the same page.

I was stoked.

We all kept going around the body designing places for access panels and then I surprised the guys when I told them that I wanted the solar panels to have a motorized mount that was controlled by photovoltaic cells that can automatically track the movement of the sun and keep the panels facing the sun when the vehicle was parked.

"That was what I wanted to do! Sam shouted out.

"So, I guess, that we now have Sam's vote!"

I said and everyone laughed as Sam blushed. And got embarrassed at his exuberant outburst.

We didn't notice but, the pizzas had arrived and no one paid attention to the delivery driver.

So, I walked up to him and paid for the pizzas so he could leave and I asked the receptionist to make an announcement to let everyone know that the pizzas were in the conference room..

So, she did, and everyone reluctantly wandered into the conference room.

We ate and drank sodas standing up and when we finished eating a slice or two we went

right back out and started hashing out some more ideas.

I was totally impressed.

These guys work just like I do.

On the fly with a sense of almost total consumption

And they seem to all work together in almost complete synchronicity

These guys were in essence , "My dream team" and I had no idea how or why we were all together. But, there we all were.

And everyone was happy to be there and all of us brought something special to the project.

This was no longer "Just one man against the rude, hard, unfeeling and uncaring world

This was a loose affiliation of like minded individuals working together for the same common goal.

I was just glad that Dick wasn't hear to see the grand unveiling. Because, it might have made him lose confidence in me.

Because of the difference in the prototypes.

Mine looked like the 1907 Wright flyer and compared to theirs,

which was looking like a formula one stealth fighter.

"Would you like to sit in it?" asked Bill.

"I thought you would never ask. "

"I didn't know if it could actually hold someone." I replied.

"Sure it can,"

" It's body is stronger than steel and a third as heavy.

" So, Please, have a seat. "

They all encouraged me, I had no other choice but, to obey.

So, Dylan showed me how to open the cockpit and I slid down in the seat and stared and the mocked up gauges and felt the control surfaces.

And believe me when I say that it was an awesome experience.

I kept telling myself.

"This is really happening."

" This is real."

" I am sitting in the cockpit of my human electric hybrid that I dreamed up. "

"Something I invented that was brought forth to 3 dimensional reality."

" And I was sitting inside it."

"We designed air scoops that suck up air from in front of the car and pressure forces the air into the cockpit for ventilation.

There is also a camelback water bottle in the back of the seat with the tube mounted to the side of the head rest.

So it's easy to get to when you become dehydrated." Added Dylan.

"Wow, you guys sure thought of everything" I said.

"Well, thank yourself."

" You had notes for these things on your plans,"

"We just figured out how to incorporate them into the design." Sam said.

I saw flash bulbs going off and I saw both Bill and Toby snap a few pictures of me inside the cockpit with the canopy up.

And one with the canopy down and me hunched down in the seat with my feet on the pedals.

"We made the crank position adjustable."

"So, different sized riders with short or longer legs can still ride comfortably." Dylan added

Well, I was experiencing sensory overload.

And I was stunned to see that it was 10:00PM

We all had been there for almost eleven hours just talking about the designs and the changes and all of the modifications and it seemed that hardly anyone noticed.

We were all feverishly working our butts off and I believe that we were making

tremendous progress and we were getting exciting results.

But, it was getting late and I needed to get Toby and Earl back to the house.

So, I asked Bill if we can set up another time for another meeting.

Bill said that would not be a problem

he jotted down some notes on his desk calendar.

"OK, I put us down for next Tuesday afternoon around 1:00PM. Is that OK?"

"Yes, Bill, that would be fine" I replied.

"OK then, you are the boss." Bill replied,

As he walked out to the project room

"May I have your attention please?" Bill shouted.

Everyone stopped taking and turned to listen to Bill.

"Everyone, we have mad a great deal of progress on the prototype so far, and we have made a number of notations and design changes based on your feed back and contributions.

So, we will plan on meeting again next Tuesday afternoon at One o'clock.

Please, let either me or Andy know if you will have any problems with the time so we can make changes if we need to.

Thank you all for being here today…, I mean this evening."

"And I hope to see you all next week." Bill Concluded

I walked up to the front of the room where Bill was standing and I began to speak.

"Once again, I would like to thank everyone here for all that you have done for me and the project".

"I am truly moved by everything you have accomplished in such a short time."

"I am extremely impressed with all the work you have done. "

"Everything looks fantastic."

"You guys rock!,"

"And I look forward to meeting with you guys again next week"

" Have a wonderful "What's left of the week,"

"And if you need to talk to me"

"Please, do not hesitate to call me at my house,"

Or you can call and speak to Bill.

"We shall see you later."

" Have a great week. Thank you again." I concluded.

And Toby and Earl bade them all good night.

We piled into the truck and headed back to Toby's house.

Toby and Earl were talking about stackable experimental batteries that Earl's company had been working on for electric vehicles to save weight and provide more space for more batteries and longer range.

It sounded like the ideal batteries for my project.

But, I didn't say a word.

I just listened to them talk about energy requirements and outputs.

I offered to buy them dinner,

But, they were just too tired and wanted to go to their respected houses and head to bed,

So, I dropped them off at Toby's house and I headed for home.

I was also tired.

But, I was so excited about everything that I seriously doubted that I would be able to sleep even if I was laying down in bed.

Just too much adrenaline.

I got home and Rhonda was already fast asleep.

I came onto the house as quietly as I possibly could.

I was very careful not set off the dog.

Thor came up to me with a sleepy look on his face.

I must have just woke him up too.

He came up to me and nuzzled his nose under my hand ,

So, I gave him a good scratch between the ears.

And petted the top of his head.

Then he walked over to his corner of the living room and he went back to bed.

I got undressed and laid down next to her.

And I was surprised.

I closed my eyes just for a moment and when I reopened them it was morning.

I hate nights like that.

I always wake up feeling like I haven't been to sleep at all.

So, I got up and I took a shower and I got dressed and I started coffee for Rhonda.

Then, I went back over the designs and started making additional changes.

I changed the way the windscreen is mounted to the body from the sliding canopy.

that would crash into the solar panels to one that is hinged at the front to allow for more room for the solar panels and improved battery space.

By changing the mounting brackets we are able to give a little more width to the sides of the body.

That would allow us to carry one more row of the advanced stackable batteries

that Earl and his company were developing.

Then, I took a look at the front suspension and the way the wheels are mounted to the spindles.

I decided that the upper and lower control arms looked to be too thin to be able to carry the vehicle and two people,

And still be able to absorb a lot of shocks and bumps.

What we will really need is a test run.

And I think that piloting the slingshot down to Houston would allow us the opportunity to develop and redesign anything that fails on us along the way.

This will allow us to test out all of the electric wheels and the generator and the solar panels and the human powered generator under actual road conditions .

And to load the frame down with the equivalent weights to simulate the weight of two people riding in the vehicle.

And see how it performs under a full load before we would load it on the container for the journey to Australia.

This way we are much closer to our resources and design teams and we can get the replacement parts made and installed prior to shipping.

So, I decided that the test ride should be in October right after the Chili Cook off.

I worked on other design aspects of the new prototype,

I think I shall call it "Project III"

Since it bares very little resemblance to the original "Slingshot Mk. II".

I'm still having trouble with accepting that the project has grown way larger than I had

ever anticipated.

I was thinking that with all the people working on the project,

maybe my plan to be the pilot and chief designer were going to be at odds with themselves.

The next couple of days were spent with me at the carbon fiber place

Working with the guys.

And assembling the vehicle.

Toby would drop by after work and help us installing the wheels to the frame and the spindles.

And we worked hard installing the rotors to the back side of the wheels.

We worked very hard adapting the prototype body to the actual challenges of equipment placement and hardware wiring.

Plus we started working on the cockpit layout and the positioning of the energy control modules and the motor actuators and the Holy grail of the entire project the heart of the vehicle.

The electrical control modulator..

This little bit of hardware monitors the charge of the batteries and the flow of energy from the solar panels.

The power output from the onboard ethanol powered generator.

And it sends the current to where it is needed most.

As soon as a battery drops below a certain level.

the modulator detects the drop in power and it sends the energy to where it is needed most.

And it switches the load to a different power source,

While it is recharging batteries or else supplying current to the motors from the generator.

Or from the human crank generator which is supplying the power generated by the flywheel..

Chris told me that this is a triple redundant system without the human element.

And that theoretically the vehicle would not need the human crank flywheel element at all.

I told Chris that the Human crank element is the most important part of the vehicle design

It melds the human into part of the system as opposed to treating the human as dead weight or cargo.

And that the human part of the vehicle could extend the range much further than on pure battery power alone.

Chris thinks that I am over scoping my design and that the weight savings and the simpler design without the human crank and flywheel generator would be much faster .

It would still be able to compete against the very famous "skycar" that the three lettered University had developed with the assistance of the second largest auto maker in the U.S.

Rather unsporting, I always thought.

A university being supported by an automobile corporation with almost unlimited resources and a huge pool of talent to draw from.

Not to mention vast numbers of researchers and scientists and secondary parts suppliers.

On the other hand,

Now, I too am guilty of assembling a "dream team" of automotive, aircraft and commercial battery designers and suppliers.

High tech cutting edge carbon fiber technology in the body construction and fabrication.

And custom developed and designed electric power plants, motors, solar power and an ethanol powered generator.

Not to mention specially developed power monitoring and management systems.

Somehow almost overnight, my Wright flyer had morphed into a truly modern and aerodynamic cutting edge vehicle.

And I was beside myself with joy.

We had finally mounted the wheels to the body.

On Tuesday afternoon Earl, had delivered six of the special batteries his company was developing for the automotive and the golf cart industry.

They are much flatter than standard batteries .

and their terminals stack into themselves

So there is no need to wire the batteries in series.

That can be done by removing an insulator plug (or cap I'm not sure what to call it.)

and just stack them on top of each other.

The vehicle was really coming along,

It was going to be extremely hard on me to leave right in the middle of the final assembly

So We (Rhonda and I) can go on our Anniversary trip to San Antonio.

But, I had planned this trip for a long time now and I was emotionally ready to go.

But, since we started the assembly,

I was really into finishing the vehicle up before we had to go,

but, time ran out for us,

Plus the vehicle is not quite ready for the open road.

But, I still didn't want to tear myself away,

I mean, it was just starting to come together.

I didn't want to miss that.

But, If I was ready to quit my job because they would not allow me to take my vacation,

Then I needed to retain my focus on what is important to me and my spouse.

It was our first five years of marriage and I was not going to be one of those husbands who can only remember their anniversaries "if it falls on Trash day."

I was ready for bear.

I had planned so many surprises that she would not know what to expect next.

The hardest part was to keep my mouth shut so I didn't blow any of the surprises that I had planned for her.

I was close to exploding from holding all the details inside.

I didn't have Dick around right now to dump all of my ideas on

And since I couldn't tell her or her family,

I just kept it all inside.

So, here we are It's time for me to go home and switch gears and get ready for "the anniversary of our lifetimes."

The Hybrid Diaries
Chapter Twenty Five
The Anniversary Begins

Thursday Morning is here! OK, I got up and did the coffee ritual and I began packing my suitcase.

I had dropped Thor off at the in-laws the previous evening.

So, we would not have to worry about him being locked up in our house.

With no one to let him out or to feed him while we are gone.

The in laws love to watch Thor, because, he is such a good dog.

They call him "their grandson."

I made sure I had shorts and normal clothes.

That I can have fun in.

In addition to the old suits I had acquired from my years at the thrift store.

And a couple of hats and an cashmere overcoat in case it gets cold.

The antique suits were only for the train trip.

(Maybe for a night on the town)

We wanted to look like we were from the period circa 1933 or so.

Definitely from the art Deco period.

Rhonda knew that I was planning a trip for our anniversary,

but, that was about it,

I did let her in on the need to be able to wear antique or period clothing from the nineteen thirties.

So, she had gone to my store and a couple of consignment and antique clothing places around town and purchased a few dresses and shoes and hats for her part.

I told her to pack a bathing suit and some comfy shoes for lots of walking.

She didn't sound very enthused about the lots of walking part.

But, she did do as I asked. I told her we had to be ready to go at 9:00AM sharp.

(the Shuttle bus would not be at our house until 9:30AM.

But, I wanted to make sure that we would be ready and waiting when they pulled up to our house at 9:30.)

She said that she would be ready, (I didn't believe her)

So, I left her to her own devices.

I focused on making a few calls.

I called the station to confirm the arrival of the Texas Eagle.

It was running behind by about a half hour at the time of my checking.

So, there was no problems so far.

Train on tracks and making it's way down to Dallas.

I also called and spoke with the concierge and they had all the items I had requested.

They had the cake and a couple of little bottles of white zinfandel wine and a couple of glasses on ice.

They had a six pack of Coca Cola in ten ounce bottles and six Hershey bars that they were keeping in the station refrigerator for me.

and a couple of surprises I had already arranged.

(I'll share with you as they unfolded.)

Well, much to my surprise.

Rhonda had everything packed and she was ready to go on time.

She was dressed an a stunning dress that hugged her figure and left little to the imagination.

She looked really hot standing in the doorway with a travel bag in her hands.

And a puzzled look on her face.

She was trying to remember all the little details that women love to focus on.

I told her not to worry about trying to remember everything.

And If she forgot something then I will replace it when we get to San Antonio.

This was going to be the most stress free vacation ever!.

This was my plan and I was prepared for everything.

So, I thought.

The Shuttle arrived right on time.

We had all of our stuff ready to load into the van,

I had packed the ice chest while Rhonda was getting ready.

I had packed some cokes in ten ounce bottles,

And I packed a few Hershey bars in a zip lock plastic bag to keep them dry in the ice chest..(just in case the Concierge forgets to have the cake and wine and cokes and Hershey bars already loaded on the train at the station.)

and I hid them under all the ice and other beverages..

I also packed a twelve pack of Rep-Tex Beer and a few Dr Peppers for myself and a couple of Big Red sodas.

Now Big Red Soda was originally bottled in Waco Texas.

And only available in Waco.

I used to get them when my folks bought them for me way back when I was a child.

And we would be going on vacations and road trips with my folks.

So, I had to have some for this trip as well.

(they are now being distributed nationally by the Dr. Pepper Bottling Company.)

Any how,

The shuttle arrived and the driver and I started loading the suit cases and the ice Chest.

A couple of carryon bags with the camera and the necessary film and batteries.

And tissues and aspirin and antacid.

And all the usual stuff that you would need when going on an adventure.

I reached out my hand to Rhonda.

I helped her down the steps and together

we walked hand in hand towards the van.

We climbed into the back seat and sat down right next to each other.

Holding hands Just like two school aged children on their way to a field trip.

The Driver confirmed our destination,

Downtown Dallas, Union Station.

The Driver's name was Gus.

And he was a big man with arms as thick around as my thighs.

He looked like he could pick up the van all by himself.

He had a husky voice and a stogie would look right at home in the corner of his mouth.

"I usually take people to the airport,

I don't get many fares heading for the train station.

He said.

That's OK, We are kind of unusual people. I replied.

(he blew the surprise about the train, but, that was ok with me.)

Rhonda smiled and giggled.

She was very excited about finding out that we were going to be traveling by train.

"I haven't traveled on a train ever since I was a little girl with my Grandmother."

Rhonda said.

She also commented on How romantic train travel used to be.

And how she would have loved to have been born in the twenties or the early thirties.

"Now you see why I wanted us dressed in Nineteen Thirties apparel." I said

"It fits the mood."

Rhonda and I are big fans of art deco the artwork and the influence on building design.

It Really had an impact on modern and post modern artists and architects alike.

After a brief trip,

We pulled up in front of Union Station in downtown Dallas.

Right next to the Reunion Hotel.

It's a white rather officious building looking like the U.S. Treasury building on the back of the ten dollar bill.

Except without so many columns.

We disembarked from the van

Gus went to go and find an available luggage cart

So, we could load up our luggage and wheel them into baggage check in.

Gus was successful and came back to the van with a luggage cart in tow.

We unloaded the van and loaded up the cart.

I tipped Gus, and he drove off.

And we pulled the cart into the station and on to the baggage check in window.

We got tags from the lady at the counter and we proceeded to tie the tags on all of our luggage.

Then a porter came by and loaded the bags onto a larger cart which already had some luggage neatly stacked upon it.

We hung on to the ice chest and use it like a convenient resting place to sit upon.

While we were waiting for the Texas Eagle to pull into the station.

Good thing too, because, I had checked with the luggage counter lady who let me know that the Eagle was now almost two hours behind schedule.

So, What we did to pass the time was walk across the street and get some coffee in a old diner that is across the street from the station.

The place was a very old coffee shop / diner that had been opened since the late nineteen fifties.

And it looked it.

Nothing was new in the place for a very long time.

The coffee was strong and almost made Rhonda gasps when she took her first long swallow.

" I bet that, the coffee could probably climb out of the cup and walk around all on it's own" I laughed.

"Most likely" Rhonda added as she took another sip.

This time she had steeled herself against the hot and semi bitter taste of artificially sweetened "cup O Joe".

I had a Frosty Root Beer.

Now, I hadn't seen Frosty Root Beer since I was a kid, and when I saw the empty bottle on display,

I simply had to inquire if they could still get Frosty Root Beer.

The guy behind the counter said that "they have some now."

So, I said, "Let me have one then."

He reached down into a old style ice box that was originally used in the nineteen fifties to Keep all kinds of bottled drinks cold.

Big metal boxes that have sliding doors on top to keep the sodas cold.

And a big metal bottle opener on the front so you can open your bottles.

The opener has a compartment which easily disposes of the bottle caps for you.

the ice boxes also have logos from the drink companies that commissioned the construction of these ice boxes.

This one was an old Seven Up cooler with the original can opener still in place.

All decked out in the original green and white and red with the green bottle artistically stenciled on the side of the box.

The simulated water running down the side of the painted bottle looked so very refreshing.

And the slogan "You like it , It likes you!" under the bottle. Very retro.

They even had an old Dr. Pepper clock on the wall that was still working after all these years.

We were starting to get hungry and the smell of burgers grilling on the stove was starting to get to us, then the train finally pulled into the station.

So, We finished our drinks and together we walked hand in hand across the street again and back to the station.

We still had to wait for the passengers that were getting off in Dallas to disembark,

Before we were allowed to board the train.

So we took all the pictures of the train that we could.

And Then we were allowed to board.

I had almost forgot my ice chest "bench" that we had stashed behind the counter at the baggage drop off.

I ran back and I got it.

And together we walked to our car and the porter came out of the door and put a step stool down on the platform.

He reached out his hand and helped Rhonda into the car.

And he looked at our tickets and took us to the back of the car where our drawing room was located.

(The unofficial name for our suite was called the family car.)

"drawing room is the old name for a car like the one we were on.

Now the room was very nice and it had two chairs on the right side of the cabin

And a loveseat that took up the rest of the space.

The chairs could unfold to make a twin bed and the loveseat folds out into a full sized bed.

And there is a fold away shelf above the full sized bed that folds out into a single twin as well.

So you could sleep four to five people there in relative comfort.

Since there was only me and Rhonda there,

we just folded out the full sized bed and slid the ice chest next to the bed,

I left Rhonda there so she could get the room situated the way she wanted it to be

And I walked down the hall and looked at the bathroom,

It was tiny and believe it or not you could actually shower in it.

So, I tested out the bathroom facilities and I went to go and look for our porter.

I found him a couple of cars down talking to another porter.

I identified myself and I asked if they had spoken with the concierge?

He said no but, he was about to check with them in a little while.

I let them know that I had "some items" that were supposed to be on the train when we boarded.

He said he would look into it.

And I said OK,

I walked back towards our car.

I sat down in our room and I closed the door

I opened the ice chest and I handed Rhonda a beer.

And I opened one for myself.

We drank a couple just before the conductors voice

came over the loud speaker and he did the "all aboard" call.

Everyone who was leaving the train

Were asked to leave the train because it was getting ready to pull out of the station.

So, we sat back and prepared for the trip.

We felt the lurch and jerk of the train as it started revving up.

I could hear the growling of the big diesel engines at the front of the train

And I could feel the vibration and the hum of power as the big train slowly pulled away from the platform.

We looked out the window and saw the panoramic view of the skyline melting away

being replaced by the back side of businesses and buildings and parking lots and the occasional back yards of someone's residences.

We would pass through train crossings and see cars stopped by the big arms of a rail road signals.

Occasionally we would hit a bad spot in the track that felt like a giant was kicking the side of the train trying to knock it over on it's side.

It was just the track in some need of alignment.

We stopped in Fort Worth and we waited at the station for about a forty five minutes before we began our journey again.

We passed the time away by trying out the beds and

spending some quality time together.

They say that making love on a train can change your appreciation for the

Simpler things in life.

Like breakfast.

Our little session had made me tired and hungry and it had made

Rhonda feel as cozy as a house cat curled up on the sofa next to a fire place.

It took some coaxing to get her to get dressed and follow me to the dining car.

The announcement said that because of the delays that they were just now

ready to serve breakfast.

(This was the second of many surprises.

we got to eat breakfast and would be eating lunch and dinner on the train.)

I LOVED the dining car!,

It was old and it had the feeling of someone's old restored Victorian house.

It had carpets and overstuffed (probably with horse hair) cushions on old

antique looking chairs.

And small but solidly built tables.

They did everything with a certain air of aplomb.

They seemed like they knew what they were doing and the wait staff seemed

accustomed to the jarring and rocking of the train when it hit a bad spot on the

track.

They set the table for us without missing a beat.

They served us breakfast on linen tablecloths and they served our drinks in

crystal glasses.

We had what looked like genuine silverware.

It was all very classy and I really enjoyed breakfast with the love of my life

right in front of me.

with a "cat that just ate the canary " expression on her face.

She said that she felt that we were wearing a sign over our heads

that told everyone around us that we had just had sex.

I told her she was just imagining it and not to bother caring about what other people thought.

So we ate scrambled eggs hash browns and bacon strips and sausage patties and pancakes with maple syrup.

We drank coffee and iced tea.

And I also had a small glass of orange juice.

After breakfast we were stuffed and just wanted to sit down somewhere and just enjoy the ride.

But, I was worried about the items that were supposed to be loaded onto the train.

So, I suggested that we go and explore the train and see what it has to offer.

We walked back from the dining car to the observation car

Which had big expanses of glass windows so people can sit down and watch the fields

and houses and farms and junked rusting cars in the middle of open pastures.

The observation car was two levels

One was high and it had seats that swivel from side to side.

So you can see out of either side of the car without having to leave your seat.

I told Rhonda that I had to go to the bathroom and wash up

So I left Rhonda sitting in the observation car and I left the car to go and find the conductor.

I found him a couple of cars away in the bar car talking to the bartender.

I told him who I was and that they were supposed to have transferred my things from the station onto the train.

He said he would look into it.

I told him where we were sitting in the observation level of the observation car.

And that we had booked a family room for our trip.

He said that he commended me for my choice of the family car and said that it was a good move.

I told him thank you,

and I went back to go and sit by Rhonda.

After about thirty minutes I saw the conductor coming down the aisle with a unhappy worried look on his face.

I let him walk up to me and when he came up, and he said

"Mr. Barrientos? I need to ask you something."

I told Rhonda that I would be right back and to sit tight.

So, I got up and I escorted him out of the observation car and down towards the bar car.

The conductor then cleared his throat and swallowed hard.

And then he began…

"I apologize, but, the concierge had dropped the ball and all of my stuff was still sitting at the station and that they would have to try and replace it along the way.

I said that I was expecting something like this to happen

And in order to make everything right, I would need his help.

He said that he understood and I asked him to come back to my room.

The conductor followed me back to the car

and we walked back to where I had the ice chest.

I told him the story of when Rhonda was little.

And when she first rode on a train.

So he agreed to do this one thing for me.

I opened the ice chest and pulled out the Hershey bars

(still dry in their zip lock cocoon.)

And I took out a ice cold coca cola and I opened two bottles

And I handed them to the Conductor

Who took the candy bar and the sodas up to the observation car.

I followed.

I waited as I watched the Conductor walk up to Rhonda and hand her the cold Hershey bar and the coke.

She had this big childlike expression on her face and she started to retell the story of when she was a little girl etc..

And she thanked the conductor and took the candy bar and coke and took a long sip from the bottle.

Then he said something that I could not hear.

So, I waited until he left and then I sat back down by Rhonda.

"What did he say?" I asked

"Oh, he told me that drinks are not permitted in the observation car

And that we would have to the Bar car if we wanted to be drinking on the train."

"Oh, OK." I said and we had to get up and go to the bar car.

While we sat there in the bar car,

we watched a couple of old guys playing dominoes

while she drank her coke and ate her Hershey bar.

She looked absolutely adorable to me sitting there smiling all happy and enjoying the chocolate and the cola.

I told her that I love her and she kissed me

And then she said that she knew that I had something to do with the Coke and chocolate Bar.

because, the conductor did not ask her if she wanted anything before he brought the items in the car.

And then he did not ask her for any money to pay for the purchases.

He just wanted us to go and enjoy them somewhere other than the observation car.

Where food and drinks are not permitted.

Rhonda is so easy going and fun to be with.

I am glad that she is a part of my life.

We watched the domino game and when we were finished we continued our exploration.

We found the smoking car at the other end of the bar car and we walked into the smoking car through a huge cloud of smoke.

We found several people sitting around the car,

which unlike the rest of the train had the look of early nineteen sixties of middle to late nineteen fifties.

Lots of post atomic era green linoleum on the floor and Formica on all table top surfaces

And uncomfortable metal chairs.

Together we left the smoky confines of the smoking car to find the end of the train where

People used to stand and wave goodbye to their friends as the train is leaving the station.

Just like they used to do in those old nineteen thirties movies with Myrna Loy and William Powell, or, Katherine Hepburn, or Ruth Hussey and Cary Grant.

Or even Bugs Bunny cartoons for that matter.

We got there and there was no platform at the back of the train.

Just the end of the car and the locked door.

With a glass window in it.

I guess so we could still see what we were leaving behind.

And a flashing red light of course.

We walked back to our room and sat down and started drinking a couple more beers.

and together we continued to stare out the windows watching the world roll by.

Soon the announcement came over the intercom that the dining car would be serving lunch in about thirty minutes,

We were still kind of full from breakfast, but, there was no way that I was going to miss a meal that I had already paid for in advance.

Train time is not like time anywhere else in the world.

Ask anyone who has ever ridden the rails for a distance travel

(not a morning commuter express train, but a liner.)

Times are technically approximate and are hopeful euphemisms and are not intended to be bound by any other time frame of genuine reference.

Meaning that if the train brochure says breakfast at 9:30 and the train gets delayed by a cow on the tracks or by a mechanical breakdown then,

They will hold off on breakfast until the problem has passed or been resolved.

Then things can begin anew.

And breakfast shall commence.

So the nine thirty serving time will get pushed back an hour.

And every meal from then on will also be pushed back a little until they can finish up serving dinner at the appointed time.

Lunch was great too, and we enjoyed our hamburgers, and fries.

And washed them down with coke and iced tea.

Then We retired back to the car to take a nap.

And we slept for a couple of hours.

When we awoke it was almost dinner time.

And it was just starting to get dark outside,

We went back to the lounge car and climbed the spiral staircase and went up to watch the sunset.

It was beautifully brilliant full of pinks and purples and oranges surrounded by indigo

colored skies in the distance and the azure of blue receding from the approaching night sky.

Then the announcement was made

We went back to the dining car for dinner.

We ate prime rib. It was delicious.

And after walking around the train,

we went back to our car and I noticed that it seemed

to be taking an in ornate amount of time for the train to get to San Antonio.

When all of a sudden.

Everything shut down and the train came to a stop in a little town of Gordon.

Basically an old postal mail stop.

I could not see anything that would indicated a large community.

Just a few old buildings clustered on either side of the railroad tracks and the glow of what could have been a store like Wal-Mart or an Albertsons or maybe a Kroger.

After what seemed like an eternity the train started moving again and we got underway.

There was a knock at the flap of the door for our room,

it was the attendant ,

wanting to turn down our room.

So we stepped aside and let him do his job, and I decided to take a

quick shower, (It was just so I could say that I had bathed on a train.

Something I have never done before.

The bathroom was tiny and so was the shower

So, I made it a quick and lukewarm shower.

When I got out and dried off,

I went back to our room and everything was all nice and comfy looking.

So, I came into the room and I slid my body under the covers next to

Rhonda and I started to give her a kiss when there was another knock on the "flap"

So, I had to get back up and answer the door and it was another attendant.

They had informed me that they had something to give us,

So, I let them into the room and they brought in a cake.

Two wine glasses and a couple of cute little 375 milliliter bottles of white zinfandel

And a birthday cake with no writing on it. I never gave it much thought before,

but, in retrospect, I will bet you that the reason the train "Broke down"

Because I am thinking they had to send someone off the train to go to whichever the local store is,

and buy us a cake and some wine, and some glasses to replace the cheese cake

and wine and glasses that I had purchased for the train trip.

but was somehow accidently left off the train when we boarded. Back in Dallas.

The attendants wished us a "Happy Anniversary"

Rhonda thought that was very sweet of The railroad to do something like that for us.

She asked me how did they knew this was our anniversary?

So I pretended to be oblivious.

"I don't know HOW they knew." I said.

Before they left I asked the attendant

"when we would be pulling into San Antonio?"

He said that

"We should be pulling into San Antonio around two thirty AM."

I told him "Thank you".

And I closed the sliding door flap and I said to myself "Wow, two thirty!"

You can normally do the drive from Dallas to San Antonio in about nine hours

But, like I already said,

train time is not bound by the normal laws of physics.

With a train, it's not the destination that's important it's the journey that counts.

But, that was going to screw up my schedule that I had so carefully laid out.

We will not be at the station in time for the shuttle bus from the hotel to pick us up.

And I seriously doubt that the shuttle would be running at three o'clock AM.

So, we will have to try and think up some alternative method of transportation from the station to the hotel.

I didn't let Rhonda know about the time.

So, I opened a small bottle of wine and I poured Some of it into one of the wine glasses

and I handed it over to Rhonda, and then I poured a glass for myself.

Neither of us were very hungry so, we put the cake aside and we drank our wine until we fell asleep.

I slept well, until we were jarred out of bed by the train coming to a stop in San Antonio.

We hurriedly gathered our things together and packed them up and we got ourselves organized and got ready to disembark, or de-train.

We walked into the station and I started looking for a payphone and a list of taxi companies that were in the city.

So, I called for a cab and Rhonda and I waited for the cab to come and pick us up.

while we were waiting for our luggage to come off the train.

We were fortunate that the luggage showed up about the time the cab arrived.

So, we would not have to try and fend off any one who would try and steal our luggage from us.

We did not see any threatening people around the station,

but, you can never be too careful.

We were sitting outside of the station because the station was very crowded and there was no room to stand.

So, Rhonda and I moved right outside the door.

We were sitting on our ice chest surrounded by baggage and holding a cake in our hands.

I bet we looked pretty silly sitting out there like that,

But, we tried to keep a positive attitude.

And we smiled when the cab driver asked us if we called a cab.

I said "Yes, and he got out and opened up the doors and helped us load our stuff into the cab.

"What's the cake for?" he asked?

It's for our anniversary," Rhonda replied

"Oh, congratulations, so, How long has it been?"

"Five years" I said.

"Well, good for you, My wife and I ,

We've been married going on thirty five years now, "

"And believe it or not, I still love her".

I congratulated him on his marital success and we wished him many more

years of happiness for him and his spouse.

He thanked us and instead of trying to rip us off,

he took us straight to the hotel.

"No shortcuts"

The hotel looked as dead as the Alamo across the street

There was no bell man at the front,

 so, I had to grab my own luggage trolley and the cab driver helped me load

our luggage and ice chest onto the trolley.

And we had to wheel it into the lobby.

I identified myself and I explained that the train was running very late,

they said that someone from the station called and let them know at they were

behind schedule.

The hotel clerk was able to get someone to help us with our luggage

and to show us to our room.

We stepped inside and he placed our luggage over by the dresser and closet.

I didn't like the room that much , it looked like it wasn't as well maintained.

As the rest of the hotel.

But, it was very late and I wasn't going to complain.

Right now anyway.

We climbed into bed and tried to relax for a little while before getting

undressed and getting ready to go to bed.

I laid my head down,

The Hybrid Diaries
Chapter Twenty Six
Exploring San Antonio Five years later.

I woke up and the sun was streaming in through our bedroom windows.

I hate nights like those.

I was hoping that Rhonda would not be too mad at me falling asleep on what is

technically our second honeymoon,

but, she wasn't in the room.

I heard the sound of running water coming from the bathroom.

So I got up and I knocked on the door and Rhonda was in the tub.

"The Jacuzzi doesn't work!" she yelled at me through the bathroom door.

"I'll call room service" I said

"OK, but, wait until I get out of the tub."

I told her I would wait until she was finished.

And I sat on the window sill and I gazed down into the roof and the courtyard

of the Alamo and the skyline of San Antonio.

I could see the Hemisphere tower (it kind of looks like the tower in Seattle)

When she got dried off and dressed.

I called room service and asked to speak to the manager.

I let them know that there were deficiencies in our room,

no rose on our pillow and no Mints,

And that the Jacuzzi was broken.

They were very apologetic and they said they would relocate us to another

better room.

I told her that we came here back when they first opened up

About five years ago and the hotel seemed nicer.

And that there were yellow roses on the bed each day and chocolate

mints in the evening.

When they would turn down the beds.

And that we came back because it was our anniversary.

We wound up with one of the larger king sized bed rooms.

And it was very nice.

So we packed up our stuff and we had everything moved to the new room.

And then we went out for breakfast.

They had a continental breakfast downstairs in the bistro.

After breakfast I ordered us a rental car so we could go and do some exploring

(Plus, I had purchased a couple of tickets to SeaWorld.)

The rental car showed up right on time and I loaded the car with the ice chest

I smuggled Rhonda's swim suit and a couple of hotel bath towels out to the car.

I made sure Rhonda had a decent pair of walking shoes on.

We no longer had to wear period clothing.

We were now part of the modern world traveling a car instead of riding the rails.

So we were wearing jeans and I wore my trunks on underneath just in case we could do some swimming at Sea World.

It was a long way out to Sea World.

And it seemed like it took forever to get there.

But, we made it to the parking lot.

I rented a wagon to haul all of our junk around the park.

When we got there we saw some dolphins and we fed them fish.

Then we saw the Sea Lion and Otter show,

Rhonda really loved the otters saying how cute they looked and so on.

I noticed a complete lack of orca here.

I didn't see Shamu anywhere.

So, after the show I asked one of the animal handlers "where are the whales?"

"Oh, they are put up now."

"We always put the whales up after labor day weekend.

And they were not going to unpack them just for us."

Rhonda asked me about the whales,

I had to tell her that the whales were boxed up for the rest of the year.

And they were not going to take them out for us.

So, whale-less, we walked around the park.

We rode a few rides and I asked Rhonda

if she wanted to go swimming?"

she said no, and that she was getting tired of walking all

over the park.

So much for the water park,

Well, you can't really blame her, here it was,

the latter part of September.

The first days of autumn and the time for swimming ended

sometime after labor day.

So we got a couple of people to take a picture of us with a

fiberglass replica of Shamu and Baby Shamu.

So, we headed out from Sea World and we drove around.

I wanted to go and take a tour a the local brewery, but, we needed to

get back and go find something to fill our time.

So we went back to the hotel and parked the car

We walked across the street and down the steps to "The River walk."

Now, if you have never been to San Antonio and know nothing about the

city…

I can Explain what the River walk is so you can understand what we are

Doing The river walk is in downtown San Antonio.

The San Antonio River runs through Downtown San Antonio

And it is considered a state park.

It is supposed to be named after St. Anthony.

There are shops and hotels and restaurants lining the banks of the river walk and there are river taxis and boat rides and dining boats that ply the waters trolling for customers.

There are mariachi bands and other colorful things to see and do there.

Rhonda and I love to take a water taxi to the head of the river walk and walk back down

seeing all the sights and eating dinner and having a few drinks as we make our way back to the Hotel.

The River runs through the lobby of one of the larger hotel chains that start with "H" and it looks really neat because they put in a waterfall.

And there is lots of glass and stone or brick walkways that let you walk over the running Water.

And there are lots of plants in the lobby. Very nice.

Now across the street is the Menger Hotel.

Now, the Menger Hotel is where Theodore Roosevelt recruited his

"Rough Riders for the Spanish American War."

Very Old and we loved the architecture.

It was getting pretty late and we decided to head back to our hotel,

when I heard the sound of bagpipes.

Yes, the Scottish bagpipes.

The sound carried in the air, I could tell that it was not a recording.

We walked around the corner, and I could see him.

Some guy wearing a kilt and walking down the street playing his bagpipes.

He was standing at the intersection and playing his bag pipes to no one but himself.

We tried to pick up our pace, but, Rhonda's feet were hurting from all the walking and so were mine to be brutally honest.

There was no way that we could close the distance so we gave up the pursuit and settled back and did our best to listen to his music for as long as we could before we arrived back at our hotel.

Not something that you see everyday.

It just stood out to me.

Well, we made it back to the hotel and back up to our room where Rhonda kicked off her shoes and flopped face down on the bed.

I opened a bottle of wine so we could have a drink and relax a little bit

(before I had to make a phone call to the surrey ride people.

I had called them from Dallas before we left for San Antonio.

And I set up a date and time For the coachman to come and pick us up.)

So, I called and spoke with Charles

he Said that he would be at the hotel in about fifteen minutes.

I suggested to Rhonda that we would need to walk around the walls of the Alamo

She said that she was just not up to walking around and doing any more sight seeing.

Rhonda said that her feet still hurt and that she just wanted to sit down and relax,

So, I told her that I had left something in the rental car and I excused myself and I headed downstairs for the lobby.

Charles was as good as his word.

Because in less than fifteen minutes there was a gleaming white carriage being pulled by a white horse

in front of the hotel.

I ran outside and I approached the coach and I asked the driver who he was here to pick up and he said my name.

I told him that it was me and I had to run upstairs and get my wife, he said he would wait and I darted back into the lobby and I got on the elevator to our floor.

I ran down the hall and opened the door and Rhonda was in the process of changing her clothes to something more comfortable for sleep rather than sightseeing

but, I didn't care.

"Rhonda, you have got to come downstairs with me there is something cool downstairs

that I want to show you." I said to Rhonda.

"NOW???" she replied

"Yes, right now, I said to her.

She reluctantly got her shoes on and she begrudgingly came downstairs with me and she followed me outside and she almost walked right past the coach without saying anything.

She just looked at the coach and said,

"Now I wouldn't mind doing some sight seeing in that!."

"O.K. " I said

And our driver "Bruce" opened the door and he held out his hand to Rhonda and offered her a lift into the seat of the coach.

She looked totally surprised and a little confused when Bruce tried to help her into the coach. I told her it was alright, that I had ordered the coach for us to do a little sight seeing.

So, she let Bruce help her into the coach and I followed behind Rhonda.

I sat down next to her and she was all smiles.

I did not entirely tell Rhonda a lie, I did have to go out to the car because the ice chest

had a chilled down bottle of wine and a couple of glasses inside it

And I needed that for the carriage ride.

Bruce did not mind if we drank in the carriage and he even opened the bottle for us.

We sat down and he made a clicking sound with his mouth and the horse started walking

and the carriage started moving down the street and we sat back and enjoyed our coach

ride around the downtown area.

Down by the market and the downtown area and past the night spots.

We sat back and drank wine and watch the world going by around us.

Rhonda told me how romantic the entire trip has been so far.

And she said that she could hardly believe that we would have to go back home tomorrow. I said that "I know,

it seems that we had just got there and that we both felt that we haven't been there that long

but, we would have to leave tomorrow.

So, I suggested that we take a late night swim before we turned in.

and she agreed to it as long as there were not too many people at the pool.

"No promises" I said

Bruce did his best and we took an extra long ride in the surrey before he said that he had to take us back to the hotel.

Rhonda and I both let out a child like "Awwwwww…do we have to???"

Bruce said yes, so we agreed to go back,

besides we had just ran out of wine anyway,

and in short order, we were back in front of the Emily Morgan Hotel.

I helped Rhonda down from the coach and I tipped Bruce for his good service and we went back inside and upstairs.

So Rhonda can go upstairs to put on her swim attire.

Together we walked out of our room and down the hall to the lobby

I asked the concierge where the pool was.

She gave us a brief tour of the hotel with us in our bathing suits and showed us a stair

case that goes into the ceiling (a carryover from when they remodeled the building) and then finally, she took us to the location of where the pool was.

I could not believe what I saw through the glass door.

A stainless steel swimming pool.

Not very big, but, it looked like we would be swimming in a giant sink.

The pool has a matching hot tub also made out of stainless steel.

Together we splashed and floated around in the pool for a while until our splashing caught the attention of other hotel residents who I guess did not have enough moxy to go to the pool on their own.

So they descended on the pool area in droves and it looked like the party was starting up.

People were jumping into the already too small pool and bumping against Rhonda and me.

So, we gathered our things and towels and we headed back upstairs to get away from the crowd.

We got back into our room and we slipped out of our wet clothes and dried off with the towels and then we climbed into bed.

I poured the last of the wine that I had opened

earlier into Rhonda's glass and I laid my still damp head on my pillow.

After a while we fell asleep in each other's arms.

The Hybrid Diaries
Chapter Twenty Seven

Morning broke through our windows like a miniature H bomb going off.

The light pierced my eyelids like they were transparent.

And the light reached deep inside my skull.

Where the hangover demon was residing.

It got him out of bed and he started

banging on the inside of my skull.

"I won't mix beer scotch and wine anymore" I thought painfully to myself.

Rhonda wasn't much better off.

She woke with a ringing in her ears.

I walked over to the Mini bar area

And I made a few cups of coffee in a small little four cup coffee maker.

The coffee helped her a little.

But, I had nothing I could drink,

So, I asked if she could get dressed and I would take her downstairs and get her breakfast.

But, in order to get her to comply I had to promise to get her more coffee as well.

So, I said OK and she got up and put some clothes on her gorgeous naked body.

I helped brush out her long auburn hair and we slipped on some shoes

and we went downstairs to the restaurant for a bit of breaky.

We ate the usual breakfast fare, bacon, sausage, scrambled eggs, toast, juice, milk, fruit slices and lots of tea and lots of hot coffee and aspirin.

We enjoyed our breakfast and we walked hand in hand through the lobby and towards the elevators that would take us back to our floor and back up to our room.

We still had to hurry up and go back upstairs

And pack up our belongings and souvenirs

And then drop off the rental car and get on the train.

Dropping off the car was no big deal

because the rental agency was across the street from the train station.

I called the station to find out when the train is scheduled to arrive.

They said that there was a delay and the train would not arrive until after eleven o'clock

so, it was an hour late so far.

That gave us time to do something I wanted us to do for us.

We packed our bags and I went downstairs to go and get the rental car and pull it up front to the hotel entrance,

Then I got a luggage cart and took it upstairs with me so we wouldn't have to schlep our baggage downstairs all by ourselves.

We had to hurry because check out time is at noon.

I got the cart through the door and I began to pile the luggage on the cart.

I did a couple of walk troughs' to make sure were not forgetting anything.

Or leaving anything behind..

Everything seemed to be packed away and Rhonda gathered her personal effects

We walked down the hall and onto an opened elevator and back down to the lobby.

I took the cart out to the car and I piled our things into the trunk and the back seats.

And then we had just enough time to drive around the block to the catholic church in Downtown San Antonio.

And we met the priest and he blessed our wedding bands for us

and he also blessed our marriage and he wished us many children.

I told him two would be enough for us.

And he laughed. And he wished us a safe return trip back home.

We thanked him for everything (taking time to meet with us on such short notice and blessing our rings and our marriage and all.) and we bade him farewell.

We got back in the car and drove over to the train station.

Went we got there we unloaded all of our stuff,

and piled it at the luggage check in window.

And the porters loaded our bags onto a big baggage cart with a lot of other suitcases and bags.

And we sat outside and we waited for the Texas Eagle to arrive.

At last our train pulled into the station and they began to offload passengers and began to load all of baggage on the train.

The station announcer made an announcement that it was time for boarding,

(yes, they still do the "All aboard!" thing) and Rhonda and I gathered

up our ice chest and we carried it back on the train.

And together we walked back to our family car.

We walked into our car and set the ice chest down.

It was a lot lighter since we have been hitting our internal stores pretty hard over the last couple of days.

And, I still had a couple of bottles of coke and a few Hershey bars still hermetically sealed in a zip lock bag.

So, we were ready for our trip back.

Unfortunately, the train arrived too late for us to get breakfast,

but, that was really no big deal since we had already ate breakfast at the Hotel.

But, lunch would be serving soon.

I decided to open a beer and kick back and relax.

And wait for lunch to be served.

Rhonda kissed my cheek and said that this was one of

the most wonderful and thoughtful trips that she had ever been on.

I kissed her back and I told her that

"it pales compared to the five years I have spent happily married to the best

woman in the world."

The trip back was uneventful and we were wondering

"why is the trip back always quicker than the trip out?"

It was almost time for dinner to be served when the train pulled into Fort

Worth.

So, we told the chef in the dining car that we would have to disembark when

the train pulled into Dallas,

He said that there would be no problem

He'd just put our dinners into "To Go Containers"

and we could take it with us when we get off.

So, we pulled into Dallas and we stepped onto the platform holding our

dinners in two plastic to go boxes,

We added our piles of luggage to the mix when the suitcases

were taken off the train.

But, our shuttle was no where to be found.

So, I called the shuttle service and they apologized for the over sight.

And they said that there would be a shuttle within the hour.

I told them to hurry.

They said that they would.

And I hung up the phone.

And we waited. And waited, until the shuttle van showed up.

We helped the driver load up our stuff and he took us home.

I opened the door and we plopped down on the sofa and let out a tired sigh.

"We are HOME!!! I shouted to the house."

The house seemed empty without Thor there,

So, I told Rhonda that I would go and pick him up in the morning,

and Rhonda agreed to it.

In the mean time, Rhonda called her folks and let them know that we were

back safe and sound and that she had a wonderful time with me.

She let them know that we were coming by the next day to pick up our "Son"

They said that, tomorrow would be fine with them,

they also added that Thor had been very good,

no accidents, and no chewing incidents.

We said that was good and we commented on how good a dog Thor is.

We hung up the phone and we got undressed and we laid down on the bed an

fell asleep. Tomorrow it will be back to work on the project.

I had made a conscious effort not to mention the race or the project or the trip

to Rhonda,

This was a vacation get away from everything around us at home.

And we enjoyed the experience greatly.

The Hybrid Diaries
Chapter Twenty Eight

Back to work… OK the vacation was wonderful, better than I had expected.

And Rhonda had a good time too.

Now I heard the alarm clock going off and it was time for me to roll out of bed,

get dressed and then go set up the coffee maker so Rhonda can have some coffee when she wakes up.

Time to head out to Grand Prairie and check out the project.

I was not expecting anyone to be working on the project while I was away so, you can imagine the look of surprise when I got to Global Composites of Grand Prairie.

I parked in the parking lot got out of the car and walked to the front door and I casually

opened the door and I went inside,

I noticed that the vehicle was assembled.

I could not believe my eyes.

The project was finally assembled and it looked ready to go.

I even noticed that there was dust and little pebbles and dirt on the wheels.

Like someone had taken the vehicle out on a test ride... Without me!

At first I was pissed.

How could someone else take the first ride???

I was supposed to be the one who takes it out on it's first test run. I felt betrayed. But, I kept my mouth shut about my true feelings and I looked up and I saw Bill come walking up to me.

" How was your trip?" Bill inquired

"Fine, it was fine. We really enjoyed ourselves" I replied.

"Has anybody took the vehicle out for a test ride yet?" I asked

"Well, actually, no"

"The guys wanted to take it across the street over to Hensley field"

"and ride it on the emergency runway there,"

"they managed to take it across the street and onto the tarmac"

" before I made them bring it back, "

"I stopped them before they could get inside and test it out."

" I made sure that the first ride would go to you."

"Besides, we still had to set up the video cameras and the radar guns to confirm all the speed data and then we would need to get all the paperwork together before we started to document the performance data from the vehicle."

I told Bill, that "I appreciate the thought and that I was excited that they had indeed saved me the honor of getting the first ride."

And since, It was all going to be documented on film or uh, tape.

I would be the first on record to drive the slingshot project Mark III.

This was even more exciting than when I did my very first test for the video that Rhonda originally shot for my proof of concept vehicle."

Because, this time it was for real."

The vehicle would be running on electricity alone.

Not actually being powered by the generator.

I asked Bill when he would be ready for the first series of tests

mainly seeing how long it will run on battery power only.

And then top speed and finally testing the flywheel generator and the solar array.

Bill said that " they were in the process of writing the data sheets and getting everything prepared for the first tests."

And that "they should be ready for the first ride in the morning."

I told them that would be perfect.

And I would let Toby and Earl know.

Bill said that he already called Toby and that Toby said he would call Earl.

Man, I'm away for a couple of days and they did all the work in that short amount of time.

Instead of feeling like I was left out of the loop,

I felt vindicated that I had truly selected the right group of people.

To be honest, I thought that the whole project would have come

to a screeching halt while I was away and that I would have to get the team to play catch up when I got back, but, that was not the case.

Everything was well on track and I was starting to get excited about the first ride.

So, I asked Bill "What time should I be there for the first ride?"

he said that " they should be ready around noon,"

that would be perfect for the solar panels.

The sun would be almost directly overhead and they should be making plenty of solar energy.

Too bad though, since we would not be using them on the first battery only run.

I went home and I called Toby and he said that he would be there in the morning with Earl so they can make any adjustments to either the wheels or the batteries.

Toby said that he needed to mount an ammeter on the rear wheel to measure

the amount of power it would generate while it is being pulled by the two front wheels.

I said " ok," and I told him

"I would be there and that I was going to take it out on it's maiden voyage."

Toby congratulated me and he said that "he cannot wait until tomorrow."

He is dying to see his wheels in motion.

"So are you going to call Dick and let him know about tomorrow morning?"

Toby asked

"No, I want the vehicle road tested and in working order BEFORE we give

Dick a grand tour of the vehicle." I replied.

"Oh, alright then."

Toby sounded disappointed I don't understand why,

but, I chalked it up to him being tired.

Then I told him I would see him in the morning.

Then I hung up the phone.

And let Rhonda know that the first test ride would be in the morning.

Rhonda was happy for me and she asked if I wanted to have her videotape the

first ride,

I told her

"I would love to have her with me when we did the ride."

She said that she loves me and that she wants to be there with me to video

tape the ride and offer moral support"

I thanked her for her positive attitude and I let her know

"I love her and I am proud to have her beside me"

The rest of the evening was pretty boring, with me doing some more

design drawings and me pouring over all of the instruments

and their position and layout on the dash board.

In fact I fell asleep on the sofa while I was skimming all of the data

that Bill and Dylan had assembled.

I woke up just before dawn. And I was too excited to go back to bed.

So, I stayed up and watched the morning news on TV.

The Hybrid Diaries
Chapter Twenty Nine
"Today is Tuesday September 25th, 1990. and here is the news…"

I poured myself a glass of tea and I watched the morning news on TV.

While I did my best to try and kill time before I had to go and make my initial ride.

And of course, time is not going to cooperate.

I sat down watched the TV and I looked at my watch.

That went on for well over an hour.

At last I could stand it no longer.

I jumped in the truck and sat there.

At first because I had grabbed the wrong keys

I left the truck keys still hanging up on the key holder.

So, back into the house and then back to the key ring .

I hung up the wrong keys and then picked up the right keys this time

Rhonda was standing in the doorway looking at me.

"So, Can't sleep huh?" She asked

"No, I couldn't. I want to go to the Carbon fiber place and take my ride."

"So, Let's go then". She said.

"I'll get my camera and we'll go.

"Alright then." I said

"But, we're going to get there really early. "

"The ride isn't supposed to start until 12:00." I replied

"So, why are you going now? Rhonda asked

"I am worried that the thing won't work."

"I'll be sitting there on the tarmac. "

"With the cameras rolling."

"I'll hit the accelerator and nothing is going to happen."

"I'll be sitting in that busted assed vehicle looking stupid." I confessed.

"Oh, I understand now. "

"You wanna get there before everything is all set up"

" And take the vehicle out on a test ride "Before everyone is ready

To make sure that it's working before everyone starts filming."

"There's nothing wrong with that."

"No one wants to look foolish, especially in front of a lot of cameras."

Rhonda said

"Look, Here is what we'll do."

" I'll film you like I did before. "

"Just you and me and the camera and the vehicle."

"no one else."

"And if it screws up, I'll be the only one with the failure on film."

"We'll take it back to the shop and fix whatever the heck is wrong with it."

Before the camera crews show up."

"But, if it works fine"

"And then screws up for the camera crews, "

"I'll have a film record of the vehicle actually working."

"No harm, and no foul."

See? This is why I love her so much.

She made me feel better about the entire upcoming ordeal.

"OK then. That sounds like a plan." I said

And back out to the truck we go.

We got in and I drove over to the carbon fiber peoples place and we walked inside.

Everyone was surprised that we got there earlier than expected,

I walked in and said "Hi " to everyone there.

I introduced Rhonda to everyone and I told them what we were planning to do

I saw Toby and Earl, and Bill and Rich and there was Dick and ….

What the heck was Dick doing there?

I haven't called him since I got back from San Antonio.

So, I didn't invite him to this, the first test run of the formerly named Project Mk III.

Dick and I agreed that we should change the name of the Slingshot.

350

We thought that "The Spirits of Texas" sounded good to us

Because, well, the alcohol reference ("spirits") seeing as how alcohol is paying for the project.

And camaraderie. All of us working together.

And since there are so many people who are sharing my vision "Spirits would definitely apply.

But, so far since I had no running vehicle, I didn't think now was a good time to change the name, until I actually had a working prototype that works.

Well that eventuality was about to happen. One way or the other.

I spoke to Toby and He said that He had invited Dick to the test run before he had a chance to ask me about it.

Toby thought that It would be really cool to have Dick here.

I told Toby that "Yes:" is was really cool to have Dick here with us

"but, I wanted to make sure the vehicle WORKED, before I wanted to have our benefactor there to witness the successful running of the vehicle. "

"What if it doesn't work? I asked Toby

"Dick might be angry with us and pull his support if we can't get it running."

I whined.

"Oh ye of little Faith" Toby replied.

"Andy, Don't you realize that I have tested the wheels on a bench?"

"By the way, they work great."

"The first run is on battery power only,"

"So, don't stress Dude. Everything will be fine!"

I tried to smile and take a deep breath.

And I swallowed hard and I said "OK" to Toby

and I walked up to Dick and I greeted him with a smile and a hearty handshake.

"Hey there Dick!" I said

"Hey Andy, How are you? "

"Rhonda!, you look as beautiful as ever!"

"Don't tell Candace I said that." Dick Blushed

"Your secret is safe with me." Rhonda said.

"How was your anniversary trip?" Dick inquired

I said that "We had a really good time"

" Thank you for asking".

"Surprised to see me?" Dick asked

"Dick, PLEASE, don't take this in the wrong context,"

" I wanted to make sure that the vehicle worked before I called you and invited you to see something concrete and failure free."

" I just wanted you to know that We are competent and that we doing our very best to show you something that would help you validate your trust in me and my abilities."

I was almost in tears.

But, I kept a brave face and I tried to exude a overall feeling of self confidence

I did my best to hide all of the apprehension.

"Andy, please, don't be stressed about all of this."

" Whatever happens today , well, it happens."

" And I know that you will get it resolved one way or the other."

"Wow, I mean, Wow, just look at that thing."

" That is the most awesome looking vehicle I have ever seen."

" It looks like a formula one racer with the body of an Lockheed SR-71Blackbird."

"Thank you Dick, but all of that doesn't matter if it doesn't work." I said.

Dick could tell by my demeanor that I was feeling pretty stressed out.
And he might have felt that he may have made a mistake by coming here to watch the first run.
But, he didn't show any concern to me.
He seemed really excited to be here and He was genuinely happy for me and the direction the vehicle development program was going.
Bill came in and said that it was time to get the project across the street and onto the tarmac of Hensley Field.
Bill has some contacts with the General in charge of the field.
And the General had given us permission to use the emergency field that parallels the main runway.
It runs almost a full mile so.
 Christopher and Sam had set up traffic cones at the four corners of the runway.
 so that I can make a two and a half mile lap of the runway.
I was weighed and my weight was recorded in the computer.
While they were busy gathering up all of the equipment and loading it onto the truck for the journey across Jefferson Boulevard
Rhonda and Dick crossed the street and waited by the big chain link fence for the rest of the crew.
The guys wanted to push the vehicle across the street.
So, they pushed the vehicle across the busy traffic infested roadway and over to the recently unlocked gates of the emergency runway of Hensley Field.
Once there, they had assembled all of the equipment necessary for us to be able to do a series of test runs designed to determine the vehicles' capabilities.
The guys had set up a canopy and tables
And then they placed the computers and monitoring equipment on the table

tops

And slid the chairs into position.

Under what we jokingly called Command Central.

I.E. a canopy to shield the testers from the sun,

a couple of cameras to record the testers

gathering data from the car's onboard sensors.

And a couple of computers so that they can enter all of the data and a couple of gasoline powered electric generators to run everything.

On the starting line they had set up more cameras and a couple of radar guns on tripods.

It all looked very professional and I could tell that they have done this kind of work Before.

probably for the military.

They must have been there since very early in the morning.

I was starting to feel much better about everything once I saw the test track and all of the equipment that they had amassed for the first series of test runs.

I was relying on their equipment and they're design and testing expertise to produce some usable data that we can apply to improving the performance and fine tuning our driving skills with this new type of transportation.

Now if only the Slingshot/Spirit will accommodate us with a couple of successful runs.

I was still worried about failing in front of everybody,

but, it was something that has to be done one way or the other and we really needed to know if we had a winner on our hands or not.

And if there were any problems, better to detect them now instead of finding out at the starting line in Australia.

After we got the Slingshot/Spirit set up on the line.

Bill came up to me with a helmet that resembled the type of helmet that fighter pilots wear in their cockpits.

It looks way cool.

With the solar tinted sun shade that you can pull down over the visor.

And I like the way the guys were able to rig a headset for the two way radio.
Built into the helmet.

"We hooked the headset jack into the radio transceiver plug and set the tuner up to the same frequency as the base radio transceiver which was set up under the shade of the canopy.

So we can be in constant communication with you at all times." Said Bill

"We also put a camera on board so we can monitor the gauges to make sure they are properly calibrated."

"And I filled up the camel back with ice and water in case you get thirsty." said Christopher.

"Thank you Christopher. I am sure I may need it." I replied.

Dylan came up to me and he opened up the cockpit canopy. (Yes, they had hinged the canopy to open from the front instead of sliding down the back side of the body and crashing into the solar panels.)

I slid down into the drivers seat.

I strapped myself into the seat.

I put the helmet over my head and I could head Bill over the radio saying "Testing, Testing... One ...Two...Three... Andy can you hear me? Over?"

"Yes, Bill I can hear you, How am I sounding to you? Over."

I said into the radio mike.

"Good. OK, we are getting set up to do our first test run. "

"This will be a test run."

"In one direction to determine if the wheels are pulling equally."

" And to see what the top speed will be on a single run."

Dylan reached over and he thrust a clipboard into my general direction.

"Here is the pre run checklist."

" You must make sure that everything is turned on and that all of the inverters and power modules are online before you engage any of the drive wheels."

"In fact, you will have to start the vehicle using just the rear wheel."

" then once you get in motion."

"You flip this switch and engage the main drive and the front wheels will take over and the rear wheel will become a traction wheel / generator."

"If you hit the brake then the electric current reverses and the wheels become electro magnetic brakes that can be used in a process called regenerative braking."

"Plus if you need to, there is a hand brake that operates the disc brake system."

" Just in case"

Dylan and I went down the check list and we checked off everything on the list from battery output to motor testing to final start up.

And when we were finished checking off the items on the lists then I flipped the switch and I engaged the rear drive wheel.

I held down on the disc brake handle so the vehicle would not start moving on it's own,

I would not have to drain additional power from the batteries by reversing the field on the front drive wheels and using them as brakes.

To hold the vehicle in place.

I could see a small drain on the batteries and I saw the output meter on the rear drive wheel light up.

I heard a slight hum coming from the back of the vehicle.

I let off of the brakes gently and the numbers on the windshields head's up display started to climb up slowly as the vehicle started to creep forward slowly.

I guided it's direction as we moved up to the starting line.

Dylan waved his arms in the air and he yelled out "Stop"

As the nose lined up with the line on the tarmac.

Bill's voice came over the radio…" Are you ready ? over"

"As ready as I'll ever be. over" I replied I took a deep breath and I said a little prayer.

"OK then, we will do a single run down the runway at half power."

" Just to test out the system before we start trying to push the vehicle to do more. over"

"Alright I'll only move the controls half way ONLY….over"

"Good, then."

" When I say go, you can floor her at half power and let's see what she can do. Over"

"Roger" I said.

I scanned the controls that covered the dashboard and I looked down the runway.

It seemed like it went on forever.

And the butterflies were busy fluttering around in my stomach.

I felt excited and scared all at the same time.

This was far worse than the first time I had driven my prototype for Rhonda and the camera at the college's parking lot.

Man. That seems like such a long time ago.

It almost had a dreamlike quality about it.

But, now… That dream was a new reality.

And that prototype was as far removed from the new vehicle as a tricycle is to a jet fighter.

And this time everyone was watching,

Except they were counting on my success instead of waiting to watch me fail.

The vehicle sat on the starting line and I waited for a signal from Bill to

Begin the first run for my creation.

Then suddenly, the thought occurred to me…

I forgot that Rhonda was going to film me on our first run.

It's too late for me to go and tell everyone to stop filming.

My thoughtlessness and excitement got the better of me.

I just had so much on my mind that I simply forgot to tell everyone that I wanted Rhonda to film this run alone.

I didn't want to fail on camera with the entire group watching.

"Then the voice over the radio started the count down. Three, two one… Start!"

I pushed the throttle forward and the vehicle quickly lurched forward and then I could feel a smooth transition of power flowing to the rear wheel.

I could hear a slight hum coming from the back of the vehicle as the rear drive wheel came to life.

I could feel the power building as the speedometer started climbing… 10 then 12, then 15

then 20..

I cautiously steered the vehicle down the runway as it picked up speed.

I could feel the sensation of motion.

And I could feel acceleration building up and It felt like I was really starting to move.

I could hear the rush of the wind and not much else.

no sounds of a gas powered motor and no pollution coming from the back of the vehicle's tailpipe because there was NO tailpipe.

The vehicle seemed like it was flying down the tarmac.

I could hear the sounds of the wheels rolling on the asphalt. 25… 27…

"THIRTY MPH!" I shouted over the headset.

It rolled very smoothly down the runway.

I pushed the throttle forward a little more.

I flew past the ¾ mark and I was still accelerating.

I wanted to hit the brake but, I knew that would screw up the test results 35…36….39…

"FORTY!" I shouted

just as I was about to cross the finish line and I began braking.

I could hear Bill telling me to chop the power

And hit the brakes gingerly.

As I swooshed past the traffic cone that marked the stopping point at the end of the run.

I pulled back on the throttle and I grabbed the brake handle firmly.

I gave it a gradual squeeze.

I felt the front wheels shutter as the electrical field was reversed .

My head and body started to come forward under negative G's.

but, the seat belts held me back into the seat as the vehicle started to slow down.

The brakes began to slow the vehicle back down to a stop.

Which amazingly took less than one hundred feet to accomplish.

I slowed down and drove around a traffic cone and I coasted the vehicle around .

I turned around 180 degrees so I would be facing the other direction facing the command center.

And I came to a stop.

I turned off the inverters and the modulator and I cut off the power to the vehicle.

Chris and Dylan were waiting at the finish line for me. and they were laughing t me

about something.

I heard Bill say over the radio.

"That was a good first run there but, you forgot to switch the drive from the rear wheel to The front drive wheels after you got rolling."

" So that run was on the rear wheel motor only."

"We will need to do the run all over again."

But, we can still keep the data that was generated."

"You hit an actual speed of 37.6 miles an hour."

" On just one wheel, that's pretty good.".

I couldn't believe it.

"It really felt like I was going much faster than 38 miles and hour"

I said.

It must be because of how close I am sitting to the road.

And I guess that they'll need to recalibrate the speedometer it was three miles per hour off.

So, Dylan and Chris and I went down the checklist one more time and reset the controls

And then we were ready.

So, they closed the canopy again and I waited.

While I was Waiting in the cockpit,

360

I went over flipping the main drive wheels on as soon as I got moving, over and over in my head.

Determined to not make the same mistake twice.

Bill counted down. "Three,…two… one….Go!"

And I once again engaged the rear drive wheel and this time I remembered to hit the switch to engage the main drive wheels once I got moving and zoom!

I felt the front wheels bite into the tarmac and they made a little screech as they fought for traction.

I felt the surge of power but I heard no engine noises or smelled no exhaust fumes.

And the sudden burst of power pushed me back into the seat.

And before I knew it I was really starting to pick up speed.

I was really moving now.

And the finish line was looming up ahead and coming up fast.

Then I heard Bill say "Cut the power and begin to brake. "

I said "Roger" into the radio and I chopped the power and I hit the brakes.

Once again I could feel my head and my body starting to come forward under the pull of negative G's. but, the seat belts once again held me back into the seat as the vehicle started coming to a stop.

This time I stopped right on the starting line.

Bill read through all of the data.

"Hmmm. Not bad, you had a top speed of 59 miles per hour on the primary drive wheels.

Very impressive."

I saw Toby give me a big thumbs up. Grinning ear to ear.

And Dick was high fiving everyone who would put up their hands.

The guys at the table remained focused on gathering data while anyone who wasn't collecting data was whooping and hollering and cheering me on my first successful run.

Rhonda gave me a Thumbs up and she blew me a kiss.

She was standing there beside Dick with a huge smile on her face, Rhonda was still holding our video camera in her hand.

I climbed out of the cockpit and I walked over to Rhonda and I gave her a big kiss on her lips.

I hugged her and she hugged me back.

And she smiled up at me and she said that

"she got the runs on tape".

I kissed her again and I told her that I love her great big.

She said, " I love you too even if you abandon me for your other wife."

Referring to the vehicle as my other wife.

I knew that she was only partly joking.

It's been my obsession for the better part of a year now.

She told me how proud she was of me and my accomplishments.

I told her "Thank you" again.

And I kissed her once more, and then Bill hollered over to us and said that "We were burning daylight and we needed to make several more runs as long as the batteries still had juice in them."

So, I told Rhonda "bye," and I ran back over to the vehicle

And I climbed back in and closed the canopy and we went over the checklists again.

The next series of tests were to see how far we could travel on our left over power after two test runs.

So, we would be driving around the runway like an oval race course with traffic cones posted at all four corners of the turns.

And we began our tests to see how many miles we could run on battery power alone.

We ran all day and into the evening.

And we were driving around the runway as a Lockheed C-130 Hercules was approaching the landing strip next to the emergency runway

we were driving by on the emergency runway that parallels the main runway. That plane looks HUGE compared to our little car and the prop wash made the vehicle shake and buffet as the big cargo plane flew past us.

As the rear landing gear wheels impacted on the runway and the plane started to touch down,

it's speed was starting to fall away and the plane began to slow down before the front wheel touched down in a loud screech and a puff of blue gray smoke.

Then the massive cargo plane revved it's engines and dropped it's flaps and it slowed down about halfway down the runway and it started taxiing to the main terminal.

(It's not as impressive as the Grumman F-14 Tomcats flown by the Naval air Reserve squadron that normally operates out of Hensley Field or the F-16s that are flown by the Texas Air National Guard or by Lockheed, also flying out of Hensley field,

But, it was still a big plane) compared to our little fighter looking, formula one, hybrid racer.

The prop wash from the cargo plane as it went shooting past us almost lifted the front of the vehicle off the tarmac and I was buffeted by the turbulence that was being generated by the combination of prop wash and the sheer bulk of such a large plane

moving at over one hundred miles per hour flying by at nearly head top level.

I was able to regain control and keep it on the runway.

363

I started to wonder how the vehicle would handle being passed by an eighteen wheeler barreling down the freeway at over seventy miles per hour,
Would it be blown off the roadway or would it get sucked up in the vortices created by the passing of a large trailer moving at high speed?
If an eighteen wheeler can suck a old Chevrolet Corvair or a Volkswagen Beetle off the road,
Just think what it could do to a vehicle that is way more aerodynamic and about half the weight.
There is even a bigger threat to us driving on the freeways of Australia, Road Trains.
Now a Road Train is a regular Large tractor trailer truck, like a Mack, Kenworth, or a
White or even a Peterbilt
Pulling not one, not two, but, three sixty foot long trailers set up in "Piggy Back" fashion.
Being pulled at an average speed of sixty miles per hour,
On what we would call "narrow country roads"
These mammoth cargo haulers are needed to ply the vast distances between urban centers and big cities scattered along the coast and in the bush.
Sometimes thousands of miles apart.
They carry cargos of freight food, clothing and all the supplies a modern city needs to keep it's populations well supplied and fed on a daily basis.
The Australian Road-Trains are the longest trucks in the world.
They have 3 or more trailers and are 174 feet long,
Hauling somewhere around 115 metric tons (253,531 pounds) of cargo.
The race brochure warns us to Keep an eye open for the Super Road-Trains.
they can have as many as 7 trailers and hauls up to 200 metric tons at freeway

speeds.

These things cannot stop very quickly that's for sure and as a rule they demand and command the respect of all the other smaller vehicles on the road and they as a rule receive the right of way.

They are equipped with huge metal grille and bumper guards or push rails on the front of the cab looking very much like the Mack truck driven by Mel Gibson in the movie "The Road Warrior" they call them "bull bars" they're for cattle.

They kind of resemble snow plows because they are designed so that they can crash into unfortunate cattle that may have escaped from their enclosures.

The stations (or ranches as we call them in the states, do not always have fences like we do in Texas,

so their herds are generally free to wander and some cattle do end up wandering down the side of the roadway.) Or anything else that might somehow accidently end up in front of the truck at the wrong time.

Like a emu, or dingoes or a suicidal kangaroo.

The roads in the outback are very narrow and there are usually no shoulders to speak of. Just two lane blacktops.

A whole lot like west Texas, a whole bunch of wide open nothing.

rocks, sand , cactus and desert. No hills to speak of.

If there is a wreck, it could shut down the entire roadway.

And if there is an accident where the truck does roll over

then the emergency crews would have to travel maybe

hundreds of miles to the crash site,

Then unload the truck trailers by hand.

load the cargo onto another road train and send it on down the line while a tow crew tries to upright the truck.

Break up the trailers and hook them up to other trucks if at all possible.

and haul the truck cab or what they call a **pri-mover (Pronounced Pry-mover)** to a repair facility "close by" that could mean several hundred miles away.

And it could take between twenty four to forty eight hours to get the vehicle cleared off

unloaded and then towed back to Melbourne or Adelaide, or Sidney.

Thousands of miles away.

So this is a very dangerous obstacle to be respected and avoided at all costs.

Along with the animals that also wander onto the Highway.

Like the emus or wallaby's or dingoes , or kangaroos.

I pulled past the finish line and I took the vehicle for another lap.

The gauges indicated that I had less than ten minutes of power left.

so, I took her for one final lap before the power completely ran out.

As I rounded the last turn and headed back to the finish line the energy gave out.

I coasted to the finish line . I pulled over and I opened the canopy.

I got out an removed my helmet.

I set it down it the seat then I stood up and stretched.

Rhonda came running up to me and she leapt up into the air threw her legs around my hips and her arms around my neck.

I gave her a big kiss and I gave her a big hug back and I gazed into her big blue eyes and it was like I could feel her excitement

I could see how truly excited she was for me.

I could feel that she too was proud of what I was able to accomplish.

And I smiled at her and I told her "Hi" and then I apologized for forgetting to have her record the first runs by herself.

She said that "it was no big deal and no harm done. "

I kissed her and I told her just how happy I was for having her there with me.

Inside my brain, my mind was screaming,

"It works, It works It' really, really works!

I can't believe it! I was driving a car that I invented around a test track!

This is so Freaking COOL!"

On the outside I was acting like it was business as usual.

Then Dick came up to me and he gave me a big pat on the back and he shook my hand very firmly and forcibly.

And he congratulated me on the first successful runs and he apologized to me

And to Rhonda saying that Candace wanted to come but, she was in the process of organizing the logistics necessary to supply a group of over two hundred chili cook off chefs with beer and barbeque and some prizes for the silent auction.

Rhonda said that she more than understood.

And that she too was working on getting all of the registered cooks their cooks package

t-shirts and aprons and also working with assigning spots for the cooks to set up their cook sites.

.Toby and Earl ran up to us and they was excited about the results and he said that the figures bore out all of his calculations.

And the vehicle performed very well.

Everything seems to be on track and going according to designed expectations.

Earl was excited.

It was the first tests for the new battery design and they actually did better than even he expected.

So, now we know how far we can go on one charge.

With just the batteries only.

The next tests will be on the solar cells or the generators or the flywheel generator.

Or the gas/ethanol powered generator.

But, soon we shall find out what our total capabilities are.

I shook hands with the crew and they congratulated me on a successful ride.

And I Thanked Bill, Dylan, Sam and Christopher for all of their hard work.

Rhonda and I were in the mood to celebrate.

So, we decided that we would invite everyone out to dinner and they all said that after we packed up everything in the truck and carted it all back across the street and locked everything back up at the carbon fiber place that they would be more than happy to go out and celebrate our success with us.

So, We decided that Mexican food would be ideal along with several pitchers of beer and margaritas both frozen and on the rocks.

I called the restaurant and made reservations for fourteen people.

And I made sure that they would have Strawberry and lemon / Lime and Mango margaritas all mixed up and ready for pouring when we get there.

They said that would not be a problem,

And so, after we packed up everything and trucked it all over to the shop.

And we locked the slingshot in the store room.

Bill locked up the shop and we all headed out to the restaurant.

I told Rhonda to climb into the cab of the truck .

I shut the door for her.

And we waited until everyone was ready to go.

And we caravanned in convoy fashion down to the Mexican family Restaurant not too far down the road..

The Hybrid Diaries
Chapter Thirty

After the meal and drinks were all consumed and the guys went home Rhonda, Dick and I were sitting at the table and since it was kind of late, I asked Dick if he was planning on going back to Austin tonight?.
He said that he had flown in on Southwest Airlines early this morning and rented a car from a car rental place.
And that he was thinking about waiting at a hotel for Candace to fly in the following day.
"Didn't Rhonda mention that Candace was flying in tomorrow so they can go over the logistics for the Chili cook off with her boss at the Leukemia Society?"
"Ex boss, I'm only a volunteer now." Rhonda interjected.
"No, It must have slipped her mind."
" This is the first I heard about it." I said looking over at Rhonda and giving her the withering glance of an angry Margaret Hamilton.
(she was the actor who played the wicked witch of the West in MGM's version of the "Wizard of Oz.")
 Rhonda blushed and she said that "she was "sure" that she mentioned

Candace's upcoming visit to me."

" Maybe I just had too much on my mind". she said

I let her go ahead and assess blame upon me and my absent mind although I am certain that she did not remember to tell me.

"Well, no harm done". I said

"Hey Dick, So, where are you planning on staying tonight?"

I asked.

"Well, I was going to rent a room at a hotel until Candace arrives in the morning"

"Did you already get a room?"

" Because, if you didn't you are more than welcome to spent the night at our house."

I replied.

"Oh no, I couldn't do that, I wouldn't want to impose." said Dick

"Dick, it isn't an imposition at all."

" We're glad to have you and Candace stay with us while you are up here"

Rhonda added.

"Well, thank you guys, I just thought that when Andy didn't send me an invitation to the trials that you guys were not in the mood for company."

"Sure Dick, Make me feel worse than I already do." I said.

"Dick, I am very sorry for not sending you an invite."

" it's just that I wanted to be sure that everything was working before I brought you up here to witness our first "Successful runs"

"Toby had much more confidence in the vehicle than I did.

"I was afraid that you would be mad if you got all the way up here and the stupid vehicle

failed to work."

"Especially after everything you have done for us,"

" I just wanted you to see that everything was progressing according to my plan."

" I just wanted for everything to be perfect when you arrived. "

"I am just delighted that everything worked as planned."

" And that you didn't lose your faith in me."

" And that everything worked like it should." I apologized again.

"Well, let me tell you something, I was totally impressed by the vehicle."

"The way it looks and the way it drives and all of the testing equipment and radar guns and the computers."

" It almost reminded me of NASA or NASCAR."

"It looks really cool Andy."

" It kind of looks like an Indy Car With the front wings and with the needle like nose and the fighter plane type canopy. It looks really fast."

"So, have you given any thought to who you want to drive the vehicle in the actual race?"

Dick quipped.

At first I didn't really understand what Dick meant. I always thought that I would be the driver.

Being my vehicle and all.

And I wanted to be the one who would be the one who beat out all of the major Universities.

I was going to be the "little David beating out all of the mighty Goliaths of the world.

With my little electric vehicle."

"No Dick, I just thought that I was going to pilot the vehicle."

"Well The reason why I was asking is that we need someone who can talk to the media and explain what the vehicle was designed to do."

"And give them the excitement and enthusiasm that a project like this requires."

"We need someone to talk technical to the media and to show off the vehicle while the cameras are rolling and the reporters are "Ooooo ing and ahhhing over the vehicle while

we are in the process of test driving the vehicle around the track.

"And nobody knows more about the project than you."

" And to be sure, you need to be available to answer questions about the project to the hoards of News and magazine reporters and photographers that will be coming to talk to you and taking pictures of the project. "

"You know, basically the same thing you and Rhonda did back when you were trying to get a sponsor. Remember?" Dick smiled coyly.

"What hoards of photographers?"

" And what news reporters are you referring to Dick?"

"When are these hoards suppose to be showing up?" I had to ask.

"Tomorrow?" Dick Replied.

"What do you mean tomorrow? Dick ,

"What did you DO?"

"Well after the first couple of successful runs.

" I called the TV news crews of all the major networks here in Dallas and invited them to come out to Hensley field and report on the vehicle and interview you."

".I also contacted the local newspapers. I hope you don't mind."

"No Dick I don't mind, I am just trying to get a handle on things."

"I wasn't ready to notify the media yet."

" I look like crap, I need a haircut and I need to get my notes together and figure out what "I am going to say to these people."

"I need to look professional so that they will think that we are a serious contender for the race."

"I do not want anyone to think of us as crackpots or weirdoes. "

"I want to get everything looking as nice as possible for the cameras."

"I was stunned by the revelation that Dick had already invited the media to the test track to record our vehicle going through it paces.

I was thinking about having the vehicle painted in a bright color scheme.

But, I don't know. I mean, I like the way it looks now all dark gray with a grid like pattern to the surface of the carbon fiber body.

It looks real cool and it shows off our mastery and competence in working with exotic materials.

I was hoping for some kind of mural or something.

like the company truck except I wasn't going to paint the vehicle in aqua.

I was thinking about an anti glare anti radiation white finish with a mural or the name of the vehicle or something to personalize the appearance.

I wanted to let the World know that we are here ."

"And we are from Texas and we have come to kick some solar powered alternative energy asses. "

I told Dick "Well at least we now know that the vehicle works."

" And by the way, Yes, I have been thinking about recruiting a couple of riders other than myself. "

"Because, there might be an accident, God forbid, or else someone might get a cramp in their calf muscles or else they might get tired or suffer from heat exhaustion.

So, I was thinking about maybe using either Dylan or Chris as a back up rider."

" we can always use more alternate drivers."

"OK, well, What about Willy?" asked Dick.

"Willy who?" I didn't have a clue who he was talking about.

"Our janitor at the brewery,"

" Willy Werkee I believe he helped carry you out of the brewery "

"The last time you were there. "

"Short, Little Filipino guy?" small but really strong?" I asked.

Yep, that's the one. Dick replied.

"That guy can ride eighty miles in one day"

" And he does it all the time."

" I don't know why he doesn't drive a car."

" He just likes to ride his bicycle."

"And he does it every day."

He gets up at five and then he shows up at the brewery at around 7:30 AM.

And then he rides it all the way back home after work every day."

"The guy's got to have pretty good legs and he's bound to have decent endurance."

"Well I know that he's a wiry little guy".

Maybe you can have Candace bring him up with her tomorrow

And we can give him a try."

" I don't have any problems with it." I said.

"Good, then I'll call Candace and ask her to bring him up when she flies into Love Field in the morning."

"Dick, did you bring up any promotional material with you?"

"No, I didn't have time to bring anything up with me."

" I was hoping that you might have something besides the truck with the name of the brewery on it,"

" All I brought with me is my business cards." Dick said

"OK then, Dick I think you need to call Candace and let her know that she will need to drive the Suburban up here and she needs to load it down with any banners or posters or signage that you guys may have for advertising purposes.

" I think that we will also need a couple of Kegs of beer and plastic cups. "

"Rhonda, I need for you to print up a sign that says

"Responsible beer drinkers drink responsibly"

And I want to make up a sign that says

"Press conference at 4:00PM."

"That should give us enough time to get everything ready to show off to the media."

"I will need to get my Rolodex and Call Bill.

I need to let him know about tomorrow so they can clean up around the shop. And clean up the conference room."

" So we can hold the press conference in there"

"And then we can bring everyone out to the test track and present the vehicle to the news world."

"Crap, We'll need to call Toby and Have him call Earl and let them know about tomorrow., I hope Toby can make it." I added

So that being said,

We paid the check and we left the restaurant and headed back to the house

We made a stop at the grocery store because I was in the mood for something dessert like

.Like a chocolate cake and ice cream either vanilla or chocolate chip.

My favorite ice cream flavor was a ice cream that was made by Baskin Robbins called "Burgundy Cherry."

But, they have since stopped making that flavor so, I guess I had to settle for

Vanilla.

"Well we did not find a chocolate cake and I didn't want to buy just ice cream

so, I bought some store baked brownies and I got vanilla ice cream, Hershey's chocolate syrup and whipped cream and a small jar of cherries and some chopped nuts (for Rhonda, and Dick and maybe for Candace and Willy when they get here in the morning.

(You see I have an allergy to walnuts, pecans and bananas due to their high fructose.

They cause blisters to appear on the tip and sides of my tongue and they make my ears itch.)

I try to avoid the nuts like a plague.

Although I used to be able to eat them when I was a child.

My grandmother used to give me sliced bananas in a bowl with sugar and milk poured over them for breakfast.

Funnier still, I can eat cashews, Brazil nuts and macadamia nuts and even peanuts.

Oh well, life is strange.

So, we got to the house and unloaded the car and put the groceries away.

I got started calling people and telling them that the press is supposed to be there at the shop tomorrow.

When I spoke with Bill he said that he had something to talk to me about.

(God, I hate that sentence. "We need to talk". Almost nothing good ever follows those four words.) I said OK and I asked Bill what was on his mind.

"Well Andy since we built the prototype vehicle, they have been making some alterations to the vehicle to improve aerodynamics and make the design more efficient. Also The current vehicle that we are testing is just a proof of concept vehicle just like your original Slingshot was not designed to compete but, to prove that the concept worked."

And that Bill thought that they can have the actual race vehicle built and completed in a couple of weeks."

And that the concept vehicle should be used as a promotional display designed to stir up interests in the project."

I told Bill, "That sounds like a wonderful idea and that I agreed that we should build more than one vehicle just in case of accidents or in case of a break down or some other kind of mechanical failure to the proof of concept vehicle.

We need to be flexible and having another back up vehicle would be just what we are needing,

Should we need to have the vehicle be in more than one place at a time. "
said Bill.

I told him that I agree and that I was thinking about getting the prototype painted a white color with some kind of mural on the side.

I also mentioned that I needed some decals or stickers with the company name on them.

so we can start plastering them on them on the side of the vehicle.

For sponsorship purposes.

He said that they had some

but, he would need to locate them.

I told him that I will need a couple of stickers for the side of the truck and for the sides and back of the trailer that I am going to purchase to haul the vehicle

So, I let Bill get off the phone and Dick called Candace next and he had a lists of demands that we threw together at the restaurant.

Candace was Delighted to learn that she wasn't flying in tomorrow morning but, now she will have to load up the Suburban,

with all kinds of posters and banners and signs and toss in a couple of kegs and taps and a couple of carbon dioxide tanks and several bags of ice and a plastic trashcan to set the keg up in.

Not to mention plastic cups for drinking beer.

and neoprene beer can wraps for keeping cans of beer cold.

(They were give away promotional items for the reporters to keep.

And to take back to their offices or to their homes.)

And Key chains that have bottle openers on them.

Basically whatever they had lying around the brewery

That also had the name of the beer and or the name of the brewery on it.

Something that they could give to the news people.

Yes Candace was so happy Dick.

You could hear her screaming with excitement at Dick.

who was holding the receiver away from his ears

He could still hear Candace's voice very clearly even from about two feet away.

Rhonda came up to Dick and she asked him for the phone,

which he more than gladly handed over to Rhonda.

"HI Candace? Hi, This is Rhonda, How are you?"

"Good, glad to hear it,"

I know.. Yes, they are insensitive aren't they?"

"Great, OK can you also bring up a couple of evening dresses?"

" You know, the backless strapless kind that Models wear at auto shows?

"Yeah, what you do?"

"Oscar De La Renta? Oooooo…"

" and Bob Mackie? Ooooooooooooo"

"Yeah, Bring those.. Ok I'll see you tomorrow."

"OK, Bye."

"Wow, How did you do that?" Dick asked

"Do what?" Rhonda inquired

"How did you get Candace to calm down so quick?"

"Oh, that was easy,"

" Candace isn't angry at me!" Laughed Rhonda and she left the room,

I was at the dining room table desperately writing copy for the girls to say to

the reporters.It started with the a welcoming greeting…

"Ladies and Gentlemen of the media, Thank you coming, Today

"You are going to witness the future in automobile technology.

"Vehicles that do not run primarily on gasoline or diesel fuel

"But, electricity."

" And solar power and human powered flywheel technology."

" Producing zero emissions and yielding well over 80 miles to a

gallon of 80% ethanol and 20% methanol fuel mix."

" Blah, blah, blah…."

I kept on writing and rewriting all through the night.

and I eventually had a script that could be read through in about four minutes

at a normal speaking rate.

It ended with the catch phrase…"Tomorrow's future begins today"

And then I went upstairs into my attic,

Where I dragged down some Halloween special effects items that we try to

use every year.

Like our smoke/Fog machine, and a couple of strobe lights and my small wind machine.

Basically a high speed fan.

And an ice chest that I had modified to cool the smoke from the fog machine. When the fog smoke is cooled it hangs on to the ground instead of floating up into the air.

Holy Crap. I just realized, I don't have a name for the project yet!

It used to be,

"The Slingshot Mark II project",

but, that title died a long time ago when We started building the advanced model of the project.

I didn't want to call it the Slingshot Mark III,

So, My mind went back to my drunken pronouncement

"The Spirits of Texas."

Referring of course to the minds and talents of the people that had gathered to combine their expertise and skills to produce this wonderful craft.

And plus there is that alcohol reference as well .

And seeing that alcohol was paying for the project I thought to be very appropriate

So, I decided to christen it

"The Spirits of Texas.

In the very same vein as The plane, which was officially named Ryan NY-P , aka "The *Spirit of St. Louis*," was named in honor of Lindbergh's financial backers who were several men from St.Louis

They were the ones who paid for the building of the airplane.

The NYP stands for New York–Paris.

And since all of us are Texas residents

(The majority are from the DFW area including myself.)

I really think the name suits the project to a tee.

So after working all night on copy and quickly designing a platform built of

boxes and cases and covered by a big piece of black satin like fabric.

I laid out where the smoke machine should be placed.

And where to put the strobe lights so that they would flash on

the vehicle and not give the photographers, cameramen and reporters seizures.

I wish I had time to build a turntable like at the auto show,

But, you gotta make do with what you have.

So you can do it. With what you've got….

The Hybrid Diaries
Chapter Thirty One
Wednesday, September 26, 1990.

The Sun was coming up over the horizon and I could see light shining through my windows and playing shadow games on my floor.

I packed up everything I could think of,

The black fabric, the display stuff.

An American flag and a Texas flag,

Some fishing line and a few eight foot wooden dowel rods.

I woke up Dick and asked him if he wanted to come with me and help set up "The Show"

Dick said that Yes, He would LOVE to come with me and help set everything up.

It turns out that Dick was actually feeling guilty about not helping out on the project and he was dying for an opportunity to help out with the project.

I told him that his position of sponsor and benefactor more than made up for any perceived absence on his part.

And that just having him here was help enough.

I woke Up Rhonda and I told her that Dick and I were going on ahead and that she can stay home and wait for Candace and Willy to show up.

That way she can show Candace where the test site is and where we would be.

So, I called Bill to see if anyone would be at the shop.

And if not, when would someone be there so we can start cleaning up the place and get everything set up and ready for the news crews to show up.

Bill laughed and said that the guys hve been there all night and they have been busy getting the shop in order .

I told him "Thank you" and I asked if they had a turn table?"

He said that they didn't have one

but, they might be able to build a platform out of boxes

and left over pallets if I could bring something to throw over it to give the pile of pallets and boxes the illusion of being a real platform.

I let him know that we were bringing and that we were planning on doing a press conference in the conference room for the reporters and the photographers and the news crews.

Bill said that the shop was at our disposal and that they would do anything we need them to if it would get some publicity for their company and to our cause.

I thanked Bill for everything and hung up the phone.

"Well Bill has everything under control at the shop.

And the guys are cleaning and getting everything ready for the press.."

"We are lucky that the truck is relatively clean and that there is no junk in the bed.

But, we will need to bring a change of clothes

 so I can wear something more presentable for the presentation." I sighed.

Well after we got everything packed into the truck,

we headed back out to Grand Prairie and drove to the carbon fiber shop.

There, we unloaded everything and started to construct the platform out of pallets.

 We were fortunate that they had enough pallets stacked up in the back of the property .

We carried the pallets inside and we started to build the platform in the conference room until one of the guys looked out the window.

It was almost 10:00AM and Dick had told the press that we were doing the presentation at 11:30 and they were already starting to arrive.

Big vans with collapsible transmitter antennas on their roofs.

So far, there were only three that were parked up front and another van driver was more brazen, by parking and setting up right in the company parking lot.

I didn't have time to look for long,

So far the turn out looks like it was going to be much bigger than either Dick or I had anticipated.

So, we had to scrap the plan of setting up in the conference room.

And we decided that it would be better for everyone if we set everything up on the production floor.

So, we cleared a space near the center of the room.

I hung the American flag and the Texas flag from the wooden dowels that I had brought from home.

I used the fishing line so that they looked like they were hanging in mid air.

They hung together side by side from the ceiling so they could hang in front of the "Stage" and serve as a curtain that we could raise

and lower from the side of the Stage.

Bill also had a HUGE American flag that normally hangs from an inside wall

It would be used as a back drop for the platform "AKA" the "stage" and the vehicle sitting on top of it.

Then we dragged chairs from every room of the offices.

Desk chairs big leather chairs all the chairs from the conference room

and any folding chairs that someone found stacked in a closet in some corner of the shop. and set them in front of the flags and the stage.

I called my house.

Rhonda said that Candace and Willy had arrived there and they were getting ready to leave.

And head out to the shop,

So, I reminded Rhonda that they would need to pick up some ice for the kegs and the fog machine before they get here.

Because there will be no time to do it after they get here.

Rhonda said that they promised that they would remember

and that they would be there soon.

So, I hung up the phone and I finished stacking the pallets and covering them with the black satin fabric that I had brought from home.

Then, I got the strobe lights out and I placed the fog machine and the fog cooler behind the pallets so they were not immediately visible to the press crews.

I set the wind machine up in front of the platform so that the breeze would blow on the girls and make their hair blow in the wind.

It would be a nice touch.

I borrowed the podium out of the conference room and set it up next to the platform

Christopher helped me hook up the loud speaker system to the mike.

And we tested the system and adjusted the volume so that we can be clearly heard in the back of the production floor.

Rhonda and Candace and Willy arrived at the shop in good time.

And they did indeed remember to bring lots of ice.

Candace had plastered some magnetic signs for the brewery on the doors of the Suburban

Willy got out and he started unloading the kegs from the back.

Rhonda and Candace came in carrying two dress bags

and a couple of bags of clothes and make up and everything they would need to look gorgeous.

Dylan and Christopher ran outside to help Willy bring the kegs and the plastic glasses and the ice inside.

Suddenly a couple of the news crew came walking over to the Suburban and they started yelling…

"Hey!, is there beer in those kegs?"

"Are those for us?"

" Need some help moving those heavy kegs?"

" We can help you lighten your load…"

We told them the beer is for after the conference and demonstration.

But, PLEASE, take lots of pictures of the trucks and the brewery signs.

And the outside of the company.

We wanted all of that on video.

They said that they would be more than happy to film anything we wanted as long as they can get a taste or two of that beer.

We said that was the initial intention .

And that was why the kegs were there in the first place.

Well, in spite of the many offers,

we were able to get everything into the building without too much of a problem.

We did decide to set the kegs up in the conference room

Since we could have greater control over who would have access to the taps.

Rhonda and Candace threw posters on the walls

And hung the banners in open door ways.

Once they were finished hanging promotional posters for the Brewery they disappeared

into the Ladies bathroom with Molly, the receptionist.

They were in there for a bit when Molly came out looking very nice, she had her war paint on and she looked pretty.

Then Rhonda and Candace came out.

Talk about "Wow factor" these girls looked off the hook.

Very nice indeed.

The girls had worked out their parts for the presentation.

Rhonda would read my copy and Candace would stand next to the project

And she would do her Carole Merrill "Price is Right" waving motion over the vehicle

And make "Oooooing and awwwwing " gestures or what I like to call poohing and pawing over the vehicle.

Just like the Super models do at the auto shows.

Funny thing about irony…

Rhonda at one time wanted to be a show model back when we first started dating,

She even worked the Dallas Auto Show for Toyota.

Doing the big smile and hand waving thing over a 1986 Toyota Supra.

They both looked very nice together

Now, she will be reading my copy that I wrote to a hopefully eager audience.

OK now things are starting to look ridiculous outside.

There are now several news trucks outside.

Parked both in the parking lot of Global Composites.

And also across the street along the fence line next to the end of the runway facing Jefferson. .

There they are with their broadcast masts up.

And then I heard someone Knocking on the door to the shop

Wanting to come in.

So, I walked over to the door to tell whoever it was that We were almost ready.

So, I opened the door and there being pushed against the door was

Our old pal John Procktor from Channel 7 news.

He was getting squashed by the other reporters trying to bully their way into the offices.

I yelled at the hoard of news people

That we were just about ready to invite everyone inside and show them what hey have been waiting patiently to see.

And I asked them to give us just a few more minutes to finish setting everything up.

That made them settle down a bit.

long enough for me to open the door a little wider and yank John in by his arm and quickly close the door behind him.

John Thanked me for saving his life,

because they (the reporters) actually had him pinned into the door

and he could see his breath on the door glass.

He was really mashed up against the glass door pretty good.

So, I poured him a beer and asked him to sit down and catch his breath and I would be with him in a moment.

John said thank you as he took the glass of beer from my hand.

And he took a long drink.

He was pretty frazzled when he took a drink,

and after a bit, he was starting to look more like I remembered him looking.

Rhonda and Candace said that they were ready,

Bill and the boys also said that they were ready

Everyone said that they were all ready and Dick and I told them that we were ready to open the doors.

So, Molly the receptionist walked over to the door

and opened it wide and invited everyone inside and she showed them to

where the conference was to take place.

Naturally the first seats to go were the high back conference chairs and then

the office chairs went next leaving only the metal fold up chairs.

But, soon they too were filled and the rest of the reports and newspaper

columnists were left with standing room only.

You could hear the din of about thirty five reporters and news paper people

jostling for elbow room.

Bill walked up to the podium, cleared the throat and he began the

presentation.

I won't go into details,

but, Bill was very eloquent and he thanked everyone for showing up and he

explained

what they do here at Global Composites of Grand Prairie.

And then he kindly introduced everyone to the production floor staff who

were standing

on the platform next to a tarp covered lump in the middle of the "Stage"

Then Bill said "so, now without any further ado,"

" We would like to introduce you to The Project and it's inventor."

" Mr. Andy Barrientos" I swallowed hard and walked out.

Cheers went up and I came out and I thanked everyone who was in

attendance.

I explained the concept of the vehicle and our goal to race in the International

Energy Challenge.

And then I asked for the curtains to be raised.

Up came the Texas and American flags

And there stood Candace looking very attractive.

Smiling and waving to the crowd of cat calling and whistling reporters.

Then Rhonda came out and she walked up to the podium and she began to read the copy that I wrote..

And when she said "The future begins with today" Dylan and Sam pulled the tarp away and the crowd gasped I could actually hear them say Awwwwww!

And there it stood, the "Spirits of Texas" still in it's carbon fiber gray material color.

But, still looking like a formula one race car and jet somehow molded together into one really cool looking vehicle.

Flash bulbs were going off like crazy.

I could hear the reporters yelling out questions for me to field.

Questions involving how much the vehicle costs went unanswered and other questions about how fast was the vehicle.

And how long could it run on a charge.

I deferred to those to Toby.

We gave the photographers about fifteen minutes of photo time.

Then the guys pushed the vehicle across the street and over to our test track.

They opened the gate and the crowd flooded into the runway taxi area

And over to Command Central

where our canopy tables and equipment was already set up.

They crowed around the testing equipment.

And they watched the readings pop up on the monitors as all of the systems were coming up online.

Now unbeknownst to us, about twenty five thousand feet up was a Texas Air National Guard F-15 strike Eagle that had just taken off from Carswell Air

Force base in Ft. Worth.

And it was climbing to about fifty thousand feet when the starboard engine jet intake

swallowed up a Canadian Goose on it migratory flight to south Texas and Mexico.

(No, the goose was not flying at 50,000 feet.

The plane was climbing to that altitude)

This is not a good situation for a jet engine.

They are not designed to ingest something that large

The impact of the poor goose completely tore out some of the turbine blades out of the starboard engine.

And the engine blew to pieces. Leaving the plane in a very serious dilemma.

Bird strikes are not uncommon but, to have one now.

At this very moment while the pilot was trying to climb to altitude was not good for us either.

I had just told the assembled multitude that I was going to do a demonstration run to the far end of the runway and then back to the starting line.

To show the media that the vehicle does indeed do what we said it can do.

So, I started down the runway and about half power.

Just as I got to the end of the runway,

I made my turn and was starting to head back when all of a sudden the runway lights came on and I could hear a Claxton blowing in the distance.

"This cannot be good." I thought to myself.

I bet the lights look really cool at night.

But, believe me they were very frightening if you are still on the runway when A big plane is trying to land.

I was betting the farm that The pilot of the F- 15 and I both said "OH crap " at about the same time..

My mind kept going back to when the C-130 landed on the runway next to the emergency runway,

And all the turbulence the plane generated when it passed me on the tarmac.

So, I immediately assumed that there was an emergency situation going on around me.

The bright Yellow fire trucks and the ambulance all with their flashing lights and sirens blaring and the Staff car rushing towards the emergency runway was also a good clue

that I was about to be in a world of hurt if I didn't do something quick.

So, not even thinking,

I pushed the control stick to full power and I slammed the throttle all the way full.

And the vehicle shuttered as all three of the wheels spooled up to full power.

I shot off down the runway as fast as the little car could go.

The heads up display said 78 miles per hour.

If there was a plane landing behind me it would be moving at around one hundred and Sixty five miles per hour.

So, if I don't have enough distance between me and the landing airplane then there is a possibility that the plane could come down right on top of me.

I held the throttle at full power and crossed my fingers and did a little praying.

As the starting line at our end of the runway was looming ahead of me.

Without the sounds of an internal combustion engine in my ears.

I could clearly hear the sound of an approaching jet,

It's one jet engine running at close to full power in a desperate bid to keep the plane aloft

long enough for the pilot to get the wheels on the ground.

And that engine sounded like thunder,

It was very loud and it sounded very close to where I was.

It drowned out the sound of the wheels humming and the tires rolling on the tarmac,

and the rush of the air around the vehicle went unnoticed

I heard the squeal of the planes wheels touching down on the runway very close behind me.

I was thinking to myself, "so, this is what a cockroach must feel like."

As I scurried away from the landing plane for the safety of the blast wall.

I got off the runway as fast as I could and turned the vehicle to the left as fast as I could

so I could see exactly where the F-15 was in relation to me.

The F 15 was almost on top of me just a few hundred feet behind.

It was still billowing smoke from its right engine and I could see the heat rising in the air behind the plane from the left engine's jet exhaust,

the heat was rising up and distorting everything behind the twin massive tails as it's wheels touched down on the runway not far behind me.

I continued to speed off the tarmac and try to get the vehicle behind the blast wall at the far end of the runway because,

Well, do the math, A F-15 engines normally produce about 17,450 lbs of thrust per engine.

The F-15 has TWO of these monster motors

And The Slingshot only weighs about sixteen hundred pounds including the driver.

If I didn't get the slingshot off the runway and fast,

the jet wash could literally blow my vehicle off the runway.

And me along with it.

The press photographers snapped some really cool pictures of the Slingshot (with me
behind the wheel) whizzing by the cameras.

And a huge F-15 taxiing right behind my vehicle.

I bet they looked really cool.

I shut down the motors and coasted to a stop behind the blast wall and I got out.

My hands were shaking, and my heart was going a hundred and fifty beats a minute.

I took off my helmet and I got out.

I was greeted by a throng of press and media people congratulating me on a great show

and they were impressed at the speed of the vehicle and it's styling..

I tried to keep everything together, and I tried my best to play along.

I tried to look as composed as possible.

Let them think that this was all planned out in advance and that this was all part of the show.

I asked some of the photographers if they got any decent pictures of me or the vehicle

Because I wanted to get some copies for my office and for press purposes.

And the photographers said that they'd be happy to give me copies of the pictures.

And then, at that very moment.

the General's staff car rolled up to where we were and the General got out and he walked over to us.

His name was General J.R. Simpson.

He was the base commander (and also the one who gave Bill permission for us to use the emergency runway in the first place.)

General Simpson said that he was happy that no one was hurt and that everything worked out for the best.

No casualties and the pilot was able to land safely on one engine.

And I was able to get out of his way in time.

But, he was a little bit on the angry side.

Because, no one told him that we were having a press conference and that the media was out there in force to cover the event.

So, the General told everyone there, that

"He would be asking for the cooperation of everyone there."

He said that the press could cover the story about the experimental vehicle being tested for the Challenge in Australia or they could cover the emergency landing of a Air Force F-15, but, not both events at the same time.

And if they chose not to cooperate

He could have his Military Police arrest everyone there for trespassing on Government property without expressed written permission.

He didn't mean us, he was mainly concerned with the camera trucks parked on the other side of the fence and the blast wall who were indeed parked on Government property without any permission.

from either the base, or Bill or from us.

So, everyone agreed, and we all retired back to the Composite place and we all drank a beer while the guys brought everything back over to the shop.

I hob knobbed with the press,

Offering them glasses of beer and telling them more about the vehicle and about the Challenge.

Dick came running up to me and he slapped me on the back and he said,

"That was freaking awesome!"

"Are you alright?

"MAN, Andy, that car can really fly!"

"Wow, you were really moving when you got back here to the start line."

" Man. That was exciting!"

"That plane was almost on top of you!"

" Are you alright?"

At least Dick was concerned for my well being.

I got to talk to John Procktor after things settled down a bit and We talked about the race,

and he commented on how amazed he was at the progress I had managed to make since the first time he met me at my house.

John commented that The vehicle didn't look anything like the original prototype that he had first viewed.

He said that the vehicle looked really futuristic and that it was really moving when I was Driving it back to the starting line.

I told John "that was because the vehicle was actually running on fear."

We both let out a hearty laugh and then John told me how truly proud he was and how

happy he was to get the call from the news desk to come out to the presentation.

I thanked him for coming out and I handed John a glass of beer.

And I said to him that the beer is courtesy of one of my sponsors.

He took a long drink and said that the beer tasted good and that he thought that I had indeed found a good sponsor with a great product to back me on my project.

John said that he would cover both stories.

And that he would do his best to make sure that we would get proper coverage.

Once again I thanked him for everything,

And John and his camera man and the majority of the news crews were heading out and going back to their stations or offices to write about the events that unfolded this morning.

And hopefully they'll write volumes about a little experimental vehicle that was Designed and built totally in Texas.

And about its designer who has it's sights on taking on the entire world.

For the sake of advancing scientific knowledge.

And hopefully benefitting all of mankind.

I also wanted John to mention the little brewery in Paluxy who was sponsoring me with the delicious beer made with only pure spring water, and the freshest hops and barley malt.

Hmmmm Hmmmm tasty..

John promised me he would do his best.

So I walked John out to the front door and I shook his hand and thanked him again for coming out. John got into the van and waved good bye.

And then he was gone.

I watched his Channel 7 news van go down the road heading back towards Dallas.

I fell hard against the door frame and I leaned against it to hold myself up.

I felt my energy all but leave my body.

Because after everything was all said and done, I was exhausted.

Everything hit me all at once.

I felt weak in the knees and I really wanted to sit down and rest a little while,

The realization of what had just happened was starting to set in.

And so, I went back inside chased some straggling media people away from the kegs and poured myself a glass of beer.

Then I poured myself into a chair and tried to relax.

Rhonda and Candace came up to me and I told them how proud I was of them and how good a job the girls did.

And also, how beautiful they all looked.

Then I shared what happened at the test track with the F-15.

And almost getting landed on

And then I told them what the General said to the press and that they could either cover our story or else the could cover the emergency landing at Hensley field.

But, not both stories in the same story.

Or else they would be arrested.

And possibly prosecuted.

"Well, I wonder how much press we'll be getting from all of this?" Candace asked.

I let them know that I have no idea right now who will give us the best coverage.

But, I also let them know that John promised me that he'd give us a good story.

So, we gathered everything together and the guys got what was left all put away and we packed up the Suburban.

And Willy helped put the kegs inside the back of the truck.

Then I called Willy over and asked him if he would like to drive the vehicle.

Willy immediately said "No" because he doesn't have a driver's license.

And plus, he rides a bicycle.

Not because he can't afford a car, but, because, when he first came to America all he could afford was a bike, and he bought his with his own money.

He was very proud of that accomplishment.

Because it was the first major purchase Willy made when he first came to this country.

I told Willy that "It's OK, the vehicle is really a three wheeled bicycle.

And that you have to crank the pedals to get it to move.

That seemed to pique his interests somewhat..

So, with everything packed up and put away,

we caravanned back to the house to watch the news on TV. Rhonda drove the crap

mobile home while Candace and Willy followed in Candace's Suburban.

And Dick rode with me in the "Company truck".

And just like I was expecting ,

The majority of the TV stations did indeed carry the story about the F-15

flying out of Carswell Air force base in Ft. Worth suffering a bird strike,

being disabled and having to make an emergency landing at Hensley field.

Amazingly you could still see me and the Slingshot leading the F-15 towards the taxi way.

But, there was no mention of me or the tests or the fact that I could have been killed if that plane had landed on top of me.

So I guess the stern warning from General Simpson was more than enough incentive for the media to not cover me and the project.

Candace and Rhonda were watching the 5:00PM news from the channel 2 news room on

the TV in the bedroom while Dick and I were watching other channels in the living room.

And when the news came on they would yell at us when the news story was being aired.

And we in turn would run into the bedroom and watch the news in there or else when the story came on the channel that we were watching ,

Then, We would holler at the girls and then they would come running in and watch the news on that channel in the living room.

So when the story was over, we'd change the channel and channel surf some more until

The 5:30 news and then repeat the process, until we viewed all the channels and we were

able to match up the channels with the news trucks that showed up earlier on in the day..

Our old friend John Procktor of Channel 7 news was the only one who covered our story

and he did such a good job,

He made the whole thing seem like it was a huge publicity stunt designed from the get go

as a media grabbing spectacle.

And that the jet behind the vehicle was there as a prop to emphasize the sleek design of

our vehicle compared to the deadly efficient lines of the fighter jet lines of the F-15.

And it was quite a story.

You could see the stunned and scared expression on my face if you get up close to the Television screen and you took a good look at the video.

Dick and Candace looked a little shocked when they saw the F-15 rolling up behind the slingshot at a fast rate of speed.

And you could almost get a glimpse of how close the plane came to touching down on my windscreen..

Then the story switched to the interview that John had with me after the debacle.

And he said that the vehicle looked very different from the prototype that he had originally covered way back when.

Sometimes it does seem like a lifetime ago, but, it was only a few months ago. I told John that was because I had found some like minded sponsors who shared my same

vision of clean electric powered vehicles replacing the pollution generating gas guzzling cars of today,

Utilizing space age materials and advanced electric motors and the latest in solar power technology to produce one of the world's first human powered electric hybrid vehicles.

Rhonda sat there white faced and quiet while she was watching the news stories,

which is not like Rhonda..

She had no idea what had happened earlier because she was with Candace and they stayed in the shop. because, well. You don't wear strapless evening gowns on a airport tarmac.

Or like the girls said, "you can't go out there in a Bob Mackie or an Oscar."

So, when Rhonda saw the newscast and the F-15 taxiing off the runway as I drove the vehicle back behind the blast wall.

She took a deep breath and she looked me straight in the eye, and she told me "Do not EVER do anything that stupid again!"

Then she got up and stormed out of the room.

And out the back door into the back yard.

I got up and I went after her and I yelled after her

" It's not like I planned that to happen. It was an emergency landing! "

"I had no control over what had happened." I said

"Exactly!" you had no idea that was going to happen and you were lucky to have gotten away with your life."

" Do you know, that I can't live without you, you big Goof!"

Then she started crying. I tried to give her a hug but she yanked her arm away and she went back inside.

Leaving me outside by myself standing next to the burned out remains of what as once my garage.

The landlord still had not rebuilt or replaced the garage yet.

I quite frankly believe that he isn't going to repair the garage until we leave or else someone else wants to move in and fixing the garage is part of the deal.

A couple of minutes later Dick came outside and he asked me
if I could help him get a keg down from the back of the Suburban.

I was more than amiable to the idea and together we took one keg down
and one trash can and a bottle of CO_2.. then we set it up on the back porch.

Dick and I went out the back gate and walked across the parking lot
and around the building to the front of the little convenience store behind our rent house.

We went in and bought a couple of bags of ice for the keg and I bought Rhonda a milky way candy bar.

(one of her favorites) to try and patch things up with her.

I've been with Rhonda long enough to know that what she was feeling was fear.

Fear of an accident taking me away from her.

Thankfully, She had not given much thought to the possibility of me being injured or killed in a vehicle accident while testing or actually racing the Slingshot

And to be perfectly honest, neither did I.

I had to spend the next hour in trying to reassure her that I was doing the entire project

"The Right way".

Not taking unnecessary chances on the design or the fabrication of the machine and not taking any shortcuts.

I was doing my best to try and engineer some level of safety in the overall design.

Which did little to ally her fears of doom.

Once those fears get into her mind it is almost impossible to exorcise those demons from her thought processes.

Even if she said "OK I feel safer about your chances,"

I know that somewhere in the back of her mind she was still dwelling on those old fears.

And She was just telling me what I wanted to hear.

Well, we all decided to order some pizzas for Dinner.

After we ate, Dick and I walked back over to the convenience store behind my house and

we snagged some copies of the Dallas Morning News.

The newspaper was very kind in their coverage of the event.

And they placed the story on page one of the Metro section of the newspaper with a

picture of me and the vehicle with the F-15 coming up behind me,

Very closely I might add.

The camera angles do make it look far more dangerous than it actually was,

And Rhonda was right, I WAS VERY lucky to have not been killed.

The pilot was an extremely good pilot to land a damaged airplane and avoid running me over like that.

Willy was in the house with the girls.

I could tell that Willy was nervous about testing the vehicle.

Maybe him seeing the whole episode live and on TV might have discouraged him or at least put the fear of God in him,

Either way, I know that he would do almost anything the Wigglers asked of him.

He is a very kind and polite man.

I called Toby and asked him if he was watching the news.

He was in an excited state when he answered the phone, .

"Heck yeah, I saw the news story on Channel 7."

" Man, that plane almost squashed you!"

I was glad Rhonda wasn't hearing this, it would just fire her right back up again.

We talked about the project and Toby was telling me that he felt that we might need

back up replacement wheels "just in case".

I told him that Bill wanted to build the final prototype and we would need to have some

wheels to hold the vehicle up, So that would be an extremely good idea.

I hung up the phone and since it was getting late and we were all pretty worn out by the day's events, we all decided to turn in early tonight.

I got the guest room all set up for the Wigglers and I got the sofa ready for Willy.

My dog Thor was all over Willy.

Sitting down next to Willy and almost imploring him to pet his head and then curling up at the foot of the sofa just barely out of arms reach.

Just in case Willy decided that he desperately needed to pet the dog again.

Well, I had Willy settled in on the sofa.

Dick and Candace in the guest room and Rhonda and I in our bed.

Before I had retired I told Dick that I was going back to Toby's shop in the morning and

He and Willy can either come with me or else they can meet me at the Carbon fiber place

And we can see about getting Willy suited up trained and ready to do some test runs.

Dick said that was a good idea and that he'd like to ride with me and Willy. I told him

that sounded like a good idea and I bade him and Candace "Good night"

Then, I went to bed.

I scooted Rhonda over gently to her side of the bed.

As I laid my head down and I tried to clear my mind and focus on trying to relax .

soon I was asleep and I began to dream about the days events.

And in my dreams I saw a Huge Red tailed hawk circling over my head.

And then it came down beside me and it spoke to me. it said

" Everything has a certain amount of danger to it."

" Watch out for the clouds on the ground."

" They will try to grab you as they go past."

Why do dreams never come out and tell you what they are really trying to say?

Why is it always this cryptic nonsensical business with abstract information?

Why can't they just say, Look out for that train?

Or watch out for the hole in the road?

Why is it always some cryptic crap?

I woke up, dismissed it all and rolled back over and went back to sleep.

The Hybrid Diaries
Chapter Thirty Two
Friday September 28th, 1990
The day after....

OK when we got back to Grand Prairie.

We walked in the door of the carbon fiber place,

Bill asked us to come into his office.

We walked over and sat down and Bill let us know some new safety policies that he had worked out with General Simpson.

All of our radios will be tuned to the tower's radio frequency and they will monitor our transmissions and notify us by radio if any more emergency situations begin to unfold

while we are testing.

The General did not revoke our permission to use the runway.

Instead, we now have a plan in case to hopefully prevent any more future "incidents".

Naturally, we agreed to abide by the General's wishes seeing as he was doing us a tremendous service by allowing us on the base in the first place.

Dick took Willy out to the shop floor to ask the guys if they had something smaller

than the cover alls I was wearing when I was testing the vehicle, they said that they would check,

In the mean time I was still talking to Bill and he was now on item two of his check list. Next on Bill's agenda… There was a knock on his office door, and Bill told me, that

There was someone who was wanting to meet me.

Bill got up out of his chair walked out

from behind his desk and he reached out his arm and turned the door knob and opened the door.

And there stood Major Dwight D. Smiley from the Texas Air National Guard.

Major Smiley reached out his hand to shake my own hand.

He introduced himself and he gave me a firm handshake.

Major Smiley whose call sign is "Reaper" congratulated me on my vehicle and my

ability to get out of his way in time.

I told him "thank you again for no landing on top of me and crushing me" and I told him

that "I was truly happy that he was able to get back down to Earth in time. And that I was happy that we were both able to walk away non the worse for wear and tear."

Major Smiley said that when he was landing and he first spotted my vehicle on the runway in front of his plane,

he had no idea what the heck that thing was speeding down the runway at a "high rate of speed"

He thought at first that it was a race car until he got a look at the cockpit.

Then he thought that GD (General Dynamics) was testing some new type of craft.

So, it was only natural that he would want to see the thing close up and personally.

I took him over to the production floor.

And I showed him "the Slingshot aka The Spirits of Texas"

He stared at it for quite a long time running his hand over the canopy and staring at the

solar panels that were unfolded and tracking sunlight

(well the fluorescent lights and the sunlight streaming in through the open sliding doors

in front of the shop) all by themselves.

"Right now, it's generating electricity all by itself." I said

"Man this is one cool piece of technology." Said Major Smiley.

"Can I have a ride in it?" He asked.

"Sure you can" I replied, if I can get a ride from you one day." I answered

"Well maybe, after I get my engine replaced. I'll let you know."

He told me that he isn't actually based out of Carswell.

 he was flying across country and had to make a scheduled stop at the base before heading back to California.

I told him that would be fine with me.

I opened the canopy and I let him sit inside the cockpit.

 And he was amazed when I turned on the power switch and the heads up display came up.

"wow, you have a HUD! That is awesome!

It looks just like a fighter plane in the wind shield.

"That is freaking amazing." Major Smiley was all grins at the technology the vehicle possessed and had rolled into it's little carbon fiber body.

So much like his own cockpit.

With the control sticks and no steering wheel.

It almost looks like a "Fly–by-wire system.

Except the control sticks on my car actually move back and forth.

While on a F-16 they are fixed in place and the grip of the pilots hand inputs steering information to the on board control systems.

Not sure if the F-15 has a fly by wire system.

 I'll have to find out somehow.

Major Smiley climbed out of the cockpit and he walked all around the vehicle, just like a judge walks around a show car.

Doing his best to view the design from every angle careful not to miss any details.

Kneeling down on one knee so he could see the vehicle from ground level.

Something he didn't have a chance to do when he was almost rolling over me on the tarmac.

He asked me about the forward wings / solar panels on the nose of the car.

And I told him that they were for generating down force and also electricity to recharge the batteries as well.

He asked me where the drive motor was and I opened the hood and I showed him the flywheel generator instead.

And I think he mistook the flywheel generator as the motor.

I was glad because even though everyone in the shop knew where the drive motors were located.

 that was as far as I wanted the information to go.

I still have my personal obligation to protect Toby's invention and it's my duty to try and help him find his "Million dollars"

I am sure that Major Smiley recognized some of the parts for the new pre production

racer that Bill and the guys were constructing as the final preproduction race vehicle.

Major Smiley once again shook my hand and said that it was a pleasure to meet me and that he would get back to me on the plane ride later.

I told him "No worries" just hook up with me when you can."

"You know where you can find me."

I smiled at Major Smiley and he walked out of the shop and across the street back towards the base.

All the while I saw thinking to myself…

" what a nice guy, and, I sure am glad that he didn't kill me…"

Back at the shop.

Dick and Willy were able to get a smaller sized jumpsuit for Willy and

They were able to use my helmet for Willy's head.

I could tell he was still very nervous about driving,

I let him know that there was nothing to worry about. And although

I did my best to reassure him and ally his fears,

I had to take the first couple of test rides with him sitting behind me in tandem.

So, on the third run, We had decided to test the flywheel generation system.

That was where Willy's legs come into play.

We set the vehicle on some blocks to keep the wheels off the ground and I had Willy sit down in the seat and I told him to start pedaling.

That part, Willy had no problems with.

That part of the test and soon he was spooling up the flywheel and the gauges started sending data back to the command center computers.

The test results were very encouraging and Willy was starting to get the hang of all the controls and soon he would be ready to take the vehicle for his first test runs.

We showed him how to work the radio.

And we did several "tests" using the radio and us giving him commands and instructions

and Willy would always answer us with "Yes, OK, very good!"

We were going to take the vehicle across the street for a few test runs.

but, the skies were very cloudy and there was a 30 percent chance of rain.

Normally when the weatherman says "thirty percent chance of rain" that usually means

that there is really a seventy percent chance of rain and a thirty percent chance that it will not rain.

So, we closed up early and headed back to the house

When we went home the girls were no where to be found,

so, we sat down and played Super Mario Brothers on my Super Nintendo system.

We played and drank some beer and listened to music on my stereo.

I played some albums.

At last the girls showed up.

They had been in meetings with the director of the Leukemia society and with Jack Coakley who was a member of the Margarita Society.

They are major contributors to the Chili cook off.

Along with the representative from the ICS (international Chili Society")

They had been planning and scheming and now they had talked themselves into the position of agreeing to supply the beer for the event, PLUS, donating some items for the silent auction, and providing free beer for the Cooks party on the Friday night before the cook off.

"It sounds like a whole lot of fun" I said to the girls.

Rhonda said, "No, it sounds like a whole lot of work!"

"But, it's all for charity!" Candace said as she came up behind Dick sitting down in a chair and trying to focus on his level and trying to avoid being killed by a koopa.

She bent down behind him and wrapped her arms around his neck and then she kissed his

ear and "Damn!"

He lost his life. "Luigi died."

I had invited Toby and Bill to come by my house for a couple of beers and surprisingly

they both showed up.

It was the first time Bill had ever come by my house and he sat down on the couch and was pretty quiet for a while.

Toby on the other hand was all smiles and in a mood to celebrate.

He had a few beers and then he began congratulating everyone on the success of the vehicle and of his wheels, generators and motors.

Then he slapped me on the back and he shook my hand and he said

"I can't believe it, we reinvented the wheel!"

Truer words were never spoken.

We had in fact re invented the wheel.

Our approach to the challenge was a truly novel one.

And We were all grateful for all of the hard work in bringing my designs to life and then making them work exactly as I had envisioned them to.

That in itself is no small feat.

Bill looked a little uncomfortable on the sofa and it seemed to me that we needed some

other form of entertainment to arouse the interest of Mr. Cabumba.

So, I asked everyone

if they would like to play a little game of cards?

"What kind of game?" asked Bill hopefully

"Poker or Blackjack. I've got chips." I replied. And that was all it took.

We were all sitting around the dining room table passing out chips and dealing cards and drinking beer.

That seemed to make Bill feel right at home.

The good thing about playing cards and drinking beer is that when the drunk level goes up then the inhibitions come down somewhat and people can work up enough courage to say what is on their minds.

Bills biggest concern was regarding the shipping and the voyage to Australia.

I finally worked up enough courage to ask Bill if he was thinking about accompanying us on the trip.

Bill said "No" at first, and then he said that he hadn't given it much thought. He had been thinking about sending a couple of the guys along with the racer for technical support but, he hadn't made any decisions as of yet.

Dick and Candace and I had explained the process of getting the project to Australia on a container ship that we had one picked out a vessel to ship the project over on.

And Dick said that while we were busy playing with the project, he had been on the phone and sending out letters to several beer distributors in Australia. And he was also shipping a couple of refrigerated containers filled with beer to try and impress the distributors into buying and selling his beer in some of their retail outlets.

The night wore on and we all shared with each other our personal motivations about the project and our feelings about the race and it was very heartening to learn that we were all now intrinsically involved in the project.

And together, we can accomplish what other people once considered impossible.

Toby tried to tell me that I was a genius. And I told him that I wasn't the genius.

I told Toby that HE was the genius.

No one could build the wheels the way I wanted them built unless they have a tremendous amount of technical know how to bring an idea like mine into fruition

Over cards we hammered out our plans for the trip and together,

we had decided that we would indeed need two ships.

A container ship that Candace had found to transport the racers, monitoring equipment

and spare parts not to mention the other two containers

414

And the beer and the "Pit crew" whose main job is to keep an eye on our containers and make sure that they get to our destination in a safe and timely manner,

And a cruise ship for the Wigglers and Rhonda and I and Toby and his wife and Bill and his spouse.

Everyone liked the idea except for Toby who was more afraid of technical spying industrial espionage.

He was consumed by the fear of someone stealing one of our power units or energy

regulators. Or possibly someone swiping one of our wheels off of our racer,

So, Toby very prudently had decided that he and his wife should and probably wanted to ride on the container ship to protect his interests.

Which is no major problem.

because the ship that Candace had picked out is staffed by a world class chef and other

atypical amenities not normally found on container ships..

Bill said that he wasn't sure if he or his wife can get away from the shop long enough for the trip and the challenge.

but, he said that he would look into it. And get back to me.

Funny thing about decisive people… When someone says, "OK, I'll get back to you…blah blah, blah…"

That usually means "a snowball's chance in Hell that anything like that's going to happen.

But if a decisive person says that, then you just never know.

Judging by what I have learned about Bill, you just can't tell.

He has a good poker face, and I lost quite a few chips to his card playing (a good host always tries to be a gracious loser and does their best to let the guests win, even if you

COULD easily beat their hand, It's always better to fold.

Unless this is a real game among serious players.) We were just playing

around to kill time and there was no money involved.

Just good fun. Rhonda was playing footsy with me under the table,

which didn't do much for my concentration.

I think that was her plan all along.

Just to keep me off balance and off my game.

We got finished playing very early in to morning.

And we could see the changing hues of a quickly arriving sun rise,

I decided that I was too tired from hosting the poker game.

And I wanted to get some well deserved sleep.

So, I told everyone involved that I was going to go to bed and get a few hours of shut eye.

And I bade our guests to drive safely and be careful and I asked if anyone needed a cab Ride.

Everyone said that they were fine and able to make their respective trips home.

I was sound asleep and dreaming about racing the new slingshot

With its new more powerful motors and what to call this machine.

The Hybrid Diaries
Chapter Thirty Three
Friday September 29th 1990
Andy earns his wings

Then the phone went off ring, ring, ring…

I called out to Rhonda to answer the phone but, no one answered, she must have gone somewhere with Candace.

I got up and Bill was on the phone…

"Did I wake you?" Bill inquired.

"Oh no, I was just getting up " I lied…

"OH, Ok then, Do you want to come down?

I have something I need you to check out"

I rubbed my eyes with my hands and I tried to get my mind to work…

"Oh, OK, Then, I'll come down in a little while."

"OK, about what time will you be here?" Bill asked.

"I don't know.. About, an hour? I've got to take a quick shower before I head Out.

" I smell pretty ripe."

"Ok, well get cleaned up and head on down here." Bill said and then he hung up the phone.

I did what Bill requested and I got showered

(to help me wake up more than to clean my body..)

and I got dressed and I went out to the truck and I got in and started it up..

Did I ever mention that I really love to drive that truck.

It's unrefined and brutal to drive

but, it's got more than enough power to pull stumps.

And I know it can haul almost any sized trailer..

It's comfortable and not as refined as a modern truck but, it's got old school styling and I really like the way it looks.

So anyway I climbed unto the cab closed the door, buckled up and I started the motor backed out of the driveway and headed back to Grand Prairie.

When I got to the shop, I pulled in to the parking lot and parked and I went inside.

MUCH to my surprise,

I saw Bill who motioned me into his office.

And standing next to Bill was Major Smiley.

And he was dressed like he was flying out of here.

I walked into the office and I shook Both Bill's and Major Smiley's hand

Major Smiley had been talking to General Simpson and together.

They were able to arrange for a trainer version of the F-16 (an F-16 D.)

To be all fueled up

and on the flight line and ready for a training flight.

So, Major Smiley asked me

" if I had ever seen a F-16 close up?"

"No," I said

"Would you like to?" asked Maj. Smiley.

"Sure I would" I replied

It was pretty hard for me to contain my excitement

but I did my best.

Bill watched me and Major Smiley as we drove across the street to the air field and we walked up to the flight line.

And there on the tarmac was a flight ready F-16D. It was fully fueled and had tags on key parts of the fuselage that have to be removed by the pilot who does this while he is performing his pre flight check lists.

So, Major Smiley took me over to the plane and I got a look at the cockpit.

And I noticed that the layout was very reminiscent to the layout that I am using for the cockpit of the slingshot.. including the control sticks on either side of the seat.

I walked around and looked up the tailpipe of the engine and then I ran my hand over the leading edge of the wings.

Then Major Smiley asked me to follow him into the building that houses the pilots briefing rooms and pilot locker rooms.

There I met up with General Simpson.

And he signed a pad that Major Smiley carried with him.

And then they took me into the locker room.

When I got inside they instructed me to follow Major Smiley to a locker where his pilot equipment was stored.

There next to his locker was a locker with My name written on masking tape stuck to the locker door.

I opened the locker door and there was a duffle bag loaded with pilot gear,

A pressure suit, boots and gloves.

" I got your clothes sizes and weight from Bill"

" I hope that these fit you ok."

Explained Major Smiley with a smile on his face.

I opened the duffle bag and I took out everything and laid it on the bench beside me.

And Dwight told me what each item of apparel was and how to use it.

The first Item I pulled out was a funny looking pair of pants Dwight said that it was a PRG CSU 13 anti G suit.

Then I pulled out a vest looking thing that Dwight said was a CSU 17 anti G vest

And then I took out what looked like a body harness, and Dwight corrected me and he told me that it was a PCU – 15 Torso Harness.

So I took the items and set them aside and I pulled out a flight suit and some pressure socks and a pair of underwear, I assume that I might need them just in case of … well.. you know.

We then walked over to a rack that had shelves that in turn were holding Helmets and steel toed boots.

I grabbed a pair of boots size 12 and we found a helmet that seemed to fit my head snuggly but not too. Firmly.

And it had a breathing mask attached to the side of the helmet that snaps and fastens into place.

Dwight said that it was a HGU-33 flight helmet, and it looks way cool.

With the gray colored face mask attached.

"Man this stuff is heavy. What's it all used for?" I had to ask

"To keep you from blacking out or redding out or passing out and crashing and dying."

"The anti G suits keep the blood from pooling up in your legs during high G maneuvers"

"The air hoses connect to a fitting in the cockpit During High G maneuvers Pressurized air is forced air into the hoses and they in turn connect to bladders that squeeze your legs and stomach, and torso area and force the blood back into your head preventing you from blacking out. "

"Or during negative G maneuvers the suit keeps blood in the lower extremities to prevent you from getting too much blood in your head."

" When that happens it's called redding out."

" It's very important if you are flying a fighter jet in combat conditions."

"Also important is breathing through your mouth and blowing out rapidly. "

"That helps you grunt and strain to keep your focus and it also helps to keep you from passing out." He explained.

So, He helped me get suited up and even though I have never worn anything like this before, at least I looked like a pilot.

Dwight had already had his flight plan submitted and he was already given permission

from General Simpson for him to take me up on a "Training mission"

basically once around the sky and then back down to earth and maybe a beer afterwards.

Dwight said that the plane we were going to go up in belongs to the 301st fighter wing,

which is based at Carswell air force base in Ft. Worth.

And the plane was being serviced at the General Dynamics facility in Grand Prairie.

Dwight said that he originally flew the F-16 before he moved up to the F-15.

The General had made some special arrangements to "Borrow " the plane for testing."

There were lot's of Naval aircraft, P3 Orions and F-14 tomcats and A-6 intruders there on the flight line.

And the Air Force plane looked a little out of place.

But, I was excited about going up.

The F-16 is much smaller than the other planes parked on the flight line.

But, it is capable of reaching speeds in excess of mach 2 and handling 9G maneuvers.

So, we got suited up and checked our equipment then we went back out to the flight line

We performed the preflight check lists and pulled tags and handed the tags and signed the log book and he handed the book over to the ground maintenance personnel.

We climbed into the cockpit and they hooked up everything .

All of the hoses and the communications equipment.

We did a final check .

The seats are set back at a 30 degree angle.

Dwight said that "it is set like that so a pilot can pull more G's without blacking out."

And soon we were ready to turn the engine over.

Dwight went through the check lists and When he was ready

He said "Let's light this puppy."

He pushed the ignition button and the engine whined and then roared to life.

Soon we were taxiing to the runway.

Dwight was giving his call sign and asking the tower for permission to take off.

We had to sit on the tarmac for a few minutes for the traffic to clear before we were given the go ahead to take off.

Once we got the go ahead.

Dwight pushed the throttle forward and he took his foot off the brake.

We lurched forward building up speed very quickly.

For those of you who have never been in anything larger than a Cessna,

let me describe the differences between the F-16 and a Cessna

Flying in an F-16 is a lot like flying in a Cessna 175 without all of the falling out of the sky stuff.

When the F-16 barrels down a runway your head gets pushed back into the head rest of the seat.

Your face feels like the skin of your face is being pulled back and your eyes feel like they are being pushed into the back of your skull as you are hurtling down the runway.

Almost exactly the way that a Cessna doesn't.

You get a real sense of speed in an F-16 half the way down the runway until the nose lifts off.

And then the rear wheels come off the ground.

Then. Major Smiley started a near vertical climb. climbing up to gain altitude As quickly as possible.

My stomach fell into my shoes.

I kept my eyes on the gauges and tried not to look forward.

Because all I could see was clouds and the sky.

All around us.

I yelled over to Dwight

"Watch out for birds! We only have one engine on this little plane!"

I yelled over the Intercom.

I heard him chuckle to himself..

Then he said, "Roger that"

Once we reached the designated altitude.

We leveled off and the sensation of speed fell away.

It felt just like being on a jet airliner in the sense that there is little sense of motion.

But, unlike a Cessna…

You did not have the feeling that the engine is straining just to keep up.

Your forward momentum is there.

And you didn't feel like you are going to suddenly drop out of the sky

like you get with most private planes.

Its ike a competent sense of false security. Knowing that you're not going to

drop out of the sky. Because of a lack of thrust and forward momentum

And the wind did not feel like it was pushing against you the entire time.

And not much wind noise to speak of.

The flight was fantastic.

I asked Dwight if he ever opened the plane up over Texas before.

And he said that regulations require him to not fly supersonic over population centers.

So, we were not permitted to push the plane over subsonic speeds. Too bad.

Either way.

He did let me take a turn flying the plane for a little while.

He kept his hand on his control sticks in case there were any problems there were none.

He said that I was a natural pilot and he said that I did a good job.

"Not Bad for someone who can't fly a plane" I said.

"Remember the bird comment?" Dwight asked…

"Yeah, I mean "Roger" I replied.

Dwight didn't respond.. well, he did, just not verbally,

He took the controls and he executed several barrel rolls.

Before I even had a chance to scream.

I could feel my blood go from my head to my feet then the G suit squeezed me like a Florida Orange.

And I felt the blood forcing it's way back to my head and it did not feel good at all.

Then the blood rushed from my head down towards my feet as we were encountering negative G's and we were inverted.

"Breathe" I heard Dwight say.."

"Puff your breath out of your mouth in short intervals."..

So, I rapidly started puffing air out of my lungs and the pressure eased up a bit as I continued to strain to keep conscious and to keep from hurling.

I kept my focus on the artificial horizon until the plane started rolling again.

Then I fixed my gaze at the heads up display doing my best to focus on the data and try not to see too far ahead of my normal field of vision.

Upside down right side up upside down right side up.

Soon he eased up on the stick and we were flying level again.

"Not bad for someone who isn't supposed to be here." laughed Dwight.

"I'm just sticking to you like your shadow." I replied.

Lesson here is "Never taunt your pilot."

Trying my best to not let Dwight know.

just how close to throwing up I had come.

I asked Dwight how long would it take us to fly to San Antonio.

"Oh, about an hour, less if we could open her up."

I was trying to get my mind around the speed.

It normally takes us about nine hours to drive to San Antonio and It took us almost 13 hours by train to get there.

So, it gives you an idea on how fast we were moving.

Truly amazing.

After an all too short amount of time in the air it was time for us to go back home and get this plane back on the ground.

We circled the field three times before we got our clearance to land.

When we started our final approach I could once again sense the speed as we dropped altitude and we started our descent over Mountain Creek lake.

I could see the landing strip and the landing lights in the distance closing rapidly.

I could also see the shadow of the plane as it flew over the water and I could see the water disappearing rapidly under the shadow of the F-16.

I saw the touchdown point growing rapidly in my field of vision.

Then came the moment that we returned to Earth.

We touched down and the tires made a screeching sound as they made contact with the runway.

We slowed down and chopped power down.

Then we reversed thrust so we could taxi up to the flight line and park the plane,

The canopy came up and we were rolling opened to the outside air.

Now, you probably noticed that I didn't go into much detail about flying in a fighter jet.

It's not like flying in a typical airplane.

Yes, you do see the endless amounts of open sky and all the patchwork patterns on the ground and the incredible feeling of freedom from the rest of humanity. Far away from the mere mortals below.

Except you are flying so high you mostly see clouds far below you.

Lots of clouds.

You could most likely get the same feeling in a Piper cub or a Cessna 175 without the falling out of the sky due to the turbulence feeling

you can always expect a little bit of turbulence from a little

plane that has barely enough power to even stay in the air.

But, the most exciting thing to me was the level of technological sophistication and the sheer volume of information and data that is at your local fighter jockeys fingertips.

That to me was the most mind blowing part of the entire flight.

That and some of the high G maneuvers.

That the plane is capable of performing. Now the "test flight"

that we took was nothing special.

We kept the plane within the legal constraints of a typical test flight

And did not attempt to break any rules governing fighters flying over population centers

and in no way did anything that might be construed as "Hot dogging"

The barrel rolls were in level flight and we in no way changed our altitude or deviated from our flight path.

It was the most amazing thing that I have ever done.

The whole time (when I wasn't trying my best to keep from throwing up)

I was riding in the plane

I felt like a dog riding in a car with the windows down.

If I could

I would have hung my head out the window and let my tongue flap

in the breeze.

It was that much fun.

We pulled the plane into it's slot on the tarmac and the ground crew ran up to

the plane and started tagging things.

They brought a ladder that we could use to climb out of the cockpit

After we climbed out, .

We were met by General Simpson when we were walking

back to the "Casa" as they like to refer to the main headquarters building.

He asked me if I enjoyed the flight, and I told them both that I had a

wonderful time up there and I was thrilled to be back on the ground.

The general asked Dwight how I held up,

Major Smiley told the general that

"I could have a career as a pilot.

Because I did so well,

I felt that he was just feeding the General a line of BS.

Just because I was there.

He did tell the General that " I took eight. "

I didn't know what that meant. So I waited until the General had left and we

were walking into the locker room to change to ask him what "taking eight

meant."

"Oh, that just means that you were able to handle a eight G maneuver without

blacking

out or throwing up." He replied.

I felt kind of proud of that. Because, I am not a spinning rides kind of guy normally.

That means that I don't like the Tilt a whirl or the octopus or the tarantula or any of the other spinning rides at your local carnival or state fair.

I ALWAYS preferred the roller coaster. Or else something that appears to be going somewhere in hopefully a straight line.

Like the log ride or the run away mine train at Six Flags.

I can handle rides that drop you from high places much easier than I can handle a merry go round. or even a Ferris wheel for that matter.

So for me to have survived eight G maneuvers and not black out or barf all over the inside of the plane is quite an accomplishment even for me.

Because a couple of times I seriously thought about taking off my boot and hurling into that. I came close,

But, I was able to keep it all together and I managed to keep my manhood in the process.

Once we were back in the locker room we got out of our flight gear and I took a quick shower, and then we got dressed.

I was given a couple of boxes by the General and by Major Smiley.

The first box was a small one. It contained a pair of Air Force Pilot's wings To pin on my shirt.

 Symbolizing that I have flown in the military.

It is an honorary title.

The wings look exactly like the wings that Major Smiley is wearing.

The General pinned them on my shirt and then he snapped me a Salute.

I saluted back and I shook both of their hands.

I thanked them for this honor

I opened the Big box next.

And there was a U.S. Air force M1A flight jacket.

I LOVE these! I said I have always wanted one of these jackets.

I thanked the Major and I shook his hand again.

He told me that my call sign would be :"Shadow"

I asked Dwight what name did he use for me when he put my name on the flight manifest

He said that my name on the flight manifest was "Able Baker"

"So, can't you call me that instead?" I implored

"Sure, I'll still call you "Shadow" in private.

But, to everyone else,

You'll be known as "Able Baker" he sighed.

"Ok, that'll work for me." Shadow, is an awesome name.

but, I also liked Able Baker too.

A for able and B for Baker. Those are my initials.

I invited the General and the Major out for a few drinks afterward.

 but, they both declined because Dwight was flying out in the morning after the flight crew finishes up the repair work on his F-15.

I told the General "Thank you" again and I thanked Major Smiley one more time

I wished him a safe return flight home.

Then I excused myself (jacket in hand) and I left the General office.

Then walked back outside.

So, I went back across the street to the shop and Bill was waiting to see how I handled the flight.

He didn't ask if I had enjoyed the flight, he was more concerned with whether I survived the flight or not.

I told Bill that I loved the flight and how exhilarating the flight was.

And how amazing the technology was with all the information at your fingertips.

Bill said "he was amazed and amused by the fact that I didn't throw up."

I told Bill not to be too amazed because, "I really wanted to".

I showed Bill the jacket that Dwight the Major had given me.

And he said, Oh, I have one of those too."

And he told me what a great honor it is to be given a real M-1 flight jacket by a real fighter pilot.

He was glad that I had a good time and he was wondering if I noticed any similarities between the slingshot and the cockpit of the F-16.

I told Bill that I was amazed at how many similarities there were between the slingshot and the fighter jet.

From the one piece bubble canopy all the way down to the Heads up display..

Bill smiled and said "Hey if it's a good layout why mess with perfection?"

"Besides you could not use a control yoke between your legs when you have to pump bicycle pedals to go down the street".

"You'd be whacking you knees against the control yoke all the time, not to mention trying to steer with one. "

"Not a good idea."

Our way is much better." Bill said with a huge measure of pride.

And, I agreed with Bill.

These guys know what they are talking about.

And who am I to disagree with their wisdom and experience?

Besides the original Slingshot steered the same way.

"By the way, How is the little Filipino guy working out as a driver?" Bill asked. I told Bill that "Willy was beginning to relax and he was becoming more comfortable driving the vehicle all the time."

" And soon we were going to test out the vehicle on leg power alone.

"All he has to do is keep the flywheel spinning and keep the car on the road."

The rest should be easy." I said with a smile on my face.

I know how hard driving the slingshot hard it can be.

It's like trying to drive an exercise bicycle down the road.

You have to pay attention to the road,

And keep it between the lines and also keep pedaling.

So you do not lose electric power generation abilities.

The good thing though is that once you get the flywheel up to speed.

You do not have to keep constantly pumping the pedals.

You can pump a few times rest for maybe a minute or two and when the flywheel starts to slow down then you pump a few more times to get the RPM's back up to speed.

I looked at the time and I had not realized how late it had gotten.

The days were getting darker much sooner now that winter is approaching.

So I bade Farewell to Bill and I took my jacket home with me in the truck.

When I got back home I pulled into the driveway.

I parked he truck and I was surprised.

I got there and went inside, there was no one to be found except for Willy.

I walked right past him and I said "Hi". to Willy,

Like I had done so almost automatically

"Oh, Hello Mr. Andy. How was your day, Sir?" Willy inquired

I told him it was a good day then I looked around the house.

Willy had cleaned the house and he was cooking dinner when I walked in the door.

What a surprise.

I had totally forgot about Willy being at the house this morning.

He was asleep in the den which I did not have time to go into this morning.

I looked around and apparently the Wiggler's and Rhonda also must have forgotten that he was there as well.

Or then again, maybe not.

I didn't remember to buy food earlier this week and I thought that the fridge should have been empty.

But whatever Willy was cooking, it sure smelled good.

"So, Where is everybody?" I asked.

Willy said that they had to go to the Leukemia Society and discuss sponsorship duties with the director of promotions.

Or something to that effect.

I went to lay down in my bed,

I closed my eyes I could still feel the springy, spongy feelings in my arms and legs and my head started spinning again.

My chest hurt and so did my stomach and my ankles.

The pressure suit really squeezes the heck out of your body.

I guess it has to in order to squeeze the blood back into your skull.

I also noticed how my hands still felt like they were asleep.

All tingly with little feeling.

I closed my eyes and I could still see the sky and the instruments in the cockpit.

It was like it was burned into my soul.

Then finally I drifted off to sleep.

I woke up being kissed on the mouth by my wife.

Rhonda and the Wigglers had indeed left Willy behind.

because they didn't know where I was or how long I was going to be gone.

And they had to go up to the offices of the North Texas Chapter, and meet with Rhonda's old boss Mr. Robert MacReddy.

Now, Rob was really an OK boss as far as bosses go.

He always treated Rhonda with the upmost respect.

And he was attentive to her ideas and input on company matters.

Whenever Rhonda had an idea for fund raising or had an idea to streamline a process or just figuring out the company's software.

Rhonda was always there for him.

And when she left.

she told him that she would still like to be able to volunteer her time and skills at her discression .

whenever she has time to help out with events and meetings.

Which Robert readily accepted.

She also got me involved too.

I would do the "Ride for Life" which is a bike ride that is sponsored by The Leukemia Society,

local bike shops and energy food companies and a few local beer distributors.

Unlike most of the bike rides that I had participated in.

they had free beer at the end of the ride.

That's a huge bonus for me.

after all, if you have been riding forty miles in the Texas heat.

a cold one would definitely hit the spot.

That fact alone would guarantee that this ride (The Ride for Life) and the ride for Muensterfest (Muenster, Texas is a German community in north Texas close to the Oklahoma border) were and still are by far are my two most favorite bike rides

Rhonda also joined me on some of my rides.

We have a big, blue multi speed tandem bicycle, I ride in the front and do all the steering and braking.

Rhonda is my stoker.

Now for anyone who doesn't know what a stoker is in bike riding.

A stoker is the person who sits on the back seat of a tandem bicycle.

AKA, a bicycle built for two. I had rebuilt an old blue Schwinn Marathon Tandem Bicycle and I equipped it with gears.

I converted it to a twenty four speed racing bike.

She likes to ride fast too.

She's always telling me to pick up the pace and ride faster.

The Muensterfest bike ride, is unusual.

It starts out riding up hill for about thirty Five miles.

And when you finish 100 Kilometers (sixty four miles) you turn around and ride back,

well ride is figurative.

'It's downhill all the way back and on a big bike like a tandem,

You can really built up a large head of steam.

The momentum and gravity start working together and you can really get going pretty fast.

On the way back after riding forty miles out of the original sixty four miles.

The bike's speedometer said that we were flying down the hill at forty four miles an hour,

Now that may not sound like much to people driving in cars.

But, you try doing that on two wheels with your partner riding behind your back on the same bike..

I assure you that this is a mind expanding experience.

You do your upmost to keep the big bicycle upright and under control.

With the full knowledge that the life of your stoker is in your hands.

And if you just happen to be flying down a long steep hill at forty miles per hour,

and you hit a pot hole or a rock on the roadway.

then it could very well be "Cancel Christmas"

Spandex and Lycra, Styrofoam and plastic offer next to no protection against the hunger of the road god

Who frequently demands it's pound of flesh.

I have paid my dues to the road god on many occasions

Fortunately, we had not had a accident on the tandem.

I am certain that if we ever did, that would end Rhonda's tenure as my stoker.

I doubt she would ever trust me with her life on two wheels again.

Rhonda told me that the decision on whether she would ever ride again

Would depend on the amount of skin loss or blood loss or if any bones were broken. I thought "Fair Enough"

The fact that she was the architect of all this madness would be of little consequence

Since in fact it was SHE who talked me into participating in the rides

And Chili cook offs in the first place.

It was really an easy sell for her.

She loves me, and I support her in her endeavors..

And also I agreed to participate for my own personal reasons.

You see, My mother died of Leukemia

And this was something that I could do to try and help others in honor of her memory.

So, Rhonda filled me in on all of their preparations that She and Candace had agreed to undertake to do their best to make this the best Chili Cook off ever.

Besides sponsoring the event and donating the beer, she and Candace agreed

to donate some items for the silent auction .

Which is held on Friday night at the Cooks Party the night before the cook off.

Then she dropped the bomb on me.

We are not allowed to win the cook off because, that

would look like favoritism.

Like we bought the Cook off because we had raised the most money.

Or donated the most money.

Or services and that the judges handed the trophy to us for that reason.

So, no matter how good my chili tasted.

Or how delicious our beans are or how tender and juicy my brisket is

or even if the meat falls off the bones of my Danish baby back ribs.

We are not going to be able to win any event because the other cooks

might or would say that the whole cook off is rigged.

Now in all truth.

Just to be fair.

some of the other cook offs may have well been rigged,

but no one could or wanted to try to prove it.

And besides, I am not one of the people who do the cook off circuit annually.

So, I don't care about accumulating and ICF (International Chili Federation)

points.

And competing in the big Chili Cook off in Terlingua, Texas.

I am doing this for my own reasons.

And that is good enough for Rhonda, and it's good enough for me.

Dick said "They are going to let us display the vehicle at the Chili Cook-off"

"Wow, that's great news Dick.

Now people will get their first look at it.

Up close and personal."

I thanked them for all of their hard work.

The girls were in full preparation mode.

And that was just about all they could talk about.

I was so glad that Candace and Rhonda get along so well.

With all the preparations and designing and building and all the testing,

I have in all truth I feel like I have been negligent to Rhonda.

And throughout the mission so far.

She has been very patient and understanding with me and my projects.

So, I know that she too is happy to have someone other than myself to talk about the planning and the details and all the preparation.

That goes into pulling off a event such as this.

Not to mention the company of someone who isn't talking about technical things Like conservation of energy, Blah, blah,blah…

And Candace thrives on doing things like this.

Dick is like me in the sense that he knows to not try to get too involved and help or get in Candace's way when she is doing charity work.

I am the same way with Rhonda, I try to keep my head down.

I do my best to try not to talk while she is in the planning phase of any event she has been tasked to coordinate.

Otherwise, I would get caught up in the maelstrom of her organizational fury.

And what do I mean by getting caught up?

"Utilized" that's what I mean.

When Rhonda is in charge she expects all available hands on deck.

And right now I am currently unavailable.

(Thank you Candace for taking up the slack.).

Anyway, we all sat down to dinner.

Willy had made us a traditional Filipino dinner

"Like his Mother used to make."

We started the dinner with egg rolls.

Willy corrected me and he said that they were called Filipino Lumpia.

Now Lumpia is a traditional Filipino dish.

It is the Filipino version of the egg rolls.

It can be served as a side dish or as an appetizer.

Next was the chicken adobo,

Willy says that this is perhaps the easiest Filipino dish to prepare.

Basically it is chicken that has been marinated and simmered in soy sauce, vinegar and bay leaves,

Willy says that this favorite can easily turn out to be too salty or too tart.

(because of the salty nature of soy sauce.)

My adobo was just right, with the savory flavors coming together perfectly.

No over salting here.

Willy cooked us some rice.

And he even added some tortillas to dinner.

It was a wonderful dinner.

Willy is a really good cook.

Thank you Willy for dinner.

We drank water and tea with dinner and then we kicked back a couple bottles of peach flavored wine.

And I told them all about my flight in the F-16C

With the guy who almost squashed me like a bug.

I had to run out to the truck to retrieve my flight jacket.

to show it off to them.

Dick, Rhonda and Candace were sad to have missed that.

It's not much to have seen as a spectator.

The plane goes up and it's out of sight.

Then later on the plane lands and taxis away.

End of the story for someone on the ground.

I did make it sound much more exciting than that.

It's been several hours later and I am still feeling kind of funky.

After our wonderful Filipino dinner,

And a few bottles of wine we helped clear away the dinner dishes

Then we all got ready for bed.

Rhonda and I got everyone situated and we bade our guests good night.

And I climbed into bed.

Rhonda was reading some papers on her side of the bed.

So, I covered up and I rolled over and kissed her Good night.

And then she hit me with the words that almost cannot bear to hear.

"Andy, We need to talk…"

Those five words send shivers down my spine.

So, I rolled over so I could sit up and get into my

"OK we need to talk posture" and I asked her what was wrong.

Rhonda asked me very pointedly,

"Why am I still driving the Crap mobile?

"Why Can't we buy a newer car that has a floorboard that hasn't rusted out?"

"I don't like watching the road passing under the car while I drive."

"And besides, it's ugly and the A/C,"

" Andy it doesn't even have an air conditioner. "

I looked her in the eye and I told her point blank and in no uncertain terms

"I don't know."

Honestly, I haven't given it any thought at all.

I've been too focused on the project to notice little things like my wife having to drive something potentially dangerous.

I have no qualms or problems driving the Crap mobile.

I was used to it.

And since It was the only wheels we could afford at the time.

I have accepted it as part of the family stable.

But, just because, I can sleep in the dirt out in the rain.

doesn't mean my wife should have to endure suffering that she is not accustomed to. I have been so focused on the project that I didn't notice that she didn't want to drive the little yellow Datsun B-210.

Not because it was ugly or unsafe or rusting out terribly, or had no air conditioner.

But, it was not her car.

It was a clunker that I didn't mind driving.

This was not a car that I would give a stranger. .

I'd feel very guilty of taking advantage of someone's trust.

Yeah, I heard of Buyer beware.

but, c'mon. the car is a pile of rusted out metal.

It does it's one job very well in spite of it's age.

It still starts up and runs and drives ok.

But, it has seen better times.

Not with me of course,

It was a cheap means of transport.

And it was all I was expecting.

Fine for me, bad for my spouse to drive.

"So, What kind of car do you want to buy?" I asked

"I don't know, maybe a Honda CRX or a Mazda RX-7?" she replied hopefully.

Both were sporty non family type cars.

They are both two seaters and have just enough room for two people to sit in comfort.

Fine for most couples, unless you have a couple of friends who want to ride along.

Then it would become an exercise in yoga and contortionism.

Since it would be her car,

I really didn't mind whatever car she really wants,

She will be the one who will end up driving it.

I told her that she and Candace could start looking for whatever car she wants after the Chili Cook off and before we leave for Australia.

Rhonda said that would be fine with her.

Money was no longer the issue.

I had plenty of cash to buy her the car outright.

if I wanted to.

but, I wanted her to pick out her own car.

And that way I know that she is buying a car that she really wanted.

So, I left it at that.

She kissed me good night and we both fell off to sleep.

The Hybrid Diaries
Chapter Thirty Four
Wednesday through Friday October 17th 18th and 19th 1990

So, let's fast forward a couple of weeks.

Nothing terribly exciting happened, we did get our passports back,

(God, I cannot believe I look like that.)

I think Passport pictures and Driver's license photos are the worse picture

you can take of yourself.

But, Rhonda looks great.

We continued testing the vehicle and training our drivers.

We completed our testing on the Spirit of Texas car

The vehicle came through in flying colors.

All of our tests were completed.

and the vehicle performed just as it was designed to do.

We worked with Willy and we taught him the ins and outs about driving the vehicle

We explained how the flow of energy runs from one source to another then to the drive wheels or to the batteries.

And so on.

We also trained Christopher to drive the vehicle as a backup driver.

So now we have Willy as the primary driver and Christopher doing the test runs as a back up driver.

Our next goal is to get the final preproduction version of the vehicle assembled and transfer or duplicate the equipment in that we had originally installed in the original prototype.

Toby was hard at work on a new set of more powerful wheels with better efficiency,

Greater power and a higher RPM output than the previous wheels.

For a higher top speed.

Hopefully on the same amount of power.

The other guys were assembling the body for the racer.

It looks even more menacing than the original prototype.

Dick and I started calling the racer "Deguello"

meaning "no quarter" In Spanish.

We were going to Australia to compete and we planned on kicking butt and taking names.

In other words we were going out there to win and we were taking no prisoners.

Figuratively speaking of course.

The Girls have been extremely busy working with the Leukemia Society.

Planning out the cook off.

I had no idea how much effort is exerted in planning and executing an event like this.

I only know that it takes an almost super human effort to gather up all of our camping gear and cooking supplies and food logistics.

Beverage supplies, loading up my smoker and my Coleman stoves,

My lanterns and dining canopy, plus tents, 2 MIL plastic for draping on the ground

(A chili cook off requirement) extra chairs and

tables and fire wood and trash bags.

All this stuff just for one overnight cooking event.

And then having to break camp and load everything back up into the truck.

and cart it all back to the house.

A tremendous effort all for the sake of Charity.

We were able to get the City of Irving to let us use the Lake Caroline spot that the last few cook offs were held at.

This is a good spot.

It is visible from the road, and it has lots of grass and sand for the cooks to set up their cook sites on.

The site is also big enough for the main tents where the judges will sit.

And where we have the cooks party the night before the cook off.

Also this year, the City of Irving was hosting

"The Ships of Columbus" exhibit.

So, there will be a full sized working replica of "The Nina" actually on the lake there at the cook off,

The exhibit is usually on display in Corpus Christi.

but, for some reason they let the City of Irving host the exhibit.

They just sent the Nina, No Pinta or Santa Maria were present.

And for contrast, there will also be a cutting edge, high tech, human electric hybrid vehicle there on display as well. (mine of course)

The fire that claimed my garage also destroyed all of my camping gear.

The only things that survived was my lanterns and my Coleman stoves

And some of my cooking utensils and pots and pans.

So, I will have to go out and buy replacement camping gear.

Yee Haw, I am going to Wal Mart!

I think the camping gear sold at places like Wal Mart is pretty decent,

A little on the thin side and not enough fabric or materials used in the construction of the tents, and the fiberglass rods are a little too easy to break.

Requiring you to buy many more replacement rods to replace the broken ones.

And the nylon that the tent is made out of seems too thin to provide much in the way of protection from the elements or retention of body heat.

But, for light camping and family outings they get the job done.

So, I called Dick in Austin and asked him when he was planning on coming up to Dallas for the cook off,

He said that he'd be here by Thursday afternoon.

So, I would have to go and buy enough equipment for Rhonda and I and for Dick and Candace by myself.

First thing I decided on would be separate tents.

One for us and one for the Wigglers.

Experience has taught me that it is wise to have a place that people can crash out in And still leave your main tent unoccupied.

This is especially helpful when some of your friends have had too much to drink and they have passed out.

You can pile their drunken bloated bodies in a corner of their tent

And still have enough room to set up your cots or air mattresses in your own tent

If you need to set up their inflatable beds.

Just use your foot to move their bodies to a unused corner of their tent and you throw a blanket or sleeping bag over their heads.

And you should still be able to drink and carry on a normal conversation with your other friends and fellow cooks.

During a cook off

People are walking past your cook off booths all the time.

And the only facilities that may offer you some small measure of privacy are the "Port-O-Cans" that have been set up at the site for temporary bathrooms.

And those can become very filthy, dirty places at best to change your clothes.

That's why I set up an extra tent that I call "the supply tent".

It's not for sleeping it is for storing extra food and drinks.

Ice chests and the boxes and plastic bags that our camping gear comes in.

And duffle bags that we store our clothing in for over nighters..

We also change our clothes in the supply tent.

And sometimes on the odd occasion we carry an over lubricated team mate out of our tents and place their mortal remains carefully on the floor of the supply tent.

Usually face down for safety reasons.(so they don't choke on their own vomit should they barf in their sleep.)

At that point they have just purchased a tent from me!.

So, I have to buy three tents and four large replacement ice chests and five sleeping Bags, folding tables. Folding chairs, Plastic tarps, folding canopies and assorted camping gear.

I still have most of my camping cook ware and stoves and lanterns.

My butcher is still open and I'm sure he ordered my briskets and my Danish baby back ribs and steak for my chili meat.

I had already called him to place my order a week ago,

but, I called him again to remind him that I was coming by to pick up my meat later on.

Plus additional groceries like pinto beans and seasonings and peppers and tortillas and paper plates and plastic disposable bowls and lots of plastic utensils and cups.

We will also need a replacement portable stereo so we can play our cassette tapes that we have made specifically for Chili cook offs.

A chef cannot do a good job if he doesn't have the right kind of music playing in the background.

Music helps to dictate the mood.

Now, the greatest commodity at a Chili cook has is ice.

And lots of it. The second major commodity is alcohol.

AKA beer and wine. You can never have enough personal supplies at a event such as .this.

And after the cooks party you have to rely on you own personal stocks to get you through the rest of the all night cooking sessions.

You have to cook brisket slowly in your smoker.

Usually about ten to twelve hours .

So it's nice and smoky and yet still juicy and tender.

With a good smoky flavor and it slices almost wafer thin.

The slicing is extremely important when it's time for judging.

Judges count off on that.

I had gotten some items together that I was donating to the cook offs silent auction.

The first item was a poster of one of the previous cook off that I had professionally matted and framed.

And a can of "Armadillo meat" that my wife had won at one of my company picnics The "Armadillo races" by blowing on the butt of an armadillo and making it run down a track faster than my armadillo (Yes, I too was blowing on another armadillos butt and I too, was crawling right beside her on my hands and knees .

I know that Dick was going to be bringing up some items from Austin. What he was bringing, besides the beer? I do not know.

Rhonda was a little angry that I would auction off such a desirable trinket as a can of "Armadillo meat" (It's really just a can of tomato paste with the armadillo meat label

wrapped around the can.

It's a joke gift that we have always kind of treasured over the years.)

I had to promise her that I would not lose the can at the auction. So, I promised

her that I would do everything that I could to keep the can in the family.

That included buying the can back at an extremely over inflated price.

We piled into the truck and we drove over to Wal-mart to pick up the camping gear and some groceries.

We ended up with two baskets full of Camping gear stuff and a basket load of goodies.

And groceries. Lots of groceries.

Then we headed to the butcher shop and we picked up a box of Danish baby back ribs and a couple of vacuumed wrapped briskets and my steaks that I'll chop up for chili meat, that I had ordered earlier.

I always use the same butcher and he does his best to get me the finest cuts of meat.

There is a lot of pressure on butchers and meat cutters this time of year.

Cooks know what they want and they can get very vocal if they do not get precisely what they wanted.

My butcher hasn't let me down in five years.

When we got home I started putting the groceries away and I began to remove the plastic off the briskets and I began the seasoning process.

There are two methods for preparing a brisket.

The first and most common one is the "Dry rub method."

Where the seasonings are mixed together and then rubbed into the entire brisket.

Then the brisket is wrapped in plastic until time to cook,

Then you remove the plastic and wrap the brisket in aluminum foil.

Then there is the wet rub technique where you marinade the meat in a marinade of seasonings and other liquids like apple juice or some other kind of liquid.

And your personal blend of seasonings.

Here, let me give you an example of a wet rub brisket.

I call it my Orange marmalade and honey brisket.

Take one brisket trim the excess fat off the back side of the brisket and soak in a mixture of orange juice and orange marmalade and seasonings and then cover the entire brisket in honey .

Cover the pan containing the brisket and allow it to soak over night .

I could tell you what seasonings to use but, If I did,

I would have to Kill you.

That's like asking a cook for his bank account number.

On second thought, you might be able to convince a cook to give you his bank account number before they would ever tell you their personal recipes.

Trade mark secrets.

Another secret is "what type of wood you burn to flavor your meats.

Now, the most common wood used would have to be hickory and pecan oak.

Both woods are very aromatic,

but Hickory is the overall favorite.

Sure you could use Mesquite wood or Redwood,

but, it most likely would not give you the kind of flavor your meat would need

to win a cook off.

So, stick with the basics.

Hickory is the usual way to go.

It usually takes between 10 and twelve hours to cook a brisket.

And if you do it right you should be left with a tender juicy brisket.

Blackened on the outside and tender and succulent on the inside.

The brisket should have a little red line between the meat and the

smoked outer shell.

And if the meat slices thinly and holds together just enough for slices.

You got it made in the shade.

I usually let my briskets marinade for two days before I start cooking them.

I always cook more than one so I can choose which one of the group is the

very best to submit at judging time.

My ribs are my specialty.

My wife doesn't eat them because…

Well, they look like what they are.

Ribs of an pig. Or a cow, depending on which one you prefer to cook,.

I like pork myself.

That's why I go with the Danish Baby back ribs.

When they are done right, the meat just falls off the bone.

That's the way I like them and that's how I cook them.

After I got the meat all seasoned up and placed in the fridge,

I had to find a place to pile all the camping gear.

And since I have no garage to keep all of my gear,

I had no other choice but to pile all the stuff in the dining room.

I really should invest in a trailer.

After getting all of my stuff in order,

My next item of business is to go to the sign shop in the morning.

And pick up the replacement banner for our booth.

Last year, we had a small banner stat just said "Welcome Chili Cooks."

This year I had designed something special,

The name of my Chili Cook off team is "Critter Du Jour"

Which means the critter of the day in French.

Now the big joke with Chili cooks is naming your team or your chili some god awful name that hardly does justice to the delicious food that chili cooks work their butts off to prepare.

Here is a small lists of names for Chili cook of teams that I have seen at these events.

Roadside Chili

Greasy spot chili

Cat nose Chili,

Bull dog Chili,

FORD chili (That stands for Found on the Roadside dead. Chili)

Fire pepper Chili,

"Armadillo Roadside Chili,

Opossum Chili.

Rat Meat Chili,

Rattlesnake Chili,

Butt head chili,

 Bayou Self Chili

Hog dog Chili

Hog Wild chili

Booger Red's

Three legged dog Chili

Big Moe's

Stookie's

Mad dog chili

Stink butt Chili

Road kill Chili

Dead Possum Chili

Bambi's Mother's Chili

Hobo's Road Bum's Chili

All thumbs Chili

Trash Can chili

Sick Sisters Chili

Cooking on Rebar Chili

And my favorite

It's dead already Chili!"

and so on…

So, Critter Du Jour is quite tame by comparison.

On the banner that I had designed.

I put two pictures of "Lil Critter" from the Mercer Mayer books

on either side of the name.

One little Critter is wearing a top hat the other

little Critter is mixing something in a bowl.

I drew the images just because he is so cute.

In the morning.

I will have to go by the sign shop and pick up the banner before Dick gets into town.

I bet he really misses Candace.

She's been working her butt off with Rhonda.

And they have been running all over town getting sponsors lined up and going around town and making sure that all of the donated items get picked up.

Mostly they stayed in the offices of the Leukemia Society.

Answering phones and processing all of the over two hundred cook off registration forms for all of the cooks.

And making the cooks packets

which include tickets and ribbons T shirts, aprons, and even some antacids.

Also, collecting and processing all of the registration fees and making deposits.

For the last week they have been coming home late and plopping down on the sofa and falling asleep where they sat

So, I'd have to wake them and sometimes run out to the fast food burger joint and pick up something for the girls to eat.

No one ever said that volunteer work is easy.

The experience dredged up some old bad feelings that Rhonda used to have when she used to work at the Leukemia Society.

The problem was not with the Society itself,

It had more to do with management styles.

Rhonda is a very sensitive person who always tries to take on way more that she can handle.

She is compassionate, kind and caring.

She has always worked for Nonprofit companies like the United Way or the Leukemia Society because she wants to feel that she is helping and making a contribution.

And in turn helping others.

But, when you work with people and children who are suffering with a life threatening disease like Leukemia.

Sometimes the reality can knock you down.

She would come home sometimes with red rimmed eyes because she had been entering

personal data about children who have been afflicted with leukemia.

And they were not going to survive.

And she cried. She cried A LOT.

It takes a strong person who can do their best to detach themselves from the harsh reality of death.

And especially the impending death of a child.

She internalized the grief.

Rhonda would spend nights up, unable to sleep because she was thinking about a child that has passed on.

Or a different person who has just found out that they have the disease.

That, plus the fact that the office was short staffed and she had to divide all her time

between office work and promotions coordinator or event coordinator.

The little time she spent back at her old office opened up old wounds that she would rather not deal with,

Candace had developed a healthy respect for what Rhonda does.

When Rhonda works with nonprofit organizations.

Coming from Candace, that is one heck of a compliment.

"Rhonda is a truly remarkable person,"

"She is beautiful and smart"

" I feel blessed to have someone like her who cares about other people's suffering in my life."

"She is a real trooper with a big heart filled with love."

" A real caring person."

Now for my part in the fund raising and participation process…

I was lucky this year.

When I used to work at other companies.

I would work with the Company's public Relations Departments.

And they would pay the entry fees for me.

This year, I no longer had to beg for my entry fees.

So they could provide me with a way to show the good Old corporate flag,

"And being a good corporate Neighbor."

This year I was going to get exactly what I wanted in the way of a cook site.

I got two cook sites right next to each other.

Both by the lake and under the monorail bridge.

(yes, there is a monorail that runs around Las Colinas.

In Irving where the event is to be held.

The bridge provides shade which is in high demand when you have to be out in the sun.

So I had more room to set everything up.

Which was no small feat.

It seems that everyone was coming out to the cook off this year.

And many large Corporations were also doing their public service.

Corporations like IOU and Telnext, were going to be there in force.

I can hardly wait for Dick to get here.

Back at the Carbon Fiber place.

Poor Willy has been getting the work out of his life.

but, since he and I both have a riders back ground he was hanging in there.

A little sore but not too bad to ride.

Willy even tried to convince me that he was still able to ride back to my house from Grand Prairie on a daily basis.

I told him that would not be necessary and that I would pick him up in the evening.

After the days tests were conducted.

Because, I could not risk him getting hurt in a bicycle versus car accident.

And poor Christopher who does not come from a biking background said that he needs

to go back to the gym to work out more because testing the vehicle was wearing his legs out on the human flywheel generator.

It's like I explained to Christopher.

It is like trying to drive an exercise bike.

But, it costs very little in the way of fuel.

And it is healthy and it goes pretty fast too.

And once the flywheel is up to speed you do not have to keep pumping your legs constantly.

You can stop and catch your breath while the flywheel slows down and then pump it back

up to speed and rest some more.

Willy was usually so tired when he finished the testing he would come back to my house and crash out.

So his home cooked meals would be more rare as the testing program comes up to full speed.

And since we are now on the same radio frequency as the control tower,

We have eliminated the problems with landing aircraft and runway close calls.

We can now monitor their transmissions.

and we now know when there are incoming planes.

We wait until they are safely on the ground before we begin another test run.

I didn't have any problems testing the vehicle.

Except for almost being squashed.

Or being blown off the runway during a test run.

I guess I am more excited and maybe I just didn't notice.

But, I am a very good rider and I have been doing it for many years.

Or else, it might be the adrenaline.

I'm not sure.

but, I am certain that they too will get accustomed to driving the vehicle.

Testing is pretty boring but extremely necessary.

Testing power and drag co-efficient and power to weight ratios and the like.

It also involves sitting. A lot of sitting waiting for the signal to begin.

And it's worse in the vehicle since there is no air conditioner.

You have to keep hydrated.

It is essential to keep fluids in your body at all times.

To keep you from dehydrating or passing out or cramping

Salt tablets and Gatorade and lots and lots of water are the normal tools for fighting dehydration.

So, everyone who is driving the vehicle gets their own Camel Back hydration system

And several bike ride type enclosable water bottles for use while driving

And as an added precaution one of the guys is supposed to give the rider a fresh water bottle with water and ice after every series of test runs.

We implement more procedures and their needs become more evident to us.

I have cancelled the testing for Friday and Saturday

(we do not usually do testing on Sunday)

So we can all attend the cook off and they can participate on the cooking team.

As the team Captain I am allowed to have five people on my team

And Since Rhonda and Candace are busy elsewhere with other duties,

I was able to have Myself, Dick, Dylan, Christopher and Willy on the team.

The Hybrid Diaries
Chapter Thirty Five
The Madness Begins

Well, with everything going on I will quickly move into Friday morning.

The first day of the Cook off.

I woke up around 10:00 AM and the girls were already at the cook off site.

Marking out cook sites with a chalk line marker machine.

They use them in sports to draw lines of white chalk powder on the ground to mark the field of play.

In our case the boundary lines and the number of cook off sites that will be available to the cooks.

I had the guys Dylan and Christopher come by my house to help me.

And Willy loaded the smoker and the camping gear onto the back of the truck,

so we could drop all of our supplies off at our spot.

They showed up around 11:00AM.

Our first goal was to load the smoker up in the back of the truck.

Now I built the smoker a couple of years ago after my first cook off

because those little R2D2 looking smokers are just too small to cook more than one item of anything at any given moment.

I had made the smoker out of recycled materials,

I made the smoker box out of the standard fifty five gallon oil drum.

With a smaller drum welded onto the end of the big oil drum.

that has been modified to serve as the firebox / Grill.

I made the legs out of metal chain link fence poles.

I had attached shelves to the sides of the barrel

so I had places to put my briskets, turkeys, or whatever else I may be cooking

and my

utensils and whatever I am drinking at the time.

My spices go underneath.

So, When the guys arrive I asked Dylan and Christopher to load the smoker

onto the back of the truck.

No small feat, the smoker is very heavy.

I knew the guys would normally have little trouble loading the smoker

onto the bed of the truck.

But, unbeknownst to me there was a major problem that was about to present

itself.

Just as Dylan and Christopher were picking up the smoker.

Dylan let out a loud Oooouuuuccchhhh! Followed by

"Son of a Bitch!"

"Something just stung the crap out of me!"

They dropped the smoker and ran away as fast as they could.

The smoker fell over and it's legs were sticking up into the air at an angle.

On the underside of the smoker nestled up close to the frame where the legs

attach to the Frame.

Tucked up and completely out of sight,

there was a huge Yellow Jacket nest.

That nest looked like it had been there for quite some time.

I hardly ever move the smoker

so It seems that the wasps had found themselves the perfect peaceful place to raise their brood.

That was a bad call.

I try to be as Buddhist as possible.

I don't crush crickets and I do not kill lady bugs,

Or garden snakes or Horny Toads.

But, If an insect goes on the offensive then I will fight back.

And That is exactly what I did,

I sentenced them to death, for the crimes of trespassing

And for assault with a deadly weapon.

Their stingers.

Using a can of WD-40 and a long fireplace match,

I turned the can of lubricant and rust preventer into an impromptu flame thrower.

PLEASE, do not try this at home.

I have experience doing stuff like this.

So, after incinerating the nest and more than a few yellow Jackets,

I apologized repeatedly to Dylan and I took some salt

and I sprinkled it onto a damp paper towel and I placed it on Dylan's sting

To draw the venom out and reduce the swelling.

Willy and I went out to finish loading up the smoker

After doing a final once over to make sure that we got all of the offending bugs.

I carefully removed the still smoldering nest with a stick.

We righted the smoker and Christopher and Willy and I loaded the smoker onto the truck

while Dylan took care of his injuries.

Then, once the drama had played out,

we all went into the house and grabbed the camping gear while Christopher and Willy secured the smoker to the back of the bed of the truck tying rope around the stake bed to hold the smoker in place.

Then we started piling the gear and cooking utensils underneath and around the smoker. And then the ice chests were loaded into the bed of the truck

And A LOT of firewood Yes, it was seasoned Hickory.

And kindling and lighter fluid (just to get the fire started).

In no time at all,

We had the truck loaded and ready to go.

Well, OK, so in reality it took about an hour and a half plus the amount of time to flame broil the offending wasps.

And to tend to Dylan. So, I called Dick to find out where he was,

on his mobile phone.

Dick said that he was about twenty five miles outside of Dallas and he should be there in about thirty minutes.

We decided to wait for Dick to get here.

I decided to order us Chinese food from our local take out place down the street.

And I asked Dick if he wanted anything

he said no, he had eaten earlier.

I focused on feeding the guys.

Several eggrolls and sweet and sour chicken and Broccoli beefs later,

We were fed and ready to go.

When we heard the loud air horn blare from outside.

And there was Dick

leading two big Beer delivery trucks

We ran outside and Dick climbed out of his Suburban

brimming a big smile from ear to ear.

I asked Dick about the delivery trucks

"How many of these big trucks do you have dick?"

"Oh, a couple." Dick said coyly

These were the big trucks that He uses to distribute beer through out the

Austin - San Antonio region.

He had two of his drivers with him and their job was to drive the trucks

and then help with the distribution of the kegs and the set up of those already

mentioned kegs for the people manning the kegs.

And also they have to provide security to guard the trucks and to make sure

than no one tries to walk off with a couple of "free kegs " for their own

personal usage.

We let the guys freshen up and get a soda before we caravanned out to the

cook site.

We had a convoy of five vehicles running down the freeway.

Past Texas Stadium and towards the cook site with our head lights on.

So everyone could see that we are together.

I though it looked impressive as hell.

The company truck with me behind the wheel and Willy riding shotgun, in the

lead followed by a black Chevy Suburban with magnetic signs for the brewery

on the doors,

and then another pick up that Dylan and Chris drove and two big ass beer

delivery trucks in the rear.

We rolled up to the site and I saw a couple of hot looking babes that were taking registration forms and instructing cooks on where to unload and which cook sites that they had assigned to them.

Rhonda and Candace took our forms and told us where to set up our sites.

Then we pulled out of the parking lot and then we went off road and we drove across the field and up towards our designated cook sites.

I could not believe that other cooks were already showing up.

And setting up their cook sites.

One old guy was setting up a stage on his cook site about four spaces down from us.

He had already set up benches made out of cinder block bases

with 2 X 8 X 10 feet boards for the bench seats.

Other people were there that I recognized from previous cook offs.

Former competitors.

For the most part, there wasn't that many people there yet,

And over the roof tops of Winnebago's and motor coaches and tents,

I could see the towering masts of the Nina riding majestically up and down on the surface of Lake Caroline.

With the flags of Spain, the U.S. and Texas fluttering in the breeze off the stern of the ship.

Man that ship is tiny with a Capital T!

They must have had guts to spare in crossing the Atlantic in something so small.

The entire boat measures barely 70 feet long.

And it displaces about 60 tons and it has a crew of 24.

A really small boat. calling it a ship is an exaggeration.

I also found out that the ship was there because they were building a permanent mooring for the ship

And they needed some place to keep her while they finished construction.

Since it was October and that's the time of year for Columbus day.

And since they had not finished her anchorage in Corpus Christi.

They sent her up to Irving for safe keeping

And to exhibit her to the local populace.

I was also prepping my vehicle for the exhibit as well.

The guys had gotten "Deguello"

Put together but, Toby had not finished with the new electric wheels yet,

So we "borrowed" the wheels from "Spirit" and slapped them on "Deguello".

And Chris and Dylan were going to the shop to bring her to the cook off,

after we set up the cook site.

I had made sure that I invited Toby and his wife and Bill and his spouse and

the rest of the guys at the Carbon Fiber place.

Plus Earl (the battery guy) and Rich (the solar panel guy), and Sonny

Maldonaldo from his restaurant

This promises to be a whole lot of fun.

My friends could not participate at the Cooks party,

that is reserved for cooks only

but afterwards (Usually around 11:30PM) When the party is over.

We crawl back to the cook sites and begin cooking.

And we break out our personal reserves of beer and wine and

sodas and we run on them until dawn.

But, let's get back to the set up.

We parked at our site and we began to unload everything.

It took us a couple of hours to set up the tents and the smoker and the folding

tables, the Coleman stoves and some folding camp chairs and unpack my

cookware and utensils.

Not to mention stacking the firewood in a safe place.

This stuff is like gold

because there are no places to get firewood on site.

Sometimes people will try to "borrow" a few pieces of wood from you "without your permission".

So, you have to keep a watchful eye out to make sure that your stockpiles are safe.

I also brought charcoal for grilling.

I was "so lucky" to have a cook site that was facing North under the Monorail bridge.

That way the sun is not beating down on us and in the afternoon the bridge provided us with comforting shade.

(Thank you Rhonda and Candace.)

It's right on the main pathway and everyone in attendance will have to walk past us.

The next best site was taken up by the "Rattlesnake Chili team of IOU"

(They never say that they are with IOU they prefer to remain anonymous and I respect that so I hid their actual name) a huge contributor.

And next to them was the Margarita Society (another Huge contributor)

They brought their party bus and they pass out self adhesive stickers that they slap on well any female passersby.

These Corporations make major contributions to the Leukemia Society

They are doing their part by being good corporate neighbors to the community

Although this can be considered a "Fun Event" believe me when I say that these events take an incredible amount of organizational skill and tons or sweat and hard work to pull all of these corporations and resources together.

Just look at what Dick and Candace had to do.

Bring up two big beer distribution trucks loaded with kegs and carbon dioxide tanks and keg stands and lots of ice.

And pumps and spigots and plastic cups.

Everything you'd need if you were throwing a party for approximately 5000 people.

All of that was provided free of charge by the kindness of the Wigglers because they are caring people who go all out for charitable organizations.

Dick told me that Candace is very active in the Austin Charity community.

And that this is the largest donation that they have ever made to any one organization.

They are paying the salary for the guys to drive the trucks and set up the kegs and watch over the trucks and their precious cargoes

And keep an eye out for our campsite while Dylan and Christopher went back to the shop to load up the vehicle for the trip over to the cook off site.

That left Willy to guard the campsite which he did like a pit bull.

This is all part of an all out effort by us and all of the chili cooks and their teams and sponsors.

It's not just a bunch of people getting drunk.

It represents the culmination of almost a year's worth of planning and attention to details that the public rarely (if ever) notices.

And all the proceeds are handed over to the Leukemia Society whose work is indispensible for the families of people who are stricken with this terrible disease.

I personally salute each and everyone of these people who are drawn together for this and other events like it.

These people are true heroes to their community and I consider myself lucky to have been a part of their numbers.

And I thank my wife Rhonda for getting me involved in the community and allowing me to help her make a difference no matter how small.

Rhonda I love you.

Now, let the fun and games begin…

The Hybrid Diaries
Chapter Thirty Six
Friday Evening October 19th 1990

Always keep your checkbook handy.

That was what my father used to tell me whenever you plan on anything because you never know what unexpected expenses might pop up.

With the beer truck drivers vigilant eyes focused on the trucks and on our cook site.

And with Christopher and Dylan back at our base camp with Deguello.

And Rhonda and Candace performing their duties at the Main gate,

It was time for us and the rest of the cooks

(some two hundred and nineteen teams already in attendance)

To go to the big main tent

AKA the judges tent and attend the silent auction.

We used the paddle method of bidding.

Although we were not issued paddles.

All we had to do was hold up our hands to place a bid.

This auction was very informal and the auctioneer decided that with so many cooks in attendance,

it would be better for all concerned to have a regular audible auction.

Where the auctioneer would start out with…

"OK ladies and Gentlemen"

" We have here an envelope here containing four tickets to the Dallas Cowboys vs. the Washington Redskins Game on Thanksgiving day."

" Our opening bid is 50.00 do we have fifty dollars. anyone? Anyone?"

" These are good seats ladies and gentlemen, "

"Right on the twenty yard line .."

"OK, we have fifty."

" do we have fifty five?

"fifty five Ok fifty five … do we have sixty?"

" Sixty dollars, anyone, anyone.

"Sixty Dollars. Ok, we have Sixty dollars. Do we have.

"Seventy five?".

"Ok, We have Seventy five dollars. Do we have Eighty???

"Eighty for the Dallas Cowboys tickets? They could go all the way this year…

it went on and on.

The tickets finally went for one hundred and seventy five dollars.

They had four envelopes with four tickets in each.

And Dick and Candace picked up a set.

Then came some other items that other people bid on.

I was focused on the can of armadillo meat that I had "donated" for the cause.

Now I had promised Rhonda that nothing would happen to her precious souvenir can.

Because of the sentimental value attached to it.

So when the Auctioneer was handed that particular auction item,

he chuckled.

And he said…

"OK I have a can of armadillo meat here ladies and gentlemen. "

"I don't know if this is a serious item or a joke item. "

"Or If it has actual armadillo meat inside it"

but, it's a can that says armadillo meat right here on the label."

" It also says product of Texas on the side."

"Let's start the opening bid at five dollars do we have five dollars?

I raised my hand and the first shot was fired.

I was thinking "Yeah, I'll get it back for like maybe 10.00 dollars or maybe fifteen at the most.

"OK, I have five I have five do we have six?

"Six dollars do we have seven ok we have seven dollars seven dollars

"do we have eight, eight dollars then I heard someone yell out

"Twenty!" I snapped my head around and Dick and Candace were laughing at my expressions on my face.

I almost had a coronary right there on the spot.

So, I swallowed hard and I was just about to yell twenty five when I heard a familiar voice yell out "Thirty!"

I turned around and I saw Rhonda looking at me with a evil glint in her eye.

So, I raised my hand and I yelled "forty dollars!"

The auctioneer had a stunned look on his face.

So, the auctioneer said

"OK, I have forty "

"Forty dollars do we have forty five?

"Fifty!" shouted Candace

Then before the Auctioneer could confirm her bid Dick yelled

Out

"fifty five!"

I was having fun with this.

I knew what they were doing.

Having me on at their own expense.

So, I shouted out "Seventy five!" to the gasp of everyone there no one could believe that I was willing to shell out seventy five dollars for a gag can of tomato paste.

Then someone in the crowd decided to piss me off when the shouted out 100.00 dollars.

Then Dick countered with

" One twenty five! Then Candace bid "One fifty!

Rhonda bid "one fifty five"

And she could see the look of shock and anger welling up

behind my eyeballs.

I didn't have near enough beer in me to justify spending one hundred and fifty five dollars on that stupid can.

I slammed my beer down and I shouted out

"Two Hundred Dollars!"

Everyone in the tent let out a collective gasp "ahhhhhhh" and the tent fell silent.

This wasn't cowboy tickets.

Or a autographed official Dallas Cowboy Jersey autographed by Troy Aikman… this was a stupid can of faux armadillo meat.

Not even a real can of armadillo meat.

a stupid can of tomato paste.

And I heard the auctioneer say going once going twice sold!

To the gentleman in the blue T-shirt. For two hundred dollars!

After a while I was able to buy back my framed Chili Cook off poster from a couple of years ago for only fifty bucks quite a bargain by comparison

Soon the auction was over and it was time for the caterers to start putting out dinner for all of us hungry (and slightly inebriated) chili cooks.

Now why on Earth would they serve us Barbeque for dinner is beyond me.

But there it was brisket smoked sausage and ribs and chicken with Potato salad and beans

Plus Texas toast and tea for whoever wants it.

I stuck with beer.

Until I was paid a visit by two girls in tight shorts and Tee shirts that left little to the imagination.

They were wearing cowboy hats and holding trays that held test tubes with the words commemorative on them

These were the Sauza Tequila girls and they were passing out shots of Sauza tequila free of charge to all the cooks who wanted them.

I had five.

Back at the campsite, I had instructed Willy to fire up the grill with Charcoal and when the coals got hot.

throw some hamburgers on the fire for the guys to eat.

I had ulterior motives for the fire.

besides cooking the guys burgers I'll explain later.

Rhonda and Candace and Dick and I sat down and ate dinner.

It was OK brisket, not like mine, but acceptable.

We ate and drank lots of beer and then the Sauza girls came by

Again

We all did a couple more shots.

They had kept the tequila bottles on ice and the

tequila was very smooth and it felt good going down.

Then Dick excused himself I assumed he went to relieve himself at the Port-

A-Can

I was guarding the girls when some guy came up.

Already drunk as a skunk and he started hitting on the girls.

It was the son of one of the guys who was co sponsoring the event.

Rhonda asked him "where is your Dad at?"

"Why do you want him when you got me?" he replied in a slurred speech.

Candace shut him down hard.

"Buddy, can't you see her husband is sitting right here?"

Motioning her hand to me in a presentation style move.

I did my best to look intimidating and menacing.

"Huh… sorry Dude" and he humbly staggered off toward the Sauza Tequila girls

Most likely for the umpteenth time.

Later on I saw him hunkered over a trash can halfway between the cooks tent

And the campsites and the Port-O-Cans heaving and grunting like a stuck pig

punctuated by "Oh my God!" then followed by much more grunting and

heaving and further pleas to his creator..

"Oh God!" and then more heaving. Then he fell off the side of the trash can

and he laid there with spit streaming out of the corner of his mouth and vomit residue

all over the front of his Tee shirt and he laid there. I left him.

Dick finally came back from his adventure in public urination with a box in hand.

The box was about the size of a cake box.

He and Candace were grinning when Dick handed the box to me.

"He said don't open it yet" as he handed the box to me.

So, I set it on the table in front of me and Rhonda

who also looked as surprised as I was.

Dick began his speech

"In every endeavor, there has to be a leader.

"Someone who takes charge and makes things happen.

"You sir are one of those,

"You have made things happen in such a short amount of Time..

That other people could never hope to accomplish in their entire lifetimes".

"And every team needs a Captain."

" And since this is your Chili Cook off team,"

" We,

"Candace and I think that you need to be the Captain of this here team".

"Rhonda joined in with a big "here, here."

I stood up and I thanked them both for their support and I promised that I would do my very best as the Captain of this here cook team.

And then, I sat back down.

Then Dick said

"Congratulations Andy on accepting your position."

" Now first and foremost.

" Every Captain needs a hat."

" That identifies him as the Captain of the team,"

And ...

"Since you have no hat here."

" Candace and I decided that you SHOULD have a hat here."

"So, open the box and wear your hat with pride."

Candace asked Dick "Are you finished?"

"Quite" said Dick.

"Go ahead and open it Andy", Candace implored.

I opened the box and I could not believe my eyes.

There is was, sitting in the box all stiff and brand new.

Just like I remembered it. the way it was

when I first bought myself one many years ago.

An Australian Army Akubra hat

also known as "The Aussie Slouch Hat", it has one side of it's brim turned up against the

crown, Akubra has made these hats for the Australian Army since the early

1900's. Just like the one I had lost years ago.

It even had an armadillo pin stuck in the side just like my old hat did.

I would swear it was a dead ringer for my old hat.

It even had the laced chin strap not the cheap plastic chinstraps sold currently

in the local Army Navy stores.

I lifted the hat gently out of the box like I was holding some priceless artifact

from some lost civilization.

I asked Rhonda to help me put it on.

Rhonda helped me place the hat upon my head,

I smiled and I thanked Dick and Candace for their generosity and their

thoughtful gesture

And I humbly accepted the title of Team Captain.

I proposed a toast for my fellow cook off team mates. "For the guys!"

and we drank a toast.

and As the music from a local country band started playing on the

stage of the cooks tent I asked Rhonda for this dance,

She wisely accepted.

We got up and then the tequila kicked in."

We held onto each other upright and vertical as we swayed from side to side

and around

and round more or less to the beat of the music.

We twirled around on the dance floor for what seemed like an eternity.

Mercifully the song was over and we made a bee line back to our table.

I was going to drop Rhonda off there and run to the Port-a-Can.

but, she needed to go too,

So We walked hand in hand to the long row of Port-A-Cans laid out before us.

Each of us Selected "our own space"

And After a long time of draining my body of excess fluids

I went out and I waited for Rhonda.

She too came back out and She was ready for a few

more beers.

So, back up to the cooks tent.

Where Dick and Candace still sat upright and their eyes were still wide open.

So, I went back to the Keg person and they dutifully refilled four more plastic

glasses of amber liquid and I trotted back to our table.

MY tradition at these events required me to stack every empty glass into the

previous empty until I have made a sizeable stack of empty glasses.

Nine beers and five shots of tequila.

Not a bad start. I still had to remain conscious for the rest of the night.

so I can cook the briskets and start on the ribs in the morning.

I told Rhonda that I needed to go back to the cook site to check on the guys

and make sure that they had eaten.

So, I excused myself again and I stumbled to the cook site.

Dylan, Willy and Christopher were listening to the tapes that I had recorded

just for this occasion. and eating hamburgers and Potato chips and drinking beer..

Dick had one of the Delivery guys drop off a keg at our cook site and they set it up in the supply tent.

Away from prying eyes and they seemed to be having a pretty good time on their own.

They all told me how much they liked my new hat.

I told them it was the Captain's hat, and they saluted me.

I asked Willy to keep the fire burning and add to additional coals to the fire to keep it hot.

And they promised that they would keep the home fires burning,

So, I asked Dylan what time it was, and he said close to 10:00PM.

Oh Boy, only twelve more hours to go.

It's like riding a bike up a hill. "Never look at the summit! Just try to pace yourself"

So, I told the guys that I would be back later on.

And I excused myself and I proceeded back to the cooks tent.

Man, It seems like it is getting further and further away from my Cook site.

Well, that guy who was hitting on first my wife and then the Sauza girls was now passed out face up on the ground

Somewhere between the trash can the cook sites and the main tent.

He must have really pissed the Sauza Girls off because they were drawing on the guys face with either an eye liner pencil or a Sharpie marker.

I know from watching Rhonda putting on make up that they were applying blue eye

shadow on his eyelids and red lipstick on his lips

and they drew hearts on his cheeks and colored them in with the lipstick.

They finished up their make over by writing "Dork" on his forehead.

I left him there on the ground like that after the girls left.

He looked so peaceful lying there.

I decided not to try wake him.

So did a lot of other people.

Passers by would shine a flashlight into his face and burst out laughing

But, no one even bothered to try and wake him up.

Some would check his vital signs to make sure he was still alive

and then they would walk off.

At least no one stripped him or tried to shave his head or urinated on him,

He had already beaten everyone else to the punch in that regard.

With the telltale wet spots on his pants.

I really hope that the Sauza girls were not using a sharpie marker.

I got back to Rhonda and the Wigglers and I told them of my discovery and Rhonda,

Dick and Candace burst out laughing.

And saying that "It serves him right."

And they slugged down their beers and Dick got up for another round.

The stack of empty glasses was getting taller and taller.

By the fourteenth beer it was time for last call.

11:30 PM they lock everything up and the caterers have long since gone.

The band was playing way off key and there were only a few couples still doing their very best to dance on the dance floor.

I had gathered up my auction winnings from Rhonda's old boss

And I straightened my hat. Then I helped Rhonda and Candace to their feet.

And Dick and I carried our lovely ladies back to the cook site.

At first, It was hard to judge who was drunker, Candace or Rhonda.

They both looked pretty wasted.

Dick and I got back to the camp site and Dylan took the auction items

and he placed them in the supply tent.

Dick took Candace and I took Rhonda to their respective beds inside of their tents.

No, we do not undress our team mates.

We leave them as they are Just in case they need to run to the bathroom.

It was pretty warm for an October night.

And tomorrow promised to be much hotter.

I could hear Rhonda..

She was complaining about the constant droning of a generator on some RV.

Because she knew that someone was sitting inside their RV with their A/C on.

And it was probably very comfortable inside.

She could imagine them sitting happily inside with their refrigerated air.

"I bet it's cool inside…" I wish that generator was quieter"

" It's keeping me awake…"

Soon, that generator sound would be the last of her worries.

I went inside and I kissed Rhonda's forehead and I told her

"Sleep tight My drunken Angel!"

And I took my leave.

I left the tent and I collected some logs for the smoker.

Earlier, I had taken some logs and I poured beer all over them.

Now, it was time to feed the smoker.

Willy came up to me and he told me that he did exactly what I asked him to do.

And I opened up the fire box and there was white hot charcoals burning brightly

So, I took the beer soaked logs and I placed them directly on top of the hot coals and the logs started pouring out smoke.

The coals were burning the wood without fire.

I closed the fire box and I went to one of the ice chest and I took out the briskets and put them on the racks in the smoker side of, well, the smoker.

It was just after midnight and we heard the putt putt sound of a ATV coming up the trail.

They stopped in front of our cook site.

It was what Rhonda likes to call the "Beer Sluts."

Now these Girls who ride along on Golf carts or ATVs are trolling for free beer from anyone who would give it to them.

They flirt and joke and kid around with the guys as long as they can get more beer.

And as soon as they drink up the party.

They move on down the line to the next suckers.

Dylan and Christopher poured them a few glasses of beer.

They went into their routine of talking loud and laughing and squealing and touching their victims arm or shoulders and patting them on their backs..

Then out of one of the tents I heard Rhonda yell

"Oh, c'mon Have some dignity!"

Then Candace yelled.

"Move along Girls!"

" Take your stories walking!"

They giggled and then they threw their arms around Dylan's neck and they kissed his cheek and off they went on towards the next victims.

Then, a roar came from the old guys cook site four spaces down,

He was wearing blue jean cover alls and a tank top that had the stars and strips on it.

And we was wearing an old snoopy style flying hat and goggles.

His big white beard made him look like Santa on a bender.

And around his head we wore a dirty red bandana.

He let out another "Oooooo Yeahhhh!!!!!!!!!!!! Into his PA system and blasted his voice

to all of the surrounding cook sites.

He was calling himself "Commander USA."

And he was selling his own brand of Picante sauce and salsa.

He'd take a few sips of whatever he was drinking

and he'd try to sing some old country standard like

"All my Ex's live in Texas" or "Lukenbach, Texas"

followed by another Oooooo Yeahhhh!!!!!!!!!!!!

After a while,

I threw another log into the fire box, checked the meat and repositioned the briskets around inside the smoker and closed the lid.

Then came another loud "Oooooo Yeahhhh!!!!!!!!!!!!"

Rhonda got up and she stormed out of the tent all hot and sweaty and not in the mood for his kind of crap

She was going to walk over to Commander USA and punch his lights out.

But, I grabbed her arm before she could get over to his site

She jerked her arm from my hand and rolled her sleeve back down

and she said that she had to go potty.

Candace too was up and she also needed to go and relieve herself.

So, they both headed to the bathroom while Dick Dylan Christopher and Willy and I played a few hands of poker.

And we listened to ZZ Top and Jerry Jeff Walker and Twisted Sister.

trying to drown out the diatribe coming from the not too far distance.

All this was going on while I monitored the progress the briskets were making.

The lanterns did a great job of illuminating the cook site.

but, it messes up my night vision.

I could barely make out the facial details of fellow cooks still wandering about doing

their best to try and keep awake.

And check out the competition.

Hard to imagine them having any problems staying wake with

Oooooo Yeahhhh!!!!!!!!!!!! Going on.

It was almost like clockwork.. a badly rendered song

followed by another Oooooo Yeahhhh!!!!!!!!!!!!

Me and the other guys had decided that it was Commander USA's job.

To scream into his microphone and keep the rest of us cooks awake way into the morning hours.

He took his job very seriously.

And he even had a crowd of zoned out drunken people

sitting on his impromptus benches.

While he passed out tortilla chips and handed out samples of his home made salsa.

I am sure that they were there because of the free food

and the places to sit and the well lighted area.

Whose light shone directly into Rhonda's eyes when she was lying down

trying to get

some sleep.

I noticed that I hadn't seen Rhonda or Candace for a couple of

Oooooo Yeahhhhs!!!!!!!!!!!!

I excused myself from the game and I went on a search and rescue mission.

It turned out that the rescue missing was unnecessary.

They were talking to the father of the guy with the makeup and Dork on his forehead.

They told him that they hadn't seen him since the cooks party.

Which was true.

So, I went off to look for him back at the last place where I saw him and he had somehow moved on to some other location.

As I walked back to where the girls were,

one of the sponsors friends came up and said

that they had found him and they drug his body back to their camp site.

I was extremely happy that they found him.

I was starting to feel bad about leaving him there.

Then came another "Oooooo Yeahhhh!!!!!!!!!!!!"

followed by an even louder

"OOOOOOOOOO RAHHHH!!!!!"

That was our cue to excuse ourselves and head back to our site.

The time was now closer to 2:00 AM.

The smoker needed more logs.

I poured some water from our ice chest over some more logs

and threw them into the fire box.

And I opened the foil on the briskets a little bit so more smoke can make contact with the meat.

And then I went back to the table and Dick was sitting there.

Drinking with the guys.

I came up to them and I sat down and I started making conversation.

I asked Dylan and Christopher how long they had known Bill,

and how long they had worked for the company,

Their answer surprised me and Dick.

It seems that they have known Bill their entire lives and they have worked for the company ever since it opened.

It appears that Dylan and Christopher are actually Bills own sons.

And that Dylan and Christopher are in fact brothers.

Dylan is the oldest and Christopher is his little brother.

"Little" was really not the word to describe these two guys.

They are about six feet tall.

Strong and powerfully built, and they are exceptionally nice and polite.

I felt honored that Bill would entrust his own flesh and blood to me and my project.

I decided that I would do my very best to watch over them

and try to keep them out of trouble.

We listened to some music and drank a couple more beers.

Then the guys had reached their limit.

 and they decided to say goodnight and get some shut eye.

They excused themselves from the table and went off to crash on one of the beds

That we set up in the supply tent .

Willy got up and headed out and was on his way to the bathroom.

The girls went back to their tents and they crashed.

Face down in their beds.

And in a little while we could hear some snoring.

Coming from the direction of the tents. Since women "Never snore" we had to assume that the guys were snoring from the supply tent.

"Oooooo Yeahhhh!!!!!!!!!!!!"

 Followed by some Hank Williams tunes no longer had any effect on our ladies.

As the time got closer and closer to 3:30AM.

The magic hour.

When it's time to check on the briskets and see how they are coming along.

This is normally the time when I begin my lonely vigil.

I was drunk.

but, I had a job to do and I took that responsibility very seriously.

I was happy to have Dick with me to keep me company.

I could tell that Dick was very close to falling out.

But, he hung in as best as he could.

I was surprised that I was drinking and staying toe to toe with Dick,

He may know beer.

But, he had a problem holding his tequila.

Sensing a weakness.

I went to the ice chest and pulled out a bottle of Jose Cuevro 1800,

That I had stashed "for the lonely times"

I brought it out to Dick.

And a mere four shots later,

Dick was down for the count.

Willy had been back from the bathroom for quite some time.

After a few shots.

He too had slipped into the supply tent with the guys.

I went into the supply tent.

I woke Willy back up so he could help me carry Dick to his and Candace's tent.

That being done,

I began my lonely watch for Brisket glory.

"Oooooo Yeahhhh!!!!!!!!!!!!"

Commander USA continued to bellow long into the morning hours.

His voice getting weaker and scratchier as time wore on.

He was sounding like some kind of drunken rooster.

And as the sun began to rise in the east he began to fade into the din of other chili cooks breakfast noises.

I put my coffee maker on the Coleman stove.

I made a pot for everyone to consume.

I had switched to a can of Dr. Pepper and I was preparing to cook the chili on one of the other stoves

All of my ingredients having been lovingly prepared in advance

All I had to do was cook it.

Adjust the seasonings and then serving could begin.

I had also dragged out the breakfast food that we had packed.

Eggs, bacon, sausage, and pancake mix, butter and syrup.

This is why I bring more that one Coleman stove to a cook off.

That way I can cook breakfast while still having burners open for the coffee maker and additional skillets

The cook off site always looks like someone set off an H-Bomb after a crazy night like the one before.

I was trashed.

I could barely keep my swollen eyes opened.

Bt, I have to press on.

The smell of freshly brewing coffee got Dick and Candace up first

Then Rhonda began to stir.

Finally the girls got up and they both looked ok to us men folk

But, apparently they looked awful to each other.

And since there was no running water or mirror or other niceties present.

They had to take their sweaty bodies across the street.

Where a nice doorman let them into the high rise office tower across the street.

Inside there was running water, a mirror and air conditioning

Also a comfortable place to apply their war paint.

By the time they had returned.

I already had breakfast ready.

And I was serving the guys.

Who for some reason were not that hungry.

They were very thirsty and it was a good thing

That I had packed plenty of sodas

I remembered to pack plenty of aspirin and Heartburn tablets.

Which they greedily consumed With their coffee before eating breakfast.

I had to wait for the Sun tea to make itself.

If you don't know what Sun tea is you take a large pickle jar.

the kind they use in fast food places and convenience stores.

And you fill it up with water then you take a couple of large tea bags and you immerse

them in the water and set the jar out in the sun for about an hour and voila,

The sun heats the water and, then you remove the tea bags and add sugar and there you have it. Sun tea.

I had just prepared plates for the girls and set them down before them and I had to stir the chili that was cooking on the Coleman stove.

And I had to make sure that the firebox was still hot because the briskets were almost ready to take out.

And the ribs will have to go in next,

I also had the beans on the stove.

They take several hours to cook.

So, I put them on right after the coffee.

So that they can cook for as long as possible.

Rhonda will still have plenty of time to season them,

The beans are Rhonda's forte.

And they are judged differently from the other categories.

So I give her plenty of room when she makes her way to the stove.

I got my chance to eat.

So, I left everything in Dick's charge.

I sat down and I had a couple of bites and Dick said

" I went out like a light right where I was sitting."

I rose from the sleep of the dead about thirty minutes later

And the team had taken up the slack.

Rhonda had stepped in and she took charge of cooking.

And she only woke me because she wanted me to check on the briskets.

They looked and smelled heavenly.

I sliced off a hunk to make sure that the meat was fully cooked and nice and tender.

"OH, dang it It's falling apart. And it taste horrible."

I am going to have to throw everything away, or else give it to the dog."

I joked to everyone.

So naturally everyone had to have a slice just to see how awful everything tasted.

And they all agreed that we had a winner there.

I didn't tell them that we could not win the cook off no matter how good our food was.

At least I knew that we had something I would be proud to serve anyone.

I had to go to the bathroom.

so, I excused myself and I went to the Port-O-Cans

I passed Commander USA's cook site.

And he was out like a light.

Passed out and lying down on his stage surrounded by empty bottles and tortilla chips.

Scattered on the ground before him.

I wish I had a camera with me.

that would have been priceless.

Oh well, I relieved myself and I walked back past the Rattlesnake Chili and the Margarita Society sites

They were also loud last night but around four AM they fell eerily quiet.

believe me when I say that they were up and in full force and slapping their stickers on any and all of the females who would let them do so.

This year their stickers said "I got Bit"

The Margarita Society had buttons that said

"The Margarita Society" in a lovely Script on a white background.

We had the buttons that I had made the year before they had the words

"Critter Du Jour"

Over an American flag backdrop.

And I was wearing my Aussie hat with pride.

And I was never more proud that I was at that moment.

Everyone doing things and all of them helping out with the cooking process.

Stirring the chili and watching the ribs, while I sliced the briskets.

Then the announcement came over the loud speaker for the teams to send one person to pick up the containers and the tickets for each event..

"I'll go, said Christopher.

So, I gave him one of our aprons and a button for his shirt.

We sent him on his way to the cooks tent to gather our containers.

Willy was indispensible,

He was packing all of the camping gear up that we no longer needed.

And he was busy getting us ready to bug out as soon as the cook off was over.

And as soon as Christopher returned with our containers.

I started piling seven slices of brisket into the brisket container.

It was ready to go as soon as they judges called for it.

The judges had Celebrity judges there in attendance one of them was a DJ for a radio station and four of the other judges were "The Beebles" a Beatles

cover / Tribute band that was going to be playing on the main stage later on in the day.

The judges were given water or beer and crackers and antacids..

And they were required to taste all of the submissions.

And rank them accordingly.

I am certain that antacids were extremely necessary.

OK, the call went out to all of the cook teams

The time came for us to send in the brisket submissions and Dylan and Christopher took the seven slices of brisket down to the judges tent.

Then we threw some flour and corn tortillas on a griddle that I used to cook the pancakes on

And we started "taking donations" for brisket tacos.

Usually a dollar a taco,

It comes with graded cheese and lettuce.

And we sold almost an entire brisket.

Then the call came out for ribs.. I sliced off seven ribs from the racks and I put them in the container.

and I handed it to Christopher who dutifully ran it over to the judges tent.

And then we started selling ribs to whoever wanted some.

In very short order all of the ribs were gone.

I knew that the chili would be next.

I tasted the delicious chili and I got it placed into a container.

And I left the lid off so it could cool off some before they called for it.

Now Rhonda and I had discussed this tactic at great length.

I always used to pour the chili directly from the stove right into the container and put the lid on it as soon as I was finished pouring.

Rhonda on the other hand insisted that we should pour the chili into the container and let it cool with the lid off.

So that no water condensation would form on top of the chili.

While it was waiting to be sent off to be sampled.

Then place the lid on top right before handing the container over to the judges.

I was now flexible to her suggestions,

I decided to try it her way this time. So I asked

Christopher not to cover the chili until it was time to hand it over to the judges.

He promised me that he would do exactly as I requested.

Toby and Isabel came out and they sat in the shade and drank some beer and ate some chili and chilled out and did some crowd watching.

Toby told me that he was almost finished building the second set of electric wheels and the generator assembly for the flywheel generator.

I told him "No worries" and I served them up another round..

Then Earl showed up and we sat him down next to The Calientre's table and fed him and plied him with beer as well.

No one walks away hungry from our booth.

And very soon after that, the anticipated call came in for the chili.

So, Christopher did what I asked and he left the chili uncovered until he got to the judges tent.

Then he covered it and took it inside.

And in a few minutes He emerged from the tent with a big smile on his face.

So, we started serving chili by the cup and we gave out crackers too.

We sold almost the entire stock pot in about thirty minutes.

Then came the call for the beans to be submitted.

Rhonda filled up her submission cup and handed it over to Christopher..

Christopher ran the beans over to the tent and he came back with another big smile on his face.

I asked him what was up with the smile and he said that there was a girl in the judges tent who was passing out free beer.. I asked him,

Did you forget that we still have a keg in the supply tent?"

"No, I didn't forget, but, you guys don't look near as good as she does"

All told we were able to raise almost seventy five dollars to donate to the Chili Society.

We had about an hour to kill before the judging would begin.

so, I took the opportunity to talk to the people gathered around "Deguello:"

I got to explain the purpose of the vehicle and all of it's special features to a willing crowd.

I spoke to anyone and everyone who was interested in hearing what I had to say.

I talked about the race and Representing the State of Texas and the USA.

In a world arena at great length.

And then I mentioned that the New Paluxy Brewery was my sponsor and that they were drinking their beer.

They were impressed with the beer and the vehicle and it's looks inspired comments of "Wow, and "Cool" and chants of USA, USA! from an appreciative audience.

I took a quick look around and I noticed that the margarita society was passing out tequila Shots.

So, I headed over to say "Hi". And I did a couple of shots with them.

It's kind of a tradition.

I left the question and answer session in Toby's capable hands.

Yes, I do still have a half full bottle of tequila that I was drinking on the night before, but, this one is "Free" and since they were already passing out shots, I decided to be a little opportunistic

and "Offer my services" to the group as a "tequila tester" not far from the Margarita Society's site was some guy who was passed out on the ground face down lying on a fire ant mound.

Unlike "the Dork" who was crashed out in the field next to a trash can that everyone left there,

And who was eventually rescued by fellow team mates.

Security was notified and they came out in force to move the guy to a much safer location.

while they were waiting for the EMT's to come out and evaluate his condition.

And to treat his many fire ant bites all over his body.

I was glad that they rescued him.

You know what? I hadn't seen the "Dork" guy all day.

I can't imagine why.

There was a guy who had a site set up,

493

he didn't do the cook off he was there with a karaoke machine and drunk people were singing their drunken renditions of their favorite golden oldies.

I was tempted to sing a couple of songs but, I decided not to join in.

At last it was time for the judging.

Rhonda, Candace, Dick and I, headed to the main stage and on the way over we could hear the sounds of a live band on stage.

We listened to the Beebles doing their cover versions of Beatles classic hits like "Eight days a Week" and then they covered "She loves you"

Then they did "In my life". Rhonda and I danced a slow dance to the music. Then they did a very good rendition of "Hard Days Night," and then they finished their set with I wanna hold your hand"

They were a very professional and tight sounding band that managed to capture the sound and the flavor of the Beatles.

When they finished, Rhonda's old Boss Mr. MacReady took the stage and thanked everyone for coming out.

And they gave an honorable mention to Dick and Candace Wiggler for all of their hard work and support for the Leukemia Society. (I.E. the beer.)

And then they began the judging…

Bottom line we didn't win.

Big surprise. We did come in the top twenty.

We placed 17 in Chili and Ribs and 5th in beans.

Our brisket was in the top ten.

Rhonda and I went up to the stage with our tip jar full of money.

We had raised an additional eighty five dollars,

and more importantly we had gotten rid of two briskets and a couple of racks of ribs a half a pot of beans and a whole pot of Chili.

I was hoping to do better by having no leftovers to take back home.

The real shocker was that Commander USA won the award for "Best of Show"

for his Catwallader performances that lasted all night and kept us awake well into the early morning.

He walked up to the mike to receive his prized trophy and he held it over his head and he bellowed into the mike one last: OOOOOOOOOOOOHHHHH YEEAAAHHH!!!!!!!!!

And the crowd let out a collective UuuugggghhhhhHHHH!!!!!!!!!!!! Sigh.

Apparently they were equally annoyed as we were.

Only they were not four spaces down from him.

I was just glad that he was not right next door to us.

Rhonda said "Can you believe that guy???"

"Winning best of show? C'mon!!!!"

And she wanted to roll up her sleeves and go up to him and punch him in the nose for ruining her sleep.

But, I restrained her and we walked away from the stage area

With me telling her "It's OK baby, well be leaving soon" and "Just let it go!"

And together, we went back to the cook site to finish packing and loading up the trucks for the return trip home.

Toby and Earl and their Misses decided it was time to go.

So, they told us that they had a good time and they'll help me finish the keg later.

We bade them farewell and told them to be careful driving home

(they put away a lot of beer and helped me drink more than a few shots of 1800.

They said that they would do their best and then they took their leave.

We used the water from the melted ice to rinse out the cookware enough to pack them away.

So we can wash them properly when we get home.

Dick Instructed the delivery guys to follow us back to the house.

So they can collect an empty keg and drop off a fresh one if there was any left.

Man, the majority of the cooks leave immediately after judging.

And the place looks like the final day at Woodstock.

Mud (from dumped water and emptied ice chest, and sand everywhere along with overfilled trash cans and port-a-cans laden with their cargoes of filth and stench

Litter and debris and soaked t-shirts and stickers scattering the surrounding area.)

Throughout it all the Nina (remember Columbus's ship in the lake) rode the waters of Lake Caroline like she belonged there.

A few people paid the admission fee to go aboard her.

There isn't much to see.

You can see the rigging and the decks from the outside without having to pay anything.

But, she was a nice backdrop to the festivities.

All and all, it was a successful cook off.

We helped raise money for the Society and in turn we helped out many people in need of their services.

And we did a good thing.

So, after we piled everything on the trucks we slowly made our way back to my house. In all actuality none of us were in any shape to drive.

But, you gotta do what you gotta do.

So, we carefully drove back to the house in convoy fashion so we could keep an eye out for each other.

And we had an uneventful trip back.

We pulled up in front of the house

just as the sun was beginning to set.

We let the dog out much to his relief.

And then we unloaded as much stuff as we could.

including the left over beans and the cookware and utensils.

And we packed as much as we could into the dishwasher and began to run the first load.

That's when I noticed that Rhonda's and Candace's shoulders were red and beginning to peel.

And Rhonda's face was starting to get that raccoon mask of freckles that she gets when she has a little too much sun.

So, both the girls were sun burned.

But, not too badly.

Rhonda took Candace into the bathroom where they applied Noxema Medicated Skin Cream to each other's burns.

It sounds sexy until you catch the camphor like smell coming off their bodies.

And you can see where Rhonda had applied it to the mask of freckles around her eyes.

The camphor and eucalyptus smell burned their eyes with that cool yet painful scent.

Candace was not much better off.

They just wanted to lie down in a cool and darkened room.

Where they didn't have to listen to the sounds of other people's generators or A/C

units and least of all Commander USA saying "OOOOOOOHHHHHH

YYYYEEEAAAAHHHH!!."

And all they wished for was for a chance to take a nap and sleep in the cool air of our house.

The rest of us guys had the tasks of finishing up unloading.

Thank goodness for Willy and the Delivery guys.

They took my empty keg back and they replaced it with a fresh one and a new CO_2 tank

I was surprised to discover that they still had a couple of leftover bags of ice on the truck so I used them to pour over my keg.

Unfortunately, none of us had any desire to drink it.

I yelled out to everyone…

"Hey, who wants Bar-B-Que???"

"Anyone want a beer?"

" How about tequila shots?"

I was surprised that no one wanted to have a drink of beer or eat bar-b-que for dinner.

(not really surprised just overtly amused.) not that I had any left.

Dick let out a loud "OOOOOOOOOOOOOOOOHHHHHHHHHHHHHH YYYYYYYYYYYYYEEEEEEEEEAAAAAAAAAAHHHHHHHHHH!!!!!"

And Candace yelled out from the bedroom.

"Dick I am going to strangle you!"

We got a good laugh out of it when Dick called back and said

"Sorry Hun!"

After everything was temporarily put away

We all crashed where ever we could find a place to sit down.

Poor Thor had to go and lie down on the floor in my bedroom with the girls just to keep out of the way and out from under foot.

Dylan and Christopher finished unloading their truck and we got the smoker off loaded

and got it back in the back yard.

Then they came back in and they said that they were heading back home.

I asked them both if they were OK to drive, they said that they were.

But, before they could go back home.

They first had to take Deguello back to the shop.

And lock her up at the shop before they could head back to their home.

We thanked them again before they headed out and they thanked us for a fun time.

They had earned a healthy respect for what Rhonda and I do for our charities.

I asked them to call us when they make it back home so we wouldn't worry.

Dick and I helped the delivery guys get all loaded up and ready for their long trip back to Austin.

They will get to rest on Sunday before they will have to go back to work on Monday.

I didn't envy that part. But, I was glad that they did what they did for us.

And Dick and I wished them a safe trip back home.

It was getting close to 10:00PM and I was starting to get sober and hungry.

So, I ordered us a couple of pizzas over the phone and had them delivered.

Because, I sure didn't feel like driving anywhere.

And I had no idea just how long the girls were going to sleep.

As much as they had to drink,

they could easily go all night and sleep through the rest of the evening.

Dick and I ordered us a meat lovers pizza with sausage, beef, pepperoni, and Canadian Bacon and extra cheese and we ordered the girls a cheese pizza.

Because that is what Rhonda likes and since my mind and brain were only functioning at half capacity.

That was just about all I could remember to order for them.

We waited for the pizza to show up and I decided that I wanted to wear my Aussie hat In the house.

So, I put the hat on so I would be wearing it when the pizza delivery man shows up.

And I put some tea on the stove to brew.

We watched the news and were getting ready for Saturday Night Live to come on TV.

When we heard a knock on the door.

It was our hero "Mr. Pizza Man" with our delivery.

I grabbed my wallet and went to the door.

Thor came running out of the bedroom with his hackles up and bristling and barking loudly.

I told Thor to sit, and I opened the door and paid the guy and took our food closed and locked the door and took the pizzas to the table.

And opened the boxes to make sure we got what we had ordered.

They looked great.

So, Dick and I grabbed a couple of plates and a few slices and we closed the box and sat down and started watching the show.

Soon we were laughing our heads off and our laughter woke up the ladies and Willy who was crashed out in the den on the sofa.

They were hung over big time and they came out very bleary eyed.

And they rubbed their ruffled brows with the back of their hands.

They had perfect timing, I had just finished mixing up the tea.

So, I passed out aspirin and heartburn tablets to all takers and then they all sat down to share the pizzas with us.

Then we finished the show and we all went to bed

The Hybrid Diaries
Chapter Thirty Seven
Sunday , October 21st 1990

Somehow I managed to wake up before everyone else.

So, I dragged my tired hung over butt to the kitchen counter and starting filling the pot for the coffee maker with water.

And then I poured the water into the coffee maker.

I pulled out the filter basket and grabbed a paper filter and put it in the basket,

Then I looked for the coffee can.

I can't find it!

I would not want to be anywhere near Rhonda without her first half pot of coffee in her.

So, I guess I'll have to run to the store for a food run.

I dragged my body back to the bedroom and I threw some crappy clothes on.

I hopped into the crap mobile.

And I went up to the grocery store.

I bought breakfast type stuff so we can have something on our stomachs other than last night's pizza.

I walked into the store looking like Hell warmed over.

I pushed the grocery cart looking like a street person.

Partly to gather my groceries.

Mostly to hold myself upright and walking.

I passed a display for potato chips and corn chips with a picture of Troy

Aikman in uniform doing a passing pose.

With a Dallas Cowboys game schedule underneath his picture.

Wow, we are playing a game today!

Cool beans!

Now I knew that replacement keg would indeed be necessary..

With all the doings going on I forgot that the Cowboys were playing today.

We're playing the Tampa Bay Buccaneers in Tampa, Florida.

I hope Coach Jimmy Johnson can get us a win after us getting our butts kicked by the Arizona Cardinals in the previous week.

I hope Troy Aikman can somehow manage to get more than one hundred yards passing this week or we're going to be slaughtered again.

So, We need breakfast and game time food items.

To feed anyone and everyone who might drop by.

So, I threw a roll of sausage and a package of bacon in the basket and a dozen eggs .

I added a couple of cans of biscuits and some all important coffee into the basket followed by the stuff to munch on while I would be preparing dinner.

Like, boneless chicken breasts and a couple of bell peppers.

and a couple of onions.

some limes and a couple of packages of tortillas both corn and flour.

A couple of large bags of tortilla chips, a jar of salsa, and a couple of cans of bean dip.

A large head of lettuce and three packages of crispy taco shells and a large can of

nacho cheese.

Grated mild cheddar cheese and a few tomatoes.

I like to serve Mexican food during the games because it's easy to prepare and it involves a lot of finger food.

I.E. tacos and nachos, chips and salsa and fajitas.

I also took the liberty of buying a few cans of Ranch Style Beans.

So, we would not have to make beans.

We were going to make Mexican rice to go with the rest of the food anyway.

Besides, beer was made to go with Mexican food.

Just look inside the ZZ Top album "Tres Hombres".

And you'll see a picture of Mexican food and beer from a place Called Leo's in Houston, Texas.

That picture usually makes me hungry.

I wandered up to the check out line going through my shopping list in my head .

Desperately trying to remember everything I could.

But, I always seem to forget some key item.

And I usually have to make a return trip.

But, I never leave the house unless it's halftime.

I put my purchases on the counter when it was my time to check out.

Then I paid the cashier and one of the bag boys loaded up by grocery bags.

He then placed the bags into the shopping cart.

And he helped me wheel the cart out to the crap mobile and load the groceries into the back seat and the trunk.

I tipped him a dollar for helping me and I got back into my car.

I drove back to the house.

When I arrived back at the house, I could hear voices from inside.

So, I brought the grocery bags to the front porch.

Then I knocked on the front door to let them know that I was back home.

And to come to the door and help me carry the grocery bags into the kitchen.

The knocking immediately set Thor off.

I could hear his big dog bark from inside the house.

Rhonda came to the door and moved Thor aside and she said

"I hope you bought coffee!"

As she grabbed a couple of grocery bags and carried them into the house.

I told her

"That was the reason I went to the store."

"Well, did you buy more paper towels and foil?"

"Dammit!" she did it to me again.

"No, honey, I forgot those."

I sheepishly replied trying to hide my frustration

"Well, you're going to have to go back and get us some then."

She replied in a terse voice.

I grumbled and mumbled something probably obscene under my breath.

I did my best to give her a fake smile.

She looked at my face and then she wrinkled up her sunburned

nose and made a face at me.

Then she stuck her tongue out at me.

"Well, I am NOT leaving right now." I replied.

"I've got to make breakfast."

"Not without paper towels and foil. " She said pointedly.

So, I ran out the back door and went to the little convenience store behind my house.

I paid twice the normal price for a roll of aluminum foil.

And about the same huge mark up for a twin pack of paper towels.

Because, I had just got home.

And I was not going to go back to the store for those two items.

I took my purchases and ran back to my house .

I plopped them triumphantly on the kitchen counter.

Rhonda was in the den with Dick Candace and Willy watching the pre game hype shows

She heard me come inside.

Then she hollered to me

"Don't worry about the foil or paper towels,

"I found an extra roll in the pantry and we still had some foil left over from the cook off."

She happily chirped

I grumbled to myself some more.

But, I didn't answer her.

I just put the foil and the paper towels away.

I went over to our Chamber's stove and turned on the burners and began to cook breakfast.

I lit the pilot for the oven and I got it all hot and ready then I turned off the oven

I put the biscuits in the oven and left them there while I prepared the bacon for the Microwave.

I sliced the sausage into patties and I put them in a cast iron skillet and put them on a burner on top the stove.

While they were cooking I scrambled the eggs.

Now, I remember what I forgot at the store.

Hash Brown potatoes.

If no one asked about them then I would be home free, but,

If Rhonda mentioned them I would be amiss if I didn't immediately leave and fetch some from the store that I didn't want to go back to.

The biscuits baked with the heat from the oven.

Chamber's stoves are remarkably well insulated and you do not need to keep the oven on to cook some meals or side dish items.

Fortunately for me.

Everyone was satisfied with the meal I had prepared for them

No one missed the hash browns this time.

I dodged another bullet this time.

After the breakfast dishes were cleared away.

I started marinating the chicken breasts.

I covered the pan with wax paper and placed it in the fridge For later.

We watched the pre game shows and got ready for the game.

I needed to dump the ice out of the keg barrel.

I made another run to the convenience store for a couple of bags of ice.

Then I poured it over the keg and pressurized the lines

And I went back inside and took a well deserved break.

We were all waiting for kick off.

Dick and I pondered our next moves.

He told me that they were expecting their environmental impact study any day now.

And that he had found some equipment at the soon to be shuttered old Lone Star Brewery in San Antonio.

He had been working on his own formula for a new beer

And he needed a working name for the beer.

Based on a old German family recipe (His Grandfather's)

I asked him if he had any names for the new beer.

"And that is where I am stuck." Said Dick.

"I have lots of ideas for names but, nothing sounded good to me."

"What do you think, Andy?" Dick inquired

"Holdensteiner!" I like that name!" I replied.

"I also like the name Holkensteiner" They sound German" I replied.

"Holdensteiner? Where did you come up with a name like that?" Dick looked puzzled when he asked me that question

"It's a play on words" I told him.

"Look, what do you do with your beer mug when you are drinking beer?" I asked Dick.

"Well, I am generally holding my beer mug in my hand"

"And what do you call a German Beer mug?"

"A stein" Dick said.

"OH, I get it!, Holding my stein!, That's brilliant!" Dick Liked the name!

"I knew there was a reason I hired you!"

he said as he slapped me on my back.

We both got a good laugh out of it.

Dick called out to Candace and Rhonda and Willy,

"Hey Guys, Andy has a name for our new beer!"

"Really?" Candace yelled from the den, not wanting to get up.

"Would you like to know what it's called?" Dick asked

"Uh… Sure Honey. What's it called?" asked Candace

"Holdensteiner!" replied Dick.

"What do you think?"

"Sounds good Honey!" said Candace.

"Or do you like Holkensteiner?" Dick added

"They both sound good to me Dear."

Said Candace in a "Just going along for the ride sort of way."

Almost condescending but, with a lack of malice or interest for that matter.

"They both sound good Honey!" She responded

So dick and I gave a toast to the new name for the new beer that he was going to be brewing at the new brewery in the Glen Rose area.

But we still hadn't decided on which one would be the one we choose to use.

We liked them both.

"So, here's to Holdensteiner or Holkensteiner Beer!"

"The new brew of Texas!"

Well we ate chips and bean dip and Queso dip.

And drank beer until the end of the game.

I used half time to get the grill side of the smoker all fired up.

I cooked the boneless chicken breasts slowly over hot coals

Then I grilled them to perfection.

I chopped the onions, bell peppers, and red bell peppers into thin slices.

I sautéed them in butter and natural juices and slow cooked the veggies

So, that they were ready when the chicken came off the grill.

I sliced the chicken breasts into strips.

Then I threw them all together and cooked them on top of the stove in a cast iron skillet.

I also refried beans and I put some rice on the stove.

Then I placed a couple of stacks of corn and flour tortillas in a tortilla cooker put them into the microwave oven for three minutes.

Ding, Dinner was ready to be served.

Just as the fourth quarter was getting underway.

By the time we were all too stuffed to move.

The game was over and the cowboys had beat the Tampa Bay Bucs 17 to 13.

Nott a blow out by any means, but, any victory you can walk away with is a good win.

So we continue our march to the super bowl with one more team defeated

And one less team standing in the way of a potential super bowl victory.

It was a good thing that we didn't go anywhere for dinner because we were all pretty

trashed out from the cook off and the sun and all the packing and unpacking.

It really took a toll on us.

It's a lot of work to compete in a chili cook off.

I am glad that I only do it once a year.

Dick and Candace were passed out asleep on the sofa.

Willy had gone over to them and removed their shoes.

I gave him a blanket to cover them with.

Willy thanked me and he took the blanket and draped it over the Wigglers.

Rhonda was asleep in the chair.

She nodded off very shortly after the end of the game ,

I left her there while Willy and I cleaned up the after dinner dishes.

Then when we were finished.

I found a nice comfy spot on the bed and I laid down just to rest my eyes.

I wasn't planning on going to sleep.

 I still had so much work to do.

But, I guess it would have to wait until I could open my eyes again.

I dreamed about a lot of disjointed things that didn't make a whole lot of sense to me at the time.

I dreamed about working at the thrift store.

And I saw some of my old employees just as I had remembered them.

So, I guess I was concerned that the store was doing alright.

What did I have to worry about?

Randall was running the store and he was the owner of the management company that ran the store.

So, It was (I assumed) in good hands.

Besides, I do not work there anymore.

I remembered warning Randall about the new thrift store that was opening up

just down

the street.

I had a bad feeling growing in the pit of my stomach.

I kept wondering… maybe I should just go up there and pay them a little

friendly visit. Just to see how everything was going.

I knew it was already too late to go by there this evening.

I decided that I would drop by there the next day.

Just to say "Hello."

The Hybrid Diaries
Monday October 22, 1990
The walls came tumbling down

I actually slept pretty good.

I woke up and I got the coffee ready, and then I casually left the house.

I took the crap mobile over to the thrift store.

And Boy was I surprised when I got there.

My store had been closed up and all the stuff was gone!

My former company truck was no longer in the pen where I used to keep it.

Everything was gone.

The place was empty.

And there was a message on the front door

that said something to the effect that

"We moved! Come see us at our new address"

"Blah blah blah blah. Road at where ever"

Maybe he realized that the new store down the street was bigger and had taken over the local market.

It must have siphoned off the customers that were frequenting my store.

Or else it was the leaky roof and the periodical break ins and the bad parking lot with all the pot holes.

Or the fact that the a/c didn't work properly.. or else they might have lost their lease.

What ever the reason.

I wasn't going to find out what happened there by staring into a vacant building.

I would need to call a couple of my ex employees

maybe even my old boss.

He still had the other store in Garland.

I wondered if that one was still there as well.

So, I drove back to the house.

I found Candace and Rhonda already awake and dressed to go out.

So, I kissed Rhonda and I asked her where she was going.

"Oh, Candace are going out to look at cars!" she chirped.

"Oh, ok then, you girls have fun.

Don't buy it until I get a good look at it."

"No promises!" she replied.

And her and Candace left.

Candace yelled to the bedroom

"Dick we are leaving!"

" You need to get up and go to the brewery site!"

"Uhhh?" I heard Dick grunt.

"You need to go to the brewery site and meet with the inspector at three."

"Grrrrrrr Ruff. As he rolled back over in bed.

Then the light in his brain came on.

"huh? Ooooh, Yeah, OK then!"

Dick managed to string a few syllables together I am sure

What he meant to say was something really concise and articulate.

But, since he was still half asleep he did a lot of muttering.

I called the shop and Bill said that they were busy working on a another

project for an aviation company across the street.

"There would be no testing today."

But, he assured me that they would have the second pre production vehicle ready by next week,

In time for some testing and a planned road test.

We will drive the vehicle down to Houston.

And then to the container company.

Where we will load up our containers full of beer.

The two vehicles and spare parts and equipment for the race.

So, I asked Willy if he needed to go back to Austin.

He said that he had things that he needed to do back home.

So, I asked Dick "since we did not have anything for Willy to do this week if it would be ok to let him go back home?"

Dick didn't see a problem with that.

He said that we could drop him off back in Austin after we went to the brewery

site to meet with the environmental impact inspector and pick up our construction permits.

That way Dick can begin construction on the brewery while we are on our way to Australia.

Dick had told me that he had already hired a construction company and they would be ready to start the plumbing.

Then they would be ready to begin pouring concrete.

After the wells were drilled and the foundation was all framed in and the rebar had been laid down.

We did some follow up on our expenses.

I got my weekly paperwork together.

Since we were heading back to Austin anyway.

I decided to drop off my expense reports at the office in person.

So, we all (Dick, Willy and myself) climbed into the Suburban

And We headed off towards Cleburne on Highway 67

We listened to some music cd's and I thought about what was coming in the future.

That was when I remembered.

We only had one week to prepare for our annual Halloween costume party and Ball..

Since October 31st would fall on a Wednesday,

We needed to decide pretty quickly if we were going to have the party on the 27th or on November 3rd..

We usually try to have it ON Halloween.

But, with it falling smack dab in the middle of the week

And since we haven't told our friends ANYTHING about the party.

I was thinking on having the party on the third of November.

so we would not have to compete with other Halloween parties.

We would have the best opportunity of having everyone show up.

Now our Halloween parties are famous for our costumes and all of the crazy party

behavior.

People tend to act like their Halloween personas.

And we go all out on our costumes.

Rhonda and I like to wear complementing costumes.

If we want to be "Dark Characters"

Rhonda would usually go as a witch and I would

go as the "Lord of the Nazgul" from "The Lord of The Rings."

Now if you ever saw Ralph Bakshi's movie version of J.R.R Tolkein's classic

"The lord of the Rings, or read the books,

The Nazgul show up riding black horses when the hobbits

are on their way to Rivendell.

The black riders are pretty fearsome bad guys.

They look like Death on a horse.

Except the horse isn't one of those walking skeleton horses

That skeleton warriors seem to be riding. around on.

I would have a hooded cape and a blacked out face.

Complete with light up red glowing eyes and an evil looking sword,

(unless I went out in public, then I would carry a safe non threatening plastic

sword.

That was safe to carry or fall on or drop onto peoples feet.)

I thought about where to have the party.

And what we would need to bring.

so that everyone could have a good time.

I was in the middle of some deep thought processes as the Suburban made its

way down the freeway.

Past the huge fiberglass manufacturing plant and the concrete plants that

throw up tons of dust into the local air.

Willy was stretched out over the back seat.

He was sleeping hard and sawing logs.

I took my cue from Willy just as my mind started to wander off.

Before I knew it I was asleep in my seat.

my right hand cradling my head as we motored down the highway towards the

"The Wiggler New Place".

I was dreaming about the last Halloween party we had the year before.

We held it at Kiest Park in Oak Cliff.

Because it was so easy to find.

The site that we had chosen has one of those big stone picnic areas that were made of local rock

It was covered with an impressive wooden roof which protected us from the elements.

Inside the picnic site there was a built in fireplace and about six big picnic tables that fit nicely under the roof.

These are usually rented out on a first come first served basis.

But, since Halloween was on a Tuesday and we celebrated on Halloween night

there was no one out there to offer a challenge us for the use the picnic site.

We had free reign.

We decorated the arches that support the walls and roof with fake spider webs and plenty of Jack O lanterns carved to look scary.

We hung black and orange crepe paper streamers, and hung up black, orange and clear balloons from the ceiling and walls.

Then we turned off all the lights and lit everything up with candles, a strobe light and light from the fireplace and the jack o lanterns.

The partially bare trees gave a ghostly impression to the surroundings.

They cast Spooky looking moving shadows on the ground.

We could watch the branches as they swayed gently in the breeze.

And we watched the light flicker as the branches were lit up by street lights from the main road that meanders its way through the park.

We had to conceal our alcoholic beverages because it is against the law to consume alcoholic beverages in a public place.

Like a park or a campsite in the state of Texas.

All of us used our costumes as best we could to hide our bottles and cans from the prying eyes of local law enforcement..

We had a lot of guests show up and a few uninvited costumed revelers joined the party.

We didn't mind them dropping by.

At least they too were wearing costumes.

We played Halloween music and bobbed for apples and played quarters.

And we danced around and passed out candy to any children who wandered up to the party (usually the kids of our friends)

We had pizza delivered and we all had a really good time.

No one was arrested and no one died or was lost.

Everybody woke up with all their fingers and toes intact.

Those were good times.

My elbow slipped off the arm rest and I was jolted back into consciousness.

Dick chuckled as I sat up and looked around to see where we were.

Dick said that we had about another half hour of riding before we got to the site.

"The inspector should be there within the hour" he said.

"Then we can go to Austin and drop Willy off .

And then we can go back to Dallas."

"Think the girls should be back home by then?" Dick asked

"I don't know, hopefully." I replied

I haven't given it much thought because I have been so consumed with getting the project off of the ground.

I have not had much time to do much of anything else.

I wondered what Rhonda and Candace were up to?

I knew Dick felt the same way, about Candace.

What, with him having to get all the affairs in order for the new brewery.

Plus all the permits and all the planning, surely, he must be neglecting Candace in some way.

Probably just like I have been neglecting Rhonda,

I was hoping that the new car would somehow show Rhonda that I still care about her
And that I am always thinking about her well being.

And yeah, I was worried…

Two hot looking babes strolling around a Car lot searching for exciting cars that "Look really cute".

She's a walking target to every sales guy trolling the car lots

That's my girl.

A doe in the headlights turned loose on the world of New and Used car salesmen.

"Looking out for a sale and maybe a little more. "

That was what I was worried about.

I also felt kind of worried because.

I normally go with Rhonda when we go out and make major purchases.

Like cars and refrigerators and washing machines and dryers.

She tells me what she wants and I do my best to make that happen.

That had always been our system for doing things and it seems to work well for us.

But, since I had to go with Dick to the brewery site..

Also since Candace was at home with Rhonda .

It seemed like the best way to handle all of the commitments at the same time.

After all I DID promise that Rhonda that "she could go car shopping just as soon as we finished up with The Chili Cook off."

And Candace loves to go out and shop,

I was concerned that they might show up at the house with a Rolls Royce or a

Cadillac or some outrageous car.

But, I know Rhonda, She loves little "Cool looking cars" the smaller the better.

She likes cars that look sporty and give the impression of excitement.

So, I was thinking she might go out looking at possibly a Honda CRX or maybe a Mazda RX 7 or maybe, even a Volkswagen Rabbit convertible.

It's Hard to say.

Whatever looks to her like it's saying

"Hi! I am really a cute looking car and I can go really, really fast!"

" Let's you and I go somewhere!"

That's what she'd be looking at.

So, as planned, we pulled up to the "Ole Wiggler's place."

Dick got out and unlocked the pad lock.

Untangled and removed the chain that was keeping the gate securely locked.

Then he opened the front gate.

So that we and the inspector (who should be arriving soon)

Should have no trouble gaining access to the property

And deliver to us his environmental impact study.

And "hopefully" sign off on the project.

That way we can start building on the site .

And get the actual construction into high gear.

Dick wanted the construction to begin before we have to go down to Houston

and load up everything onto the containers for our voyage to Australia,

It didn't take the inspector very long to show up.

He was a nice man and he had brought the economic feasibility study

And the environmental impact report with him.

The Inspector proceeded to unroll the studies.

And then explained that the area for the new brewery

Should be able to find a enough skilled workers in the area.

It would have a positive impact on the local tax base.

And then he covered the impact study concerns.

Number One on his lists was the high amount of water consumption.

His concerns about Dick and the brewery draining all of the available drinking water from the local aquifers.

We explained that we were also using water from local lakes and rivers and the municipal Supplies system

And that our impact on the local area would be minimal.

That seemed to ally his concerns.

He happily signed off on the entire project.

He then congratulated us on our site selection and wished us much success on our future Construction project and then he left.

We piled back into the Suburban and headed down towards Austin.

Along the way we made a pit stop.

Then we grabbed a bite to eat before continuing on our journey.

By the time we pulled into Austin.

The sun had already gone down.

Willy had fallen back asleep in the back seat.

So, we woke him up and he gathered his things and got himself ready.

We pulled into the brewery parking lot.

He thanked us for everything in his heavily accented voice.

Then he left us to go back to his home.

At first I thought that he might have to ride his bicycle home.

But, someone was waiting on him when we pulled up.

Dick said that his wife had come to collect him and take

Him back home to Round rock.

That was good news for the entire Workey family.

Their father / husband was back home after a couple of weeks testing out the prototype vehicle in Dallas.

I turned in my expense reports and Dick and I left with a couple of cases to hold us over

Once we get back home.

I was fortunate enough to find a copy of the Dallas Observer at the restaurant.

I was thumbing through the entertainment section.

When I saw that my two favorite bands were coming to the Arcadia Theater on Halloween night.

The headliners were the Ramones and the opening act was Joan Jett and the Blackhearts.

"That is where I am going to throw my Halloween party."

I yelled out loud to myself.

It stunned Dick to hear me shout out over the top of the magazine.

"What party?"

"Where???" asked Dick

So, I told him what I just read.

He wasn't nearly as excited as I was.

"I LOVE the Ramones and I think Joan Jett and the Blackhearts rock." I said.

I wasn't trying to convert Dick over to the type of music I like.

I was just excited that they were coming to town.

And that they were both on the same bill.

And they were both going to be at the Arcadia Theater.

I was stoked, and I couldn't wait to tell Rhonda.

She feels about the same as Dick does.

She's not as into the Ramones or Joan Jett as I am

I was sure that I would get a near dead pan reaction from her as well.

But, that wasn't going to dampen my enthusiasm.

I was already planning to buy as many tickets for my guests as possible.

I started trying to run through last year's guest list.

I could try and get a ballpark estimate as to how many tickets I would need for all of my guests.

I'll have to think more on this after I call the theater in the morning.

We continued our drive back home,

We were lucky because while we were driving down the freeway.

We were passed by a old Black Oldsmobile Toronado.

Which was thundering down interstate I-35N at slightly over one hundred and ten miles per hour.

Dick hit the gas and we rocketed down the freeway in an attempt to catch the Toronado and follow him.

We finally caught up with him

We passed him just as the Suburban hit one hundred and fifteen miles per hour.

The driver pulled over just outside of Waxahachie, Texas and we met him.

He turned out to be an insurance Salesman who makes the drive from Austin to Dallas weekly,

He knew where the police hot spots were.

We thanked him for being our "Front Door"

Then he went one way and we went another.

We shaved almost forty five minutes off our trip time.

It was almost nine thirty when we pulled into the driveway.

The house looked dark and the Company truck was parked in the driveway.

Nothing looked out of place,

Except Candace's car was nowhere to be seen.

So, I assumed that Rhonda and Candace were not home yet.

I was wrong.

I opened the door and Dick and I walked into the house.

I immediately let Thor out to do his business.

he's such a good dog never had any accidents in the house.

We never had to house break him.

He just knew what he was supposed to do and he did it.

Then we put the beer in the fridge.

I sat down on my chair and Dick sat on the sofa

Then turned on the TV to catch the news at ten,

Rhonda and Candace came out of the bedroom with smiles on their faces.

That worried me.

"Hi guys!" they shouted at us.

Well, ok, they startled us a bit, because, we were not expecting to find them already home.

"Hey, there you guys are!" I said

"Where is your car Candace?" Dick asked

"Oh, it's over at the fruit stand behind the house," Rhonda replied.

"What are you guys up to?" I asked suspiciously

"Oh, nothing Dear".

"How did the meeting with the inspector go?" asked Candace

"It went fine.

"The inspector signed off on the environmental study.

"We now have permission to start construction and drilling."

"We are good to go" Dick said with a big grin on his face.

"Rhonda? What is going on? " I felt like something was up.

How did the car search go?" I asked

"Oh, about that, hang on for a minute."

Rhonda said as she picked up her purse.

Then she went out the back door.

Walking past Thor in the door way.

She went out the back gate and disappeared into the darkness,

A few minutes later I see headlights pulling up front.

I walked over to the front door.

Thor is barking his head off.

While I am straining to see what just pulled up in front of the house.

It was white, and shaped like a race car.

She went with the RX-7.

"They really know how to pick them."

I said to Dick who came out to see what all the commotion was all about.

"Well, at least she looks good in it." replied Dick.

"Good thing we're going to Australia, or you'd never know where she is" said Dick with an evil grin on his face.

"Funny, really funny…" I replied in my most sarcastic voice.

Rhonda couldn't wait to throw open the passenger side door open and invite me in.

So, I slid into the seat and I closed the door.

Rhonda turned on the ignition and I could hear the hum of the rotary engine.

I gazed at the orange dash lights and the green lights of the car's aftermarket radio.

which gave the insides that "Tokyo by night" look.

It was an automatic because Rhonda doesn't like to drive a stick.

As soon as I was buckled up and closed the door.

She threw the car in gear and we rocketed down the street to the stop sign.

We made a right turn onto Jim Miller.

We drove to the stop light and then we hung a right turn

Onto Scyene Parkway then she floored it.

My head was jerked back as the engine pushed the rear wheels

which were struggling to find traction.

Next thing I knew we were cruising down the street.

Rhonda was chattering away happily in my ear about how much she loved the

car and how wonderful it handles and so on.

The only thing going through my mind was

"Gee how soft the passenger door arm rest is.

And "how nice the center console arm rest felt against my fingernails"

as I dug them deeper and deeper into the upholstery.

It only mildly reminded me of the flight in the F-16.

The seats in the RX=7 were also reclined to about 30 degrees.

Just like in the F-16.

Supposedly so, you can pull more "G's"

Except when I was pulling some "G's"

I was in a pressure suit with my socks squeezing the blood out of my feet.

And up into my skull.

Inside of Rhonda's car I had no such safety gear to rely on.

And believe me when I tell you.

Rhonda has a lead foot.

And when she is behind the wheel, she doesn't even realize it..

Here she is, just chattering away about how she and Candace found the car.

At this used car dealership.

And that it was exactly what she was looking for.

And how nice the salesman was to them

And how she and Candace struck a really good deal with him..

I glanced over to the speedometer,

I interrupted Rhonda long enough to say

"Honey? "

She stopped talking long enough to look at me.

And she said "What?"

"Seventy?" I replied.

"Seventy what?" she said with a perplexed look on her face until her eyes

went down to take a look at the car's speedometer.

By now it was closing in on eighty miles per hour.

Eighty freaking miles per hour.

In a forty five mile per hour zone.

"Oh my goodness" she had just noticed that she was almost doing twice the

speed limit and didn't even realize it.

So, she backed off the gas and slowed down to fifty.

Just as a police cruiser was passing us heading in the opposite direction.

He didn't swing his cruiser around and give pursuit.

He kept on going where he was headed, but, had he seen how

fast she was moving only moments before.

I am sure he would have swung that squad car around on a dime.

I'm also sure we'd have a huge fine to pay the City for excessive speed.

After a few more miles,

She asked me If I wanted to drive it.

I eagerly said yes.

I would feel more secure that we would make it home alive.

Of course, I didn't tell her that she drives like the famous formula one race car driver Juan Manuel De Fangio at the Argentine Gran Prix.

She was just as happy as a clam.

She had gotten what she wanted, and everything was good.

I turned the car around and we headed for home.

I did one small detour, I took her by the liquor store.

And we picked up a bottle of champagne to celebrate the new vehicles arrival.

Dick and Candace were waiting for us to return and when we pulled into the drive way.

I turned the car off and I handed the keys to Rhonda.

I got out with the bubbly.

Rhonda got out and tossed the keys to Candace.

She and Dick got in and they were off on a test drive of their own.

I asked Rhonda, "So where is the crap mobile?"

"Oh they were so nice at the dealership, they gave us 300.00 dollars for the crap mobile

And they applied that to the taxes title and registration fees." She said

I honestly felt sad that my old car was gone.

Sure the fender and the hood were a different shade of yellow from the rest of the car.

And OK, the floorboards were rusted through.

And OK, there was no A/c and the interior was all torn up from wear and tear, but, it was our salvation.

A piece of car crap that never knew when to quit and it never did.

in spite of all of its woes.

It had served us well,

And Yes, I was going to miss her.

I could not believe that I was thinking that while staring at a new to us Mazda RX-7.

Yeah, It didn't really make sense to me that I should have developed a bond with that old Car.

But, I did.

I had tied desks and book cases on top of its roof and used it like a pickup truck.

I abused the car in ways the designers at the Datsun factory would have never ever considered.

 still did its job without any complaints.

I didn't even think to say good bye to it before I left for the trip.

Don't get me wrong,

I was Happy for Rhonda getting what she wanted.

But, I was still going to be sad for my personal loss.

"Bye, bye Crap mobile," I said softly to myself.

Rhonda and I decided to go back into the house and wait for Dick and Candace to return.

I put the bottle in the freezer so it'd be all nice and cold when they got back.

I shared my day with Rhonda.

She told me what she and Candace did besides car shopping,

Mainly mannies and peddies.

I told Rhonda about the Ramones and Joan Jett coming to the Arcadia.

She was just as stoked as I expected her to be.

"That sounds nice honey."

" I don't have to go, do I?" she replied hopefully.

"No, not if you don't want to."

I told her that I wanted to have the Halloween party there at the Arcadia,

She wasn't very accepting of the concert venue for our party,

She said that "No one could hear themselves talking, and that the music, wasn't quite "Dance-able."

True, guitars that sound like chainsaws and screaming lyrics are not the stuff of conversations .

I would have to revise my plans and nix that idea.

Later on when the Wigglers returned from their test drive,

I popped the top on the champagne and we poured four fluted glasses and dropped in some sugar cubes and Dick raised his glass and made a toast.

"To the future, may it be healthy, happy, and financially carefree!"

We all said "Here, here" and we drained our glasses.

After a few more toasts,

We decided that we were all pretty tired.

I fed the dog helped the Wigglers get all set up and comfy in the guest room And we headed off to our beds for some well deserved rest.

We were all a little too tired to even want to eat dinner.

I laid down and Rhonda laid down next to me and then she rolled over and kissed me

She told me

"Thank you for the car. I just love it."

" It's the car I always wanted."

I kissed her back and told her that she was welcome.

And that I was glad to have been able to do that for her.

That made me happy just seeing the look in her eyes.

The Hybrid Diaries
Chapter Forty
Tuesday October 23rd 1990

Well not much happened today, I DID have a visit from Earl with "All Nationwide Batteries".

He had just dropped off several of the latest experimental deep cycle lithium batteries that they are developing in conjunction with my project.

We have been running the "Spirit of Texas" using a few of the regular deep cycle nickel/ cadmium Batteries that the company produces for marine applications (mostly electric trolling motors and other types of marine applications)

They finally had developed several of the new stackable batteries that I designed and they had developed for me.

They are more rugged and durable, and can hold a charge for a much longer period of time.

It has the effect of increasing the duration between charges and therefore increasing the range of the vehicle on pure battery power alone.

So we were going to load them in both test cars to see if we can get them running at peak performance.

The guys were already installing a set of the new batteries into "Deguello" And she'd be ready to run in about an hour or two. So, I went into the office and I talked to Bill and Earl about some of the details involved in the trip down to Houston.

Tthat we'd be testing the endurance and reliability of the systems and the components

of each vehicle, so that we can determine the individual characteristics of each vehicle and what their projected weaknesses could be.

The test run from here in Grand Prairie down to Houston would be going down soon.

Bill was excited about the test runs He called it our "Baptism by Fire."

I had decided that we would test BOTH vehicles to see if they both break down on the Way.

And if so, what parts failed and what parts survived and which vehicle went the furthest without any problems.

That way, we can make changes to the designs and

repair anything that broke along the way.

It would be the new vehicle's chance for them to prove themselves.

We all had a lot of time money and energy invested in these vehicles.

It is amazing. Staring at them both side by side,

Just how markedly different they both are to each other.

Spirit looks much more like my original designs,

but, Deguello reminds me of the "Batmobile."

Just because of the canard wings on the nose and all the solar panels on the back.

The windscreens give them that enclosed fighter cockpit look.

The scoops that feed the air turbine powered electric motors buried deep within the vehicles bodies make it look menacing.

"You know, you really need to be proud of your accomplishments" Bill said.

I didn't even know that he was standing behind me.

I was too focused on my vision

I didn't notice anyone walking up behind me.

"How long have you been standing behind me Bill?"

I asked him as I turned my head to see his face.

He also had a big smile on his face as well.

"Oh, I've been here long enough." Bill said coyly.

Out of the blue,

Bill asked me a question that I was sure he had simmering on the back burner of his Brain for quite a long time.

"So, How did you come up with the ideas for all of this?"

"Are you some kind of an engineering wizard?"

I told him about my dreams.

How I designed and built the prototype.

I was inspired to build the air powered turbine generator by a child playing with a pinwheel.

"Yeah, but, this is WAY more advanced than your first designs

"That you brought to the Shop."

" Sure we did some changes to the body and stuff to make it look cool."

"But, all the electrical innards, are truly fascinating."

" We did a test on the vehicle without human power."

"The vehicle ran just fine on just the generators alone ."

"it also ran fine on the batteries by themselves."

"And the solar panels recharged the batteries in a few hours."

"You really do not need the crank and flywheel generators to make it go."

"Once it is going, the air going into the intakes is enough to keep the air turbines running at the right velocity to produce electricity."

" Why did you still include the human element?"

"You really don't need it"

"I wanted the vehicle to be more organic"

" It needed the human element to have a hybrid blend of man and machinery.

"It's symbiosis"

"There hasn't a vehicle built quite like this before."

"I am just the luckiest man alive to have had the opportunity to have my dreams and vision brought to fruition."

"And yeah, I know that there are enough energy resources

"Incorporated in the design that makes the human element a little too redundant, "

"but, it's the human quality that makes it special. "

"Just don't tell anybody else." I replied.

Bill promised to keep my secret and he said that it was

"one of the finest projects that he has had the pleasure of working on."

though we never talked about costs in designing and manufacturing.

Or the materials or the use of the large auto claves that we used to heat and mold the carbon fiber panels and components.

Or even the use of his sons on the project.

I knew that we were in good hands.

All of our efforts have brought forth a treasure trove of valuable data.

Our data would in the future benefit others.

Those would wish to develop a vehicle much like mine.

Or at least help them build an electric car of their own.

Based on what technology that Bill, Toby and I choose to release to the world.

Bill did warn me that

"There was a possibility that the flywheel generator might over amp the entire electrical system",

"We would need to be extra careful when using it in conjunction with the air turbines."

Earl came out of the office and he said that he had good news,

"His Company has decided to commit to the project by.

"Supplying us with however many batteries we would need for both testing.

And for the first long range test runs down to Houston.

And for the actual race in Australia.

And that they were going to send Earl along as a sponsor and a company representative and spokesperson."

I told him that "We would be glad to have him on board."

"We both welcomed him to the "Madness" that we all called my vision."

All we had to do was decide on how Earl was going to travel.

Would he want to travel with us on the cruise ship?

Or would he want to travel on the cargo ship with the guys?

So, he could keep an eye on his Company's contribution to the project?

(the vehicles experimental batteries.)

Or maybe he'd decide to fly over on Quantas Airlines.

With less than a month to go, he'd have to make a decision soon.

I went back out to the production floor.

The guys were having trouble getting the new batteries to fit properly.

And still be able to close the compartment back up.

They decided that they would have to modify Spirit.

Deguello was not having the same problems

The batteries were designed to fit into Deguello's battery compartment from the onset.

And they had no such problems adapting.

I wanted to take Deguello out for a run.

But, that would have to wait.

The General was out of town and he would not be back until sometime after Halloween.

Sometime around the third of November.

And we needed to get the General's permission to use the emergency runway and tarmac again.

I thought about taking it over to the college parking lot where I test ran the original Slingshot Mark II.

The same college campus that Rhonda and I used to for testing and recording the original videos that Rhonda shot.

It would be sort of like going back to where it all began.

But, there was a problem there too.

We'd need to wait until the weekend.

when all of the students were out of classes.

 and the campus parking lots would be

for the most part, devoid of parked student cars.

So, I decided to leave for the day and go down to the Arcadia Theater and buy the tickets for the big concert..

I drove down I-30 headed to the Munger Street exit and then down to Greenville Avenue where the Arcadia Theater is.

As I pulled into the parking lot I noticed that there were several people already at the ticket window.

Seeing just how long that line was made me want got go back to the car and go home.

I got out and joined the people waiting in line for the ticket window to open

So, I can buy some tickets too.

I ended up buying 5 tickets.

(one extra just in case.)

I rushed home to show Rhonda my prized tickets.

She was not all that excited about the show.

In Fact she started telling me that she was getting a stiff lower back and she was starting to not feel very well .

I told her that " I was sorry that she wasn't feeling up to par".

I went to the kitchen to get her a couple of aspirin and a glass of water to wash them down with.

She was thinking that she might be getting sick with a kidney infection.

And that maybe she should go and visit

"Good Ol' Doctor Dave." (Our family doctor)

And see if he could give her something medicinal to help her get over being ill.

"I told her I thought that would be a good idea. "

I promised to take her to go see him.

After she called and set up an appointment for a visit.

Rhonda called the doctor's office and set up and appointment for the next day.

The Hybrid Diaries
Chapter Forty One
Wednesday October 24th 1990
Almost forgot about the doctor

Very early in the morning I had to go down to the shop in Grand Prairie and meet with Bill and Toby about getting the vehicles set up for a reliability run down to the Houston shipyards.

We had decided that that we'd need to break up into two teams one for each vehicle.

We'd need to form a caravan so that we can protect the vehicles from traffic.

And to offer ourselves a measure of control and safety over our chosen path.

Right down I-45 through Corsicana and then right down to Houston.

Normally this would be a three and a half hour drive from Dallas.

It's a nice big freeway with nice shoulders on the side of the roadway

In case of roadside emergencies.

We'd need to have the truck and the trailer loaded into the containers.

I told the guys that I wanted them to bring their pickup trucks

And we could load up all of the spare parts.

Then we could form a decent convoy.

that would offer the vehicles some protection from other vehicles.

Namely cars and trucks and other road obstacles on the freeway.

We were just about to finalize all the remaining details when I got a call from Rhonda.

All she said was "Come Home now."

I had almost forgotten about her doctor's appointment.

By the time I got back home she was lying down in bed almost doubled over in pain.

So, I helped her up and I guided her out to her car.

I drove her down to Oak Cliff where our doctor's office is located.

Once we got there.

We got her signed in and the doctor promptly saw her.

He began by having his nurse draw blood and checking her for fever.

It was his professional opinion that she indeed had a kidney infection.

It was a pretty bad one too.

He pumped her full of antibiotics and prescribed some meds for us to pick up from the drug store on our way home.

She wasn't supposed to drink any alcohol.

And she should stay off of her feet and try to remain as comfortable as possible.

So, No, concert, or partying for Halloween night, poor girl.

I had to call her family and let them know that she was sick.

Then I called our friends and let them all know that she was sick as well.

Fortunately with Halloween falling on a Wednesday it wouldn't be that big of a deal not to have a party in the middle of the week

And since our parties are usually followed by a hangover the following day.

And since no one likes to go to work with a hangover.

It'd be for the best to re schedule the event for the following Saturday night

The doctor said that she'd need to drink plenty of water and cranberry juice to raise her acidity levels so her body could fight off the infection.

I told the guys that we'd have a surprise party for Rhonda on November 3rd.

This would be the next Saturday night.

I had also decided on a theme.

We'd all dress up as doctors and nurses and we'd do our very best

To offer up a prescription of fun and good times to help ease her suffering.

I called Toby and Bill back and I let them know what was going on.

And that I'd be ready to finish up the plans and get all the logistics together the following week,

Most likely on Monday November 5th.

Time is fast running out on us and we needed to have everything in place so that we can have everything ready.

We'd already arranged to have two empty cargo containers waiting for us in Houston.

And we were getting the manifest and the tickets ready for the guys who were riding along with the containers to Australia.

We needed to make sure that customs would be there when we arrive so that they can do their inspections and give us the green light to have the containers loaded aboard the ship.

This wouldn't happen until after Thanksgiving.

When we were actually ready to embark on the journey..

The vehicles would have already been loaded into the containers and would sit on the docks until the OK for them to be loaded had been given.

Bill said that he understood and I hung up the phone and I stayed home and took care of "The light of my life."

Her kidney infection was a bad one.

But, the doctor was able to prescribe the proper medications.

She would be back up and around in a few days.

But, no drinking alcohol and she needed to drink lots of water and cranberry juice to wash the infection out of her system

(Well, no alcohol at least until Saturday night anyway.).

So, we decided on having a party for her on November third which was a

Saturday.

It would be a costume party,

With all of the people dressing up like doctors and nurses.

We even invited the Wiggler's to the surprise extravaganza.

Unfortunately for us.

Dick was in the process of having the contractors and the architects

(who were drawing up the plans for his new brewery)

There at his office and there was no way that they would be able to come up for the party .

Although they did send their condolences to Rhonda.

Along with and their best wishes that she would be feeling better soon.

Wednesday night came.

Although Rhonda did not want to go.

She said that I was still free attend the show and take a couple of my friends

(who would actually enjoy the concert)

Go in her stead as long as I bought her a tee shirt.

I told her that I would.

It was the most enjoyable concert that I had been to in a long time.

There they were two of "New York's finest bands on the little stage of the Arcadia.

There really isn't a bad seat there in the Arcadia.

A converted movie house that now houses live venue concerts and stage shows.

It has a great little bar .

I started Cheering when the Ramones Started playing "Blitzkreig Bop and I didn't stop until Joan Jett finished Crimson and Clover and closed the show.

My throat was sore from cheering so much.

My ears were still ringing from "I wanna be sedated" well into the next morning.

And yes, I remembered the Tee Shirt, I bought two.

So Saturday came and we celebrated Rhonda's recovery with an alcohol free (for her anyway) party .

There everyone was.

Some dressed as Doctors in white lab coats others dressed up in surgical garb.

And some of the ladies were dressed up as nurses.

The rest of them were dressed also as doctors.

I've never seen so many stethoscopes in one place short of an actual hospital.

Some of the "doctors" were covered in fake blood which made them look like their patients didn't survive their procedures.

(It was for Halloween after all.)

It went on into the night and well into the morning.

Everyone had a good time and we finally retired just as the sun was peeking over the eastern horizon.

The Hybrid Diaries
Chapter Forty Two
November 5th 1990
Heaven, Hell or Houston

OK well, it was time to take DeQuello and Spirit out for their first real road test down to

the Houston ship channel and onto the container ship that would carry it and everything

else all the way to Australia.

It was a clear crisp November morning.

The Sun was up and it promised to help us produce enough power to keep the vehicles on

the road all the way.

Four Hundred plus miles from Grand Prairie.

All the way down to the ship channel.

And out to our two waiting containers where we'd load up the trucks and supplies.

And get them ready for inspection by the Port Authority prior to use leaving for the

"Ocean voyage of a life time."

I had notified our old friend John Proktor with Channel 7 news.

He showed up with his camera man Mike Heller and the Station crew and a press

photographer to take pictures of us.

And do a human interest story on our journey down to Houston.

Earl from Nationwide Batteries had a special present for all of us.

He had gone to "Dikies Uniforms and ordered us some tan colored jumpsuits.

(tan being the color for All Nationwide Batteries).

He also had to company logos sewn onto the shirt pockets and he put patches of the state

of Texas over on the shoulder of each jumpsuit.

On the other shoulder he had sewn the American Flag.

For the bikers he had some Bicycling body suits made

(Also in Tan and set up like the jumpsuits)

A suit for myself, Christopher, and Willy along with several more for changes.

And spares "Just in Case"

He had also bought us cycling crash helmets for our trip also "just in case"

Several years earlier.

I had taken an unexpected trip over the handlebars of my racing bike.

When I was hit by a 1977 Blue Toyota Corolla Lift back driven by a hit and run driver .

Who didn't give a single thought to stop and check to see if she had killed me.

Some nice woman in a van who saw the whole incident.

She turned her van around and stopped to render aid to me.

She said that she was sorry that she could not get the license plate number

of the car that ran me over.

But, she was there to do whatever she could to help.

So, I asked her to help me load what was left of my bike into the back of her van.

And, if she didn't mind me bleeding all over her passenger seat of her van,

I had asked her if she could drive me back home.

(I was still living at home with my parents at the time)

She said that she was more than happy to take me to the hospital.

I told her that I would like my parents to do that for me.

Because I didn't have any of the insurance information I would need to give to the

Hospital Admittance staff.

So, she obligingly took me home.

She gave me a towel that she had in the van for me to put over my bleeding head.

It took twenty one stitches to close the gash in my forehead to stop the bleeding.

And since that day,

I am always aware and mindful that I have to wear a helmet.

So, I was beholding to Earl for thinking about our safety.

After talking a while.

We had decided on who would ride with who on our journey down I-45.

Dick and Candace would follow us in their Suburban loaded up with Advertising for the

Brewery.

They had ordered two additional containers for all of the beer they were going to be

exporting to Australia.

These were refrigerated containers of course, (to keep the beer at just the right

temperature.)

The beer would arrive just before we all set sail.

One of the jobs the guys would have on their trip would be making sure that the

container generators had plenty of diesel fuel.

So the beer wouldn't go flat on the long trip.

We had my truck, trailer and Bill would drive it.

Dylan and Toby would drive down in their trucks.

Earl would follow up in the All Nationwide Battery Company Van loaded up with

batteries and an inverter that he'd use to keep them charged.

In the event that the charging systems stopped working.

I'd ride in Spirit and Christopher and Willy would take turns driving DeGuello.

We'd all convoy down loop 12 to Spur 408 and then I 20 to US Highway 45 all the

way to Houston.

It all sounds easy enough. Right?

We started out around 8:00AM and we had my truck in the lead.

Spirit and then DeGuello would be next, followed by the rest of the crew.

They would drive with their flashers on to warn the on-coming traffic about what

lies ahead.

We rolled out of the carbon fiber place on one wheel motor.

Taking it easy on battery power.

Then we rolled out onto Jefferson Boulevard

heading east towards Loop 12.

No problems.

Then we turned onto Loop 12 and headed south towards Spur 408.

The next big hurdle would be what I used to call "The hills of pain."

These hills lie between "Keeneland Parkway" and "Illinois Avenue. "

They are big hills that I used to train on.

To improve my leg strength and physical endurance.

I did this when I was a younger rider on my first decent 10 speed.

Not fifteen years later on when I am out of physical shape.

These hills are murder on you.

The electric motors made short work of the long hills

This in retrospect was the reason I first though up the idea for the vehicle to begin with.

Normally, I'd be coughing up a lung trying to get up those hills.

But, with the motors driving the vehicles at 50 miles per hour.

they were no match for us.

Proving to me that I had the right idea all along.

It felt good to know that I was on the right track just a few short years ago.

when I first thought up the initial design for "Spirit".

I yelled over the two way radios when we got to the top of the 30 degree hill climb.

"Woo Hoo! We made it!" I cheered.

With the cool temperature outside (barely 72 degrees,)

I didn't even break a sweat.

A few more hills were conquered.

We made the right turn onto Spur 408 heading towards I 20.

Average speed was around fifty miles per hour,

The first battery was just now beginning to get low.

So, We switched batteries to a freshly charged bank.

And I started pedaling the crank that runs the flywheel generator.

Then I flipped the switch that engaged the other two wheels into generator modes.

Then I switched the weak battery into charge mode.

I also engaged the inverter that turned on the solar cells.

Their power would be recharging the weak battery in conjunction with the other power

sources in a few hours.

Hopefully in time to re engage it for more long distance driving.

Another big hill lay ahead.

The on ramp to I 20 has a long sweeping curve to the left and an elevated bridge.

With a big hill that is pretty high up.

Followed by a nice gentle drop down onto I 20.

No problems there.

Now our convoy was rolling on down I 20 and heading towards I 45

Then we'd be heading south towards Waxahachie and Corsicana and then Houston.

Long stretches of open road going from 4 lanes, to three lanes, and then to two lanes

with shoulders.

Not much to see except for miles and miles of fields and trees and land.

Interspersed by the occasional road side antique shop or roadside restaurant.

Maybe a gas station or two.

And plenty of signs for "Frontage Road".

In my opinion, the longest road in the world.

Oh, and plenty of horses and cows.

Lots of cows.

Normally my mind would begin to wander.

This time however, I had too much to look at on my HUD (head's up display)

To be zoned out by the road.

Being wary of the occasional 18 wheeler towing a single wide mobile home.

which can blow you right off the road is another reason to be alert.

We were passed by a convoy of big rigs that were hauling a couple of U.S. Army

M1A1 Abrams main battle tanks down to Fort Hood.

Very big and very impressive.

Their huge 120 mm smooth bore guns were pointing backwards.

As they rode majestically on their flat bed trailers.

They looked like they were pointing right at us.

Occasionally, we'd see the Channel News 7 van shoot past our convoy.

They would drive ahead of us and then park on the side of the road.

Then they would set up their cameras so that they could take pictures of us as we sped by for their news story.

Man I wish I could have seen the segments.

But, I never got a chance to see the news story on the air back at home in Dallas.

We cruised through Corsicana without a problem.

That is until the Corsicana Police department pulled us and the entire convoy over.

They wanted to know what the heck was going on and what those things were on the

road.

And if they were street legal.

We assured them that the vehicles were indeed street legal.

They had functioning lights and mirrors.

A horn, and even windshield wipers.

And the words "Prototype and experimental vehicle" prominently displayed on their

sides.

The officers decided to take us at our words and allowed us safe passage through their

fair city.

With the promise that we would not be driving the "questionable cars" on their roadways

again.

So, we agreed and they let us go.

We snapped a couple of pictures for our scrapbooks and made up funny captions for them

on the radios.

I had Willy take the controls and drive Spirit for a while.

While Willy was driving Spirit,

He saw a puff of smoke coming from the passenger side of the vehicle.

Then more smoke.

The wheel had over heated and then caught fire.

Willy had to pull over while we rushed to get him out of the burning car.

The damage was confined to the wheel and the outer wheel housing.

Fortunately the composite body didn't burn up in the fire and after

replacing the wheel with a spare.

And checking all of the other wheels to make sure that they were in no danger of

combusting.

We again set off for Houston.

We easily made our way to Harris County.

Then we stopped just outside of Huntsville (home of one of the largest state prisons in the

U.S.)

to see the huge white statue of the great Sam Houston.

The General of the Army of the Republic of Texas, First and only President of the

Republic of Texas and the first State Senator of the State of Texas.

This statue is made out of sixty tons of steel and concrete.

And stands over sixty seven feet tall.

Very impressive indeed.

We gathered under his alabaster feet and we took some pictures of "Good Old Sam".

Just as we were leaving the statue.

I did something stupid.

I tripped over a cable that was stretched between some wooden posts at ankle height.

These little posts lined the parking lot area

And stretched around the statue site.

I twisted my ankle.

Now, I can't drive the Spirit.

Oh, I can still drive it.

I just can't pump the pedals very well.

So, I asked Christopher to pilot Spirit the rest of the way.

And I'll have to ride with Rhonda in my lead truck.

Towing the parts trailer.

That way I can guide the convoy to the ship channel.

And the terminal at Turning Basin which is just outside The City of Houston

And adjacent to Galveston Bay.

We rode on through the surrounding counties before we had to pull over again.

This time Bill's prediction about spiking the main power circuit had come true.

Dylan was pedaling the crank which in turn powered the electric motor that spun the

flywheel generator.

There was a power spike and the vehicle blew the main buss fuse.

We had to stop and check DeGuello for damage.

Fortunately the circuit breaker was reset and we were able to

Continue on with the test ride through Harris County.

We passed over the huge bridge that spans the Houston ship Channel.

And we rode into the City of Houston. "The largest City in Texas."

The smell of the sea, petro chemicals and the smells of the coast fills the air.

We rested on the side of the road for a few minutes while we decided how we were going

to ride the final miles to the Port.

I was more than satisfied with the vehicles performances.

They both validated my original concepts and they proved not only to me

but, to the entire posse.

That this was going to be the new wave of the future in personal transportation.

And that you don't have to be a world class cyclist to ride one of these vehicles.

Anyone with a battery charger and good sunlight can operate one of these vehicles.

For up to an hour and a half of straight battery power.

without even having to pedal the cranks.

And Thanks to Dick's confidence in me and my ideas.

We have two proof of concept vehicles .

That more than demonstrated that the concepts were sound.

And that this was no bullshit from a crazy inventor.

But solid and practical designs which operated just like genuine vehicles.

And they could also move at an acceptable speed through traffic just like real cars.

I was proud of my design team.

And all the hard work that Bill, Dylan and Christopher and of course Willy and Toby had

done It was because of all their efforts that I was able to stand at this historic time.

We are all standing on a bridge to the future.

And before us lay a vast "City of possibilities."

And what lay before that?

"Two oceans vast and wide that we'd still have to cross before

we finally get our chance to prove to the world.

We have arrived and that we few guys from Texas are here and ready to show them all."

"We are the way to future Independence.

Less dependency on fossil fuels and greater potentials for not only ourselves.

but, for the entire planet.

Showing them that our dependence on steadily dwindling supply of fuels.

Need not spell the end for our societies.

But, they offer us a chance for a better and healthier alternative to a new and better life.

Man, I was excited.

I felt almost like Columbus when he first laid his eyes on the New World.

With four hundred miles behind us.

Our convoy pulled into Turning Basin.

And the container company that had reserved the two containers that we would be leasing

for the voyage.

We got everything stowed inside and locked the containers and took the keys.

And since most of our vehicles were now inside the cargo boxes,

We would all have to pile into The Wiggler's Suburban and the another trucks and

the TV News van for the trip back to Dallas.

On the way home.

We decided to stop at Leo's Mexican Restaurant in Houston.

You know Leos if you've ever bought a copy of the ZZ Top album "Tres Hombres".

Just open the album and look at the spread of Mexican food and beer.

That is what we wanted laid out before us.

Because, after pedaling and driving two little bumpy riding concept cars.

For four hundred miles all we wanted was a nice comfortable chair, a few

Beers and copious amounts of Tex Mex food.

We decided to stay over for the night and we got ourselves a couple of rooms at the local

Holiday Inn.

The next morning we would awaken fresh and slightly hung over and ready

for breakfast.

Ready to head back to Dallas and prepare for our trips.

On the way back home.

I asked Dick and Candace and the guys if they didn't mind making a slight detour.

They all said OK on the promise of me buying them lunch.

So, we took the Port Arthur ferry over to Baytown to see two things that I thought were

Important.

One, The San Jacinto Battlefield Monument.

And the other was the battleship BB-35 the U.S.S. Texas.

Both icons are just a stone's throw from each other.

The Monument symbolizes the Victory of Sam Houston and the Army of the Republic of

Texas.

Over the forces of Centralist Government Ruler General Santa Ana and his Mexican

Army.

The other held even more personal significance to me.

The Battleship Texas.

The Battleship is the last surviving U.S. Dreadnought.

That is a class of capital ship that were developed after 1905.

When the British Navy had launched the history making Battleship by the name

HMS Dreadnought.

This was the first Capital ship consisting of an all big gun armament of 10 12' guns.

The USS Texas main battery consists of 10 fourteen inch guns.

I got everybody on board and the girls wandered the decks looking at all the historic

Displays.

while I took Dick and Earl and Bill and the guys down below.

This was not my first time here.

In fact I like coming here because of all the historic significance.

My destination was the main engine room.

Deep in the bowels of the ship.

There within was the last pair of surviving triple expansion oil fired steam engines in the

World.

And here was the reason that I wanted to show them to the guys.

I began my history lesson.

These engines represented a "thumb in the nose" of the Westinghouse Corporation by

President Taft.

The then president of the United States.

The Texas was the second ship of the New York Class of dreadnoughts for the US Navy.

(Although she was actually completed first.)

Now the Westinghouse Corporation was the main supplier of the Parsons oil fired

Turbines That powered all New Major Battleships in the U.S. Fleet.

Westinghouse set the price for the turbines at an astronomical price that the US

Navy thought was price gouging.

Rather than caving in to the demands of their supplier,

The US Navy went ahead and had the ships constructed with the old outdated coal fired

Reciprocating steam engines that had powered all previous ships of this size.

Once the Westinghouse Corporation saw what the Navy had done.

They immediately dropped the price of the turbines to the Navy.

And all subsequent ships (except for the USS Oklahoma) were all equipped with turbines.

Many thought that the turbines had reliability issues and used too much fuel.

but, that was soon forgotten in the course of history..

It kind of represented my personal philosophy.

I too wanted to thumb my nose at the general belief in fossil fuel dependency.

And I was willing to show the world that it was no longer a necessary evil.

We can do without fossil fuels in the future.

And our dependency on the oil producing nations.

And in turn help save the environment from an impending global catastrophe.

I bought a poster of the Grand Old Battleship.

And then we had lunch on me.

And then we headed back to Dallas.

When we got back to Dallas.

Toby took me aside and then he lowered the boom on me.

"I can't go with you to Australia."

"Why?" I asked him.

It would have been a huge payoff for him and his wife to go with us.

And he'd get to see the results of all his hard work come to fruition.

He'd be missing out on all of the recognition that he had most certainly deserved.

It just didn't make sense to me.

He said that He wasn't going to be able to be away from work for that long a period

of time.

And he needed to stay home with his wife.

So, I could not argue with those reasons.

He said that we had plenty of parts and he would make us a couple of spare wheels.

Just in case.

It didn't feel right to me to leave my number two man behind.

But, Like I already said. I could not argue with his reasons.

If he says that he can't be gone that long.

Then He can't do it.

He'd had just gotten his and Isabel's passports, shots and everything he would need to

Come along with us.

It had to be a last minute decision.

I knew that he was truly looking forward to coming with us.

"Oh well. If you can't be gone that long,

Then, it is what it is." I said

"I am going to miss you terribly" I said.

I hugged him and I let him know that it was alright.

I respected his decision and that there were no hard feelings.

The Hybrid Diaries
Chapter Forty Three
November 22, 1990
Thanksgiving Day

I had woken up early to watch the Macy's Thanksgiving Day parade and then

The Adolphus Children's Medical Parade.

Not quite as good as the Macy's day parade.

But, it's what we've got here.

It's still a tradition that I have enjoyed ever since I was a kid.

All last night I had been up smoking two turkeys.

that involves Tossing a log into the smoker and drinking a beer.

Then repeating the process every hour until the turkeys are done.

I had prepared a feast for all of my former employees at the Thrift store.

(That's where the first turkey went.

The other one was for my wife's family.)

The store had been closed because there was a newer and bigger Thrift store

that had opened

Down the street from my old store.

and Randall could no longer afford to keep it open.

He was losing money and he decided that it was in his best interest to just

close up the doors and cut everyone loose.

So, there went my employees.

Out of a job and on the streets.

For that I feel responsible and truly sorry.

I had no idea that my leaving would have brought that much suffering to so many people.

We had the dinner of the now abandoned building out in the parking lot.

I had brought all the food from the house.

We brought music in the form of a battery powered radio.

It was a bittersweet reunion.

But, we all had a good meal and talked about the good times that we all had working together at The store.

Then We had to break up the party to go to our respective households to share more Thanksgiving meals with the rest of our families.

It was good to see all of my old employees one last time.

Rhonda and I did what we do every year.

We spend the early part of the day with her family watch the football game.

And since my family doesn't do anything early, we finish up Thanksgiving at my folks house.

We spent the first part over at the In laws.

The Cowboys beat the Washington Redskins 27 – 17 in the Thanksgiving Day Classic.

So we Dallas fans truly had a good Thanksgiving indeed.

After the game,

We helped clean up the kitchen.

From all of the celebrations and then we bade the in-laws a happy Thanksgiving and promised

that we'd come over for the second part of the family tradition "Left over night"

That's when we go over to their house again

And help eat all of the remaining leftovers

We usually eat more deserts.

Because we normally don't have much room for deserts after the big

Thanksgiving feast".

We reminded them that we were leaving for Australia in just a couple of days.

And that we were bringing Thor over to their house so that they can take care of him while we were away.

Then We had asked them to drop by the house and keep an eye on it while we were gone

They promised that they would.

We said that we would see them some time after

New Years. Sometime around my birthday in late January.

We dropped our house to let Thor out.

So he could do his business and brought him back inside so he could "protect the house."

Then we headed over to my parents house to see my Mom and Dad and my sisters.

We all brought different things.

My wife had brought a chocolate pie and we brought the other half of the turkey.

We didn't leave the house until almost three thirty in the morning.

After all the eating and drinking it was the hardest thing to drive back home.

The turkey had me feeling so sleepy..

But the dog needed to go out before his bladder burst and we wanted to go home

And sleep in our own bed.

And wake up whenever we were ready to get up.

I hugged my Mom and Dad and told them what a wonderful time we had,
We also reminded them that we would not be home for Christmas.
And not until sometime after New Years.
We told them that Rhonda's parents were watching the house and taking care of Thor.
But we'd appreciate it if they would drive by the house and keep an eye out on it for us.
They said that they would miss us and my Mom had to remind me that this would be the first Christmas that the family would not be all together.
I said that I was sorry about that
But, business calls and I had to do what I had to do.
With that I and Rhonda said good bye to my sisters and we left for home.
It was going to be very strange for me not to be with my family for Christmas.
I know Rhonda felt the same way I did.
We both have strong family ties and neither of us have ever not been with our families on Important holidays.
Christmas being number one on the Holiday list.
Plus, we'd be spending the holidays in a tropical climate far away from the cold and snow that sometimes falls on Dallas in the winter months.
Not to mention that we'd be on a cruise ship out in the middle of the Pacific Ocean far away from the country that we have just left behind.
Man, it's a lot to consider.
I really haven't thought this thing all the way through.
But, I guess there is nothing else to do but, hunker down and face the upcoming challenges as they come,
Maybe our horizons will be broadened.
I joked with Rhonda that this Christmas.

Santa will be wearing a Hawaiian print shirt and sandals.

The Hybrid Diaries
Chapter Forty Three
November 24th 1990
The guys leave for Australia

It was time for the guys to embark on their journey to Australia on board the Container ship.

The Customs inspector had given his blessings for our shipments to the land down under.

And the guys were getting ready to head out.

I had to stop and check in on their accommodations.

Yes, they were not as nice as the Cruise ship promised us,

But, they were nice and comfortable.

The guys liked the fact that there was a pool table on board.

I can only imagine how hard it must be playing billiards on a moving ship.

So, I brought them a few gifts to make the voyage a little more enjoyable.

I bought them a television and a Nintendo and a Super Nintendo and a Sega video game system and as many games as I could find.

So, that should keep them entertained for the trip.

The ship was huge.

It was about 850 feet long and stacked to the sky with cargo containers.

Over head a big gantry crane skillfully picked up and hoisted the cargo boxes high into the air.

They were placed deftly unto the ship by skilled crane operators picking up these boxes from the dock and off loading them onto the backs of truck trailers.

It was a well organized dance of commerce.

It was amazing to watch them load and unload these containers like little pieces of Lego

building blocks building the top portion of the ship.

The ship was called the "Aledo Sunshine Hyundai" or just the Aledo for short.

I am sure that the names held some kind of meaning for the ship owners.

 but, I couldn't figure it out.

I got to meet with the Captain.

He was from Finland.

And he talked with a thick European accent.

But, he seemed like a nice guy.

He explained to me that the ship should dock in Sidney around the 20th of December.

Give or take a day or two.

And that they would be making several ports of call along the way.

He said that people usually love the food.

And the ship has lots of things on board for the traveler to keep them from being bored.

I told him that I was grateful that they would be taking care of my friends.

And that all of the ships facilities were open to them.

I wished him a save voyage and left him.

To go and do the same with Dylan, Christopher Bill and Willy.

I had to hurry back to Dallas so that I could finish packing.

 I wanted to wish them all a safe journey and tell them that I'll be seeing them soon.

I told them to tell me what it was like to pass through the Panama Canal.

The Hybrid Diaries
Chapter Forty Four
November 26th 1990
Bon Voyage Texas. Hello Friendly Skies. Not.

Ok, Here we go. The day has come.

I hardly slept the night before.

Checking and double checking our luggage trying my best to make sure that we didn't forget anything important that we'd need.

For either on the plane.

Or on the cruise ship.

I was certain that if there was anything I missed,

I'd be able to pick up a replacement somewhere.

All of the tickets were safe in my wife's purse along with our passports and all the necessary documentation that we'd need for our flight to Long Beach.

California.

Her parents were going to drop us off at the airport.

So that we wouldn't have to leave Rhonda's little RX-7 in the parking lot for a month and a half.

It'll be safe at home.

We dropped Thor off the night before so he was all set to spend the time away.

Guarding his Grandparents house.

He really loves Jim and Betty and they love him too,

He's a good dog.

So, It's bright and early.

The sun is just coming up.

But, we can't see it due to a huge bank of storm clouds.

That rolled into Dallas and Ft. Worth overnight.

Last night Rhonda was telling me all about flying commercially.

She flew back from California before.

She explained to me that the only time there is a sense of speed

Is when you are taking off and landing.

And it would be nothing like flying in an Cessna or even in a F-16.

She said that the flight would be as smooth as glass.

And you could look out the window and not have a sense of motion.

I was excited to be flying on a big jet airliner for my first time.

We got to the airport and said good bye to the in-laws.

Rhonda hugged her parents and told them.

"That we would call them when we landed in Long Beach."

We went through the ticketing gate.

Checked our baggage and then proceeded into the waiting area at Gate five.

The windows looked like the windows of a car wash.

Water beading down the big glass panes.

The storms were intermittent and occasionally heavy.

We were right by the duty free shop.

I decided to stop in and pick up a few flight essentials.

A couple of drinks for the waiting area.

A couple of magazines for the flight and some chewing gum for our ears.

The F-16 taught me the value of chewing gum.

On take offs and landings.

When the pressure levels drop suddenly and your ears pop.

Sometimes, airlines will cancel flights.

Due to inclement weather.

I was afraid that it might happen in this instance.

But this time the airline wasn't going to let a little rain stand in the way of progress.

We saw the plane pull up to the gate and I was immediately disappointed.

The plane turned out to be a McDonnell Douglass DC-9.

One of the smallest jet planes in the American airlines fleet.

I was expecting something much larger to take us all the way to California.

I was thinking Boeing 747.

I know that would have been unrealistic, but, still, I wanted to ride in a Big plane.

Not something on the small size.

" It only holds around 80 passengers and crew." I said.

Rhonda assured me that the plane would be more than adequate for the flight.

And that I should just try and enjoy myself.

"Ok, I'll do my best." I said.

The plane docked at the gate and we were finally allowed to board.

We walked down the galley way.

We got on board and went to our assigned seats.

I wanted to sit by the window and Rhonda said that I that I could.

So, I took the window seat.

She sat beside me there was no one else in the remaining seat.

We had the aisle to ourselves.

The Stewardesses secured the doors.

We strapped into our seats.

They went over the standard instructional speech.

"Welcome to American Airlines flight 202 from Dallas D/FW to Long Beach

Municipal Airport in Long Beach, California."

She explained the emergency oxygen mask drill

Then she showed us that there are exits here , here and here, .

Then explained that the seats that we are sitting on also

double as "floatation devices. "

We sat back and the plane taxied out onto the runway.

Then waited for clearance from the tower.

The decision was given.

And sure enough I felt the sensation of speed building up as we streaked down the runway and into the air.

I had passed out gum to Rhonda so, that it would help equalize the air pressure in our heads.

And although our ears did eventually pop,

It wasn't quite as painful as it could have been.

I saw the airport fall away into the distance.

As the plane flew skywards and the clouds shortened our visibility.

Then "Whoomp" The plane felt like it had fallen several hundred feet in the air.

Then another floomp and a shutter as the plane hit pocket after pocket of turbulence.

From the massive thunderhead cloud formation that we were flying through.

I felt a sharp pain as Rhonda's nails dug deeper and deeper into my thigh and hand.

"So, where is the smooth ride you promised me?" I asked sarcastically.

"I didn't expect us to take off in a thunderstorm" she said.

"Whoomp." We hit another air pocket.

The engines felt like they were straining.

Doing their very best just trying to keep the little plane in the air against the might of mother Nature.

Then the pilot came over the intercom.

His name was Captain Johnson and he was from Longview Texas.

He had a thick Texas drawl in his voice.

It almost sounded made up.

I seldom hear people talk like that in Texas.

And I LIVE here.

He said that "we were flying through a little bit of turbulence and that the stewardesses would begin to feed us just as soon as we cleared the storm front and were in clear skies".

That took a while.

He said that we were over Brownwood when we finally got into less bumpy air currents.

It turns out that we were the last flight out.

Before the airport had to shut down due to bad weather.

The stewardesses passed out drinks and they began to serve us our in flight breakfasts.

We had some mimosa's (orange juice and Champagne).

Breakfast consisted of your choice either two pieces of bacon or two sausage links.

And a scrambled egg that looked like an omelet and a hash brown potato patty.

It was edible.

Later on we played a trivia game with the stewardesses.

And I won the grand prize.

A bottle of Champagne.

I happily stuffed that into my bag in the overhead compartment.

I would save it for when we got on board the ship.

With all the excitement from the storm.

And the odd and drink and the trivia contest.

And me staring out the window.

I barely had time to read my magazine.

At last the Captain came over the loud speaker and announced that we would be landing

In Long Beach Municipal Airport in just a few minutes.

"Now, you're going to see just how smooth the landing will be" Rhonda said.

So, I braced myself for the smooth landing.

And "Boom, Boom, Skreet!"

The wheels slammed into the tarmac with a hard thud.

And then a loud squeal as the tires

Laid a strip of rubber down the runway.

We bounced up and down in our seats.

The plane began to slow down on the runway.

We taxied up to the airport terminal and a gangway was pushed up to the side of the plane.

The stewardesses opened the plane doors so that we could de-plane.

We walked right out onto the runway and we walked on the tarmac to the terminal with our carry-on baggage in our hands.

It was like we had landed in the 1970's

After a landing like that we were more than happy to leave the plane.

"OK. The worst part of our flight is over.

And now we can focus on getting to the Port Authority and dropping off our luggage dockside.

Then we can meet up with the rest of the group on ship.

And settle down for our trip.

We had no tools or parts to work on.

And All I could conceivably do is work on my designs.

And since all the designs and blueprints for the racers are safe and sound back in Dallas,(except for some electrical wiring diagrams).

I had a clean slate and a new design book to work on.

It looks like we were set up for three weeks of pampering and eating and drinking and body massages and soaking I n the pool and lounging on the deck.

The ship offered gambling tables and slot machines.

Once we got out into international waters and the pictures in the brochure showed us

The ship even had a small workout room.

(not much bigger than a closet with an exercise bike a rowing machine and some wall mounted weights.)

I imagined that all the equipment was covered in dust due to lack of use.

he brochure also said that there was a disco onboard where we could dance the night away.

There was also shopping at the ships stores

"For that special gift for that special someone".

They also offered skeet shooting off the back of the ship.

And porpoises frequently follow the ship.

We could see seals sea lions and even whales.

We were excited thinking about all of the things that we would be doing once we got on the ship and set sail for the wild blue horizon.

Of course we still had to get to the docks.

Then on board and then set sail.

That would happen soon enough,

We still had a few hours lay oven before we could even board.

We loaded up all of our suitcases onto a taxi.

The cab driver drive us over the big green bridge.

And over to the Port Authority.

We dropped off load our luggage and have it loaded onto the ship.

There we met up with the Wigglers who took a different flight out of Austin.

We decided that we had a few hours to kill.

I talked them into coming with us over to see the Spruce Goose and the Queen Mary.

We caught another cab and in just a few minutes we were standing outside of the big Dome that houses the Hughes H-1 Hercules Flying boat.

Now commonly known as the 'Spruce Goose'

(a nickname which Hughes hated).

The aircraft has gone on to become a popular cultural artifact which has a remarkable story of sacrifice, determination, and technological development by Hughes and his team.

And is the result of reading a business contract in a totally wrong manner.

The Spruce Goose is still the biggest aircraft ever built and was decades ahead of its time

In the early 1940s.

It revolutionized jumbo flying bodies.

With its huge wingspan and large lift potential.

Shaping modern flight today.

The popular Spruce Goose is now appropriately regarded as a true American icon.

Engineers hung eight of the most powerful engines available on the huge wings

(Wingspan: 319.92' (97.54m).

They designed an enormous fuel storage and supply system to allow for long oceanic flights

(Fuselage: 219' (66-75m).

The Spruce Goose was built from an urgent national need to fly over the enemy submarines ravaging shipping lanes during World War II.

Billionaire, Howard Hughes, responded to the call with his Hughes Flying Boat.

This was to be the biggest airplane ever built.

And probably the most extraordinary aviation project of all time.

The original US Navy contract called for "Flying boats to carry 740 men in full battle gear.

It didn't exactly specify "HOW many" flying boats were to be made, jus a carrying Capacity and at a range of 1,000 miles.

The Navy proposal would have suggested that if a standard flying boats carrying capacity

was fifty fully armed men.

Then the number of flying boats necessary to carry 740 men

would be around 1450 airplanes.

Instead Howard decided to build one VERY big flying boat that could do the job in one load by itself.

The navy said that it couldn't be done.

Then they threatened to sue Hughes Aircraft for breach of contract.

And since the boat was not ready until after the war.

It was no longer needed.

And it still had not flown.

So on November 2, 1947, Howard Hughes and a small engineering crew cranked the engines up for a taxi test.

Howard Hughes took the controls and the Flying Boat lifted the plane off of the waters over Long Beach Harbor.

To an altitude of 70 feet.

It flew one mile in less than a minute.

At a top speed of 80 miles per hour.

Then came in to make a perfect landing.

The Navy decided to drop the lawsuit.

But, refused to take delivery of the huge flying boat.

This was to be the only flight it ever made.

It then sat for years in an aircraft hangar until Hughes' death in 1976.

It was donated to the Aero Club of Southern California.

It emerged from its hangar after 33 years.

Eventually to become a tourist attraction

The City of Long Beach constructed the dome to protect the huge plane from the elements.

We also saw some world war two artifacts like an M-4 Sherman tank.

We also went over and saw the Queen Mary.

One of the largest cruise ships of her time.

She is permanently docked at Long Beach Harbor and has been converted into a five star hotel.

She was a troop transport in World War Two.

But reconverted back to her former cruise ship glory after the war.

She now serves as a tourist attraction and offers guest a chance to stay in one of her 322 state rooms.

She is outfitted in splendid art deco furnishings.

I wish that we had time to spent the night on board her.

but, time was running short and we had to get back to the Authority hopefully be allowed to board Our cruise ship for our long ocean voyage.

OK, we arrived back at the ship.

The Little Princess Star.

She was anything but little at over 900 feet long.

And having 14 decks this was the biggest ship I had ever seen.

The Container ship looked almost as big.

but, most of that was cargo containers piled as high as the moon.

The hull itself was about 50 feet shorter than the Cruise liner but the cargo Ship was much wider at the beam.

At last they were finally allowing the passengers and guest on board.

We found our cabin on the Lido deck.

It was a suite on the port side of the ship.

With a sliding glass door that led out onto a breeze way and then to the upper deck.

We could open the door and look out onto the dock.

And see people coming and going.

And we saw several forklifts running in and out of huge doors on the side of the ship.

Loading fresh vegetables and cases upon cases of sodas and alcohol and other boxes of goods and laundry into the hold of the ship.

I'd joke to Rhonda.

"See that pallet (the one full of alcohol) that one's for me."

"You see that pallet? (the one with the fruits and vegetables) that one's for you!"

The ships intercom was blasting "Farewell to thee" at a pretty loud volume so it could be heard on the docks.

And it just kept it over and over on a continuous loop.

With about a thirty second delay after each time the song was finished.

We'd barely get a word in edge wise and then It started blasting out again.

"No wonder they wouldn't let us on board any earlier.

The music is enough to drive you nuts". Rhonda shouted.

On the table in our suite was a lovely fruit basket and a bottle of champagne sent by the travel agency.

There was also a pair of glasses next to the basket on a little tray.

That reminded me.

I called down to the ships steward on the ships phone.

I asked them to bring up an ice bucket full of ice.

And I took the bottle of champagne that I had won on the flight over.

and I put it on ice, so that it's be nice and cold when we pulled out of port.

The steward came in and he had the ice bucket.

He introduced himself as David.

He had a British accent

(just like the brochure said. I thought to myself).

The Wiggler's cabin was a couple of doors down from us and they were also complaining about the loud music.

Trying to find some way to turn it down.

It would not have made much difference with the doors and port holes opened on the ship.

Sound carries everywhere.

The brochure said that the cabin staff were English.

And the dining room staff and the chef were Italian

The rest of the crew were of different nationalities.

They offered breakfast, lunch, Tea at four o'clock and Dinner and then a midnight supper, plus in room catering.

In case you didn't feel like eating in the formal dining room.

And room service, and laundry services.

And each room has its own bed turn down service with a English cabin steward.

It was almost after four, so It was close to tea time.

but, since we were still at anchor and the ship still had guests on board, there would be no tea time today.

Our first meal on board the ship would be a late dinner.

We had cards to fill out letting the crew know when we would be going out to eat our meals

And when we would be out of our cabins so that the cleaning staff could prepare our rooms for us.

They would prepare our table assignments and let us know what time dinner (or breakfast) would be served.

And when we would be expected to come down to the formal dining room.

If we wanted.

We could skip the formal dining room and eat our breakfast out on the upper decks.

There was always going to be food around the ship.

And plenty to drink.

At last the whistle blew and the Captain made the announcement

" This is the Captain speaking, All ashore that's going ashore"

"All guests must be prepared to leave the ship as we are weighing anchor and preparing to embark."

Fifteen minutes later.

We heard "last call. All ashore who is going ashore." over the intercom.

And the music started blaring even louder.

I guess to build excitement.

In about 30 minutes the back of the ship began to shutter and the water at the stern of the ship began to boil.

We could feel the power of the engines 10 decks up from the engine room.

We saw streamers and confetti flying past our window and sliding door.

So I ran to open the champagne and hurriedly poured four glasses.

I ran to give one to Rhonda and then I bolted out of the room and ran down the hall knocked on the Wiggler's door and I tried to hand them two glasses.

They were way ahead of me and they had just opened the bottle that the Travel Agent had sent to too.

So, I had two more glasses to take back to our room.

The Wiggler's followed me back to see how our room looked.

They liked ours better because we were on the side of the ship and their cabin was right on the corner.

So they had an additional window but, their cabin was irregularly shaped.

The massive anchor rode upwards on its capstan and the chain came through the Hauser Holes.

The anchors were secured at the bow of the ship.

Water started boiling around the ship as a tug came out from around the dock.

It began to pull on the ship.

At first we felt a slight thud.

And the water all around the hull started boiling.

Then we started inching forward.

Almost imperceptibly at first.

Then we started to gather up some speed.

And at last we started moving away from the dock.

We screamed out to the crowd below.

"Bon Voyage!"

I threw a roll of toilet paper out of the window and watched it unfurl.

And then splash into the ocean.

"I hope that wasn't our only roll" Rhonda sighed.

Rhonda had found the streamers and confetti on the night stand.

But, she failed to mention that to me.

I wanted something to throw.

We threw the streamers and confetti over the side.

and watched them rain down on the crowd of well wishers below on the docks.

We raised a toast

"To a happy voyage."

And we clinked glasses and drained both bottles of Champagne.

We still had the one that I won on the plane sitting in ice,

So, I popped the top on that one too,

And we downed it in short order.

Dick called down for another bottle

But, we were not yet at sea

So we would have to wait until we cleared the harbor.

"Oh well, No big thing." Candace sighed.

"We have a whole ship to explore."

"And fruit to eat." I chimed in.

And at last, the music finally stopped.

Next stop Hawaii in four and a half glorious days.

2,551 miles from Long Beach to Hawaii.

The Hybrid Diaries
Chapter Forty Five
November 29th, 1990
Mele Kalikimaka!

OK, We have been at sea for almost four days.

We have done almost everything on board the ship.

The lounged in the salt water pool.

"That was funny. I didn't know it was salt water until I jumped in and I tried to open my eyes under water.

Then I got a taste of the salt water up my nose.

We did shows.

At the front of the ship they had a multi use area that the staff used to put on live shows.

During the day they hosted bingo.

And then we shot skeet off the back of the aft deck.

And we ate until over stuffed and drank ourselves into a stupor.

We played cards and gambled at night.

And then danced the night away in a bedroom sized disco.

This played all of the disco greats from the late 1970s and early 1980s.

All amid flashing lights and a disco ball the size of a world globe.

Truly tiny.

Rhonda scored big time on the slot machines.

Winning two hundred dollars.

And just as quickly feeding it all (Plus an additional fifty dollars) right back into the same machine.

I just played roulette.

And I stayed on the tables all night.

"Just as long as they kept bringing me free drinks."

I started off with forty dollars and I ended up with about sixty five by the time I left.

Rhonda would not listen to me when I said that it was time to go.

She had been hooked by the one armed bandit.

And it wasn't turning loose of her.

At last she ran out of money and she was forced to leave the casino.

I was outside of the main dining room eating food from the midnight buffet.

She and I got back together and I spent the next hour peeling boiled shrimp for her consumption.

Every night we would dine in the formal dining room.

And every time the head waiter would come to our table.

And he'd shout out

"Tonight, I make for you something special. Fettuccini Alfredo"

And he did.

Then he'd serve it to us.

Man, the food was delicious and we ate nonstop.

I could have put on at least 20 pounds on ship.

If not for the massages and jogging around the deck.

The girls were amazed by the fact that it was almost December and they were walking around the ship in two piece bikinis.

They would lay out on the sun deck and sun bathe

While we had to rub lotion and tanning butter all over their glistening bodies.

That was rough. (Not)

It was a dirty job, but, someone had to do it.

When we slept in our cabin, we would have strange dreams.

They say that when you fall asleep on the ocean it really affects your dreams.

They were right. Most of them I cannot recall.

I would sometimes think of the vehicles.

Other times I would dream about Toby and how we had made some kind of mistake.

But, I couldn't put my finger on what the dreams meant.

In another one,

Everything was gone,

I didn't understand where everything went.

That scared me a bit.

Not the fact that I had lost everything.

but, the not knowing what or how had me feeling uneasy.

I did my best to put those feelings aside.

And just try and focus on having a good time on the ship.

With all of its opulence and all the things to do on ship.

I was starting to get a little bored.

I wanted the ship to hurry up and get to Australia.

so, we could get on with the race.

At least that is something I know what to expect.

(I think)

I think that some new sights and new metal stimuli is required and soon.

Fortunately, we'll sail into the big island.

I used to think that the "Big Island was actually Oahu."

Shows what I know doesn't it?

The Island of Hawaii is actually the big Island.

And that is where we docked.

When we got off the ship we were immersed in the Hawaiian culture as much as stepping off a ship in Mexico will immerse you in the Mexican culture.

We listened to steel guitar Hawaiian music.

And saw the basic stereotypes of the culture and the lore of Hawaii.

All you see is young and old Polynesian men and women in grass skirts and wearing and passing out leis.

We got to go on a luau; we saw fire dancers and a hula festival.

I had enough alcohol in me that I built up enough courage to talk a barefoot walk on hot coals.

I didn't get burned fortunately.

I practically flew over the embers after getting my Feet all covered in ashes.

Dick and Candace were suitably impressed.

We ate fresh pineapple suckling pig and drank Kola coffee

(well I didn't drink it because, I don't ever drink coffee. I'm a tea person.)

But, Rhonda put pot after pot away.

Because the coffee tasted so good.

She was wired from all of the caffeine for a couple of days.

Here is a little history of the big island.

Legend has it that two deities the Volcano God Pele and the demi-god Kamapua'a (the latter of whom could control the weather) — struck a deal to make the vast Big Island of Hawaii's west side so dry, and its east side so wet.

The story's short version is that, after a battle.

The pair divided the island in two,

With Pele taking the western half and Kamapua'a, the eastern

Covering 4,028 square miles.

The Big Island (or the "Orchid Isle") is the youngest and The largest of the Hawaiian Islands.

Twice the size of all the other major Islands combined.

And with two of the five volcanoes that created the island still active.

It continues to grow.

Kilauea Caldera is the longest continuously erupting volcano in the world.

Its present eruptive phase dating back to 1983.

Mauna Loa, meanwhile, last erupted in March of 1984.

Sending lava to within just a few miles of East Hawaii's Hilo town.

Of the remaining three volcanoes on the island.

Mauna Kea and Kohala are extinct.

While Hualalai is considered to be dormant.

Having last erupted in 1801.

The King Kamehameha the Great.

Who unified the Hawaiian Islands under one king for the first time in 1810.

He is believed to have been born in the Big Island's North Kohala area.

US Navy Captain James Cook.

Who is widely considered to be the first European to set foot in the Hawaiian Islands.

Was killed at Kona's Kealakekua Bay in 1779.

The ship is anchored out in the middle of the bay.

The crew lowered some motor launches to ferry passengers to and from the shore.

For a day of on shore shopping and sightseeing.

We all had to wear our life jackets while the motor launches.

(which were skillfully piloted by the crew)

Just in case someone fell overboard.

It was nice that they were taking care of us.

And didn't want any fatalities on their hands.

The girls decided that they wanted to do some sightseeing without us guys.

"so that they could do some Christmas shopping."

I can't wrap my mind around the fact that it is getting closer and closer to Christmas.

It doesn't seem like it is Christmas time.

No snow and no cold weather.

No bundling up in blankets and drinking hot chocolate.

The ship was just beginning to pipe in Christmas tunes over the intercom.

That did little to make it feel like the holiday.

We men folk decided that we would try our hands at either surfing or snorkeling in the Crystal blue and pristine waters that surround the main island.

The girls gave us a friendly reminder "

"Don't forget to bring your towels!"

Snorkeling seemed like it would be the easiest thing to do.

Since we were both good swimmers.

And it doesn't take that much in the way of equipment to do.

All you need is a diving mask a snorkel and some decent swim fins.

And of course a nice bay to dive in.

Finding the bay was quite easy.

It lay just a mile or so from where our ship was docked.

We had our watches so we could keep track of the time.

We didn't want to stay out in the bay for too long.

Because we were both a little scared of a possible shark encounter.

At the bay there was a beach club that rented snorkeling gear and sold drinks and some Food (burgers and fries mainly).

We rented the gear,

Locked up our towels and stuff in a locker.

This was conveniently placed for rent there at the beach club.

The keys were on chains that we could wear around our necks

We took the keys with us.

And we headed out into the ocean like two Jacque Cousteau's.

It was very beautiful under the waves.

I found out that you can go a long way from shore

Just swimming along in shallow water using your hands to gently pull

yourself along the Coral reefs.

The water wasn't much deeper than four to six feet. So the snorkel tube

worked fine for reef diving.

At last I got into deeper water when I approached a steep drop off.

I followed the reef line and continued to pull myself along the bottom.

There were all kinds of sea life all around me.

Little yellow Tang fish blue and yellow Tangs.

And a wide myriad of other colorful fish that I have only seen in someone's

salt water aquariums.

We dived for over an hour before I cut my knee open on a sharp piece of coral

that I Didn't notice.

So, when I looked up I noticed that I was in at least thirty feet of water.

And the surface was pretty high up there.

So, I swam up for some air and got my bearings.

I had no idea I had swam out to sea that far away from shore.

Neither did Dick.

I told him that my knee is bleeding.

We need to get the heck out of the water ASAP.

We made a bee line straight for shore.

 Fortunately, no shark encounters.

The cut wasn't too bad.

The bartender at the club was more than happy to give me a

Hand towel and a few band aid bandages for the cut.

He also had plenty of alcohol for my Troubles and pain.

Dick and I ordered a couple of burgers and fries and some cold frosty

Ones to wash it all down.

Don't get me wrong,

The food on the ship is excellent.

But sometimes you just gotta have a big ol' greasy hamburger and fries.

The ship could whip up a filet Migon in no time flat,

Served on china and with wine in a crystal stemware glass

but, it lacks the simplistic charm of a beach side dive.

Where beer is served cold and in unmatched plastic beer mugs.

Well after a few more beers.

I was no longer feeling the pain from my recently rendered knee.

It was getting late.

So, we went to the lockers and got our stuff.

Dropped off the diving gear.

Got back our deposit and hailed a cab for the short trip back to the docks

Where we caught a launch back to the ship.

It was almost time for tea.

We waited for a while, but, still no girls.

So, we did tea like the gentlemen that we are

We put on a couple of suits and headed down to the main dining room.

On the way we stopped by the bar and ordered up a couple of shots of Crown

Royal

And took them under our coats into the dining area

Then we sat down at our table and got ourselves comfortable.

We covered the shot glasses with our napkins.

The waiter came by and asked us if we would be having tea and buttered scones.

We assured him that we would.

But first we wanted some iced water.

A glass of ice each.

And one can of Coke and one can of Seven up.

The waiter was accustomed to our antics by now and he hardly ever raises an eyebrow when we make dubious orders.

We also requested some extra sugar and some lemon wedges.

Soon the waiter came back with a tray full of glasses.

Two filled with ice water,

two others just filled with ice.

One can of Coca Cola and one can of Seven Up,

 a bowl of lemon wedges and some extra sugar packets.

Then he left, but, soon came back with the tea pot full of hot water.

 two cups, four tea bags and a tray of biscuits and sweets.

We took the tea bags out of their wrappers

And put them into our cups.

 And then we poured the hot water over them and covered the cups with our saucers.

We opened the sodas and poured the two shots of crown into the glasses of ice.

Then we added our sodas.

We drank a toast to ourselves and to our beautiful women who were still in abstencia.

And drank them down.

We wiped our mouths off with our pinkies held high in the air.

And we each grabbed and buttered a scone.

After we consumed the dry but buttery biscuit,

we drank down our water to about a half glass.

We took the saucers off of our cups of steeping tea.

And dunked and re dunked the tea bags several more times

to get all of the caffeinated goodness out of each bag.

Then we set the bags on the side of the saucer

And added about eight sugar packets into our cups and stirred vigorously.

Once the sugar had dissolved completely.

We poured the tea cups into our half consumed glasses of ice water.

We used our tea spoons to mix the liquids together and boom! Iced tea.

Just like we make it at home.

Dick prefers his with a twist of lemon, I like mine straight up with no lemon.

We spent two days on the big island.

And tomorrow we were heading off to Oahu to see the Pearl Harbor Naval Base.

And the Arizona memorial.

It wasn't a sure thing that the girls would want to come with us.

There is so much to do on Oahu.

Besides in just a couple of hours.

We were either going to be eating in the formal dining room

Or scarfing down suckling pig at the last luau on the big island.

We hadn't decided that either.

"Where are the girls?" I wondered out loud.

Dick heard me and he replied.

"Where else?"

" They're Out spending a lot of money!"

We both laughed until we heard a familiar voice come from behind us.

"So, what are you guys laughing about?" asked Candace.

We both turned around and there was Rhonda and Candace both of them with new purses and shoes to match.

And, both of them looking like they just came back from the beauty parlor.

Their hair and nails looked freshly coiffed

"So, how long have you two guys been here?" They asked.

"Not too long."

" Just long enough to grab a glass of iced tea."

"Andy, That is not how you are supposed to be drinking tea here."

"especially at this time."

snapped Rhonda.

"Purists!" I snapped back

She kissed me and then she wrinkled her nose at me.

And then she commented about the distinct odor of good bourbon whiskey on my breath.

"Ok, I smell something."

Candace and I are going to the bar and catch up."

They turned around and left the dining room for parts unknown.

It was good to see them and even better to know that they were both safe and sound.

I think it was Dick who made the decision to leave the ship and to attend the luau.

It didn't matter,

We made a stop by our favorite bar on the starboard side of the ship

And we did a few more shots.

While we were there we ordered a couple of bottles of wine to

be sent to our state rooms.

We asked that they be put on ice so, that when we got back we

Would have something to drink when we got there.

The ship stops selling alcohol after

11:00 PM and they stop selling wine and beer after 2:00AM.

So, we planned accordingly.

We both knew that we needed to lay in extra supplies "just in case".

It was around six thirty when I showed up back at the room.

Rhonda had successfully achieved the same level of comfort that Dick and I

currently resided in.

And the room was filled with boxes and bags of things.

I knew better than to ask what was in all of the containers.

So, I moved a few boxes out of the chair and I sat down.

She got up.

Rocked a little on her new heels.

And then she gave me a present that she said

" I could open."

So, I took the box from her and I gave her a kiss on the cheek

and I asked her again if it was OK for me to be opening this box.

Once again she said that "it would be fine if I did."

So not wanting to be wasting any more of our time.

I tore open the gift wrapped package with avarice

There inside was something I would have never purchased for myself.

A flowery Hawaiian print cotton shirt and some khaki cotton shorts

"All that's missing is a pair of sandals." I said

"They'll look good with you Aussie hat." added Rhonda.

(She was obliviously very pleased with herself.

And she thoroughly enjoyed seeing me change into these touristy trappings)

I obligingly took off my suit and tie and threw these clothes on.

I felt very silly wearing the shirt.

The shorts I could handle.

But, the flower print pattern of the shirt was causing me vertigo.

Rhonda told me that she and Candace had gone souvenir shopping and gift buying so that they can "send things home to the families for Christmas."

I told her that she was very sweet and thoughtful for thinking about our folks back home.

I realized that this was going to be hard on her as well.

She had never been away from her family for this long a time.

And she had never missed a major holiday like this before either.

And to add to it.

Being stuck on a ship, on a tropical island far away from snow.

Not to mention ice and a sharp chill in the air

That only added to her missing the company of her family.

She needed to feel like she was doing something Christmasly to get her in the mood.

I would have to think about this.

Maybe I could come up with a solution.

But, right now, I had to get ready to go out and face the world looking like a beach bum.

"Oh well, here goes nothing."

The Hybrid Diaries
Chapter Forty Six
November 30th 1990
The coming of the storm

We got back on board after a night of poi and celebrating.

We loved the hospitality of the Big Island.

And tomorrow the ship would weigh anchor and we'd be off to Oahu.

It was decided that when we got there.

Rhonda and Candace would once again split up with us,

they would spend the day taking hula lessons and working on their sun tans,

drinking Mai Tai's and going on a tour of the volcanoes.

In the mean time Dick and I would head over to Pearl Harbor Naval base and see the Arizona Monument.

then we'd rent a car and cruise around the island and check out the sights.

We'd pretend that we were police officers like in Hawaii Five O.

Dick was the obvious choice for Dan-O since his name started with a D and I got to play McGareth.

We'd spend the day looking at Buildings shot up by the Japanese and wonder why there are not many people who are in denial that the attack ever happened at all.

We got the notification when we got on board

That we would have an additional day on Oahu due to a typhoon around the Philippines.

The storm was called "The Super Typhoon Owen"

And its fury mainly centered mainly around the Marshall and Caroline Islands.

but, its fury and storm effect would stretch out many miles into the Pacific.

We immediately thought about the container ship and our friends .

They had already passed through the Panama canal and they were well ahead of us and much closer to the Philippines than we were.

Due to the fact that a container ship has a schedule to follow

And it is fairly tied to its port arrivals.

We wondered what would happen to the ship if they were in the path of such a storm.

Would they alter course or try and barrel through the storm to keep its scheduled arrival?

We said a prayer for them and hoped that their captain would do what was best for his Ship, his passengers and cargo.

Meanwhile, we were heading to Oahu.

The Hybrid Diaries
Chapter Forty Seven
December 1st 1990
Visiting Pearl Harbor

The dawn broke brilliantly into our state room, as the ship pulled into the big harbor of Oahu.

We boarded a long boat which took visiting tourists to the ship docks.

We could rendezvous with our guide who would take us out to Pearl Harbor...

Pearl Harbor is on the Island of Oahu, Hawaii.

It is a significant historical location.

It was here where the Japanese Fleet of Six Carriers under Command of

Isikuro Yamamoto

Attacked the U.S. Pacific fleet on December 7, 1941.

This led the United States to enter World War II.

His original plan was to catch all of the Battleships and the two U.S. aircraft carriers

(YorkTown and Enterprise) riding peacefully at anchor inside the harbor.

And using bombs made out of shells from the Japanese Battleship Nagato.

A and specifically modified torpodoes.

Blow the entire U.S. Battle fleet into the Stone Age.

He brought with him the cream of the Japanese Naval air arm with him on six aircraft carriers.

The Akagi, Kaga, Zuikaku, Shokaku, Hiryu and the Soryu.

Fortunately for the U.S. the two U.S. aircraft carriers were at sea at the time of the attacks.

Which came in two waves.

So, the U.S. Navy was able to fight in the Pacific.

And bring the war back to the Japanese with a raid over Tokyo six months later.

Today, Pearl Harbor (including the Arizona Memorial) is

The most-visited destination on Oahu, with more than 1.5 million visitors per year.

Our visit to Pearl Harbor and the Arizona Memorial is a solemn learning experience

The Arizona Memorial, is a platform that was built atop the sunken battleship USS Arizona.

Inside of its white walls are the names of every crew member

who was killed aboard her on that fateful day.

We were literally standing on top of the graves of 1,177 soldiers who perished aboard her.

From the platform you can see what was left of the battleship,

which lies six feet below the water's surface.

After more than 48 years after the attack,

oil is still leaking out of the battleship,

This is also called the "Black Tears of the Arizona."

Our tour guide told us

"You'll have to board a small Navy shuttle boat to reach the Memorial."

The designer of the Monument was named Alfred Preis he is the architect responsible for the memorial's design.

The structure has a sagging center and its ends are strong and vigorous.

"It commemorates "initial defeat and ultimate victory" of all lives lost on December 7, 1941"

The Guide told us

"The best time for you guys to visit The Pearl Harbor Memorial is early in the morning,

when it's not too crowded yet."

We chose to take the tour with the help of one of the local tour companies (which also included the aforementioned guide).

Once we arrived, Dick and I strolled through the Pearl Harbor Museum,

Featuring World War II memorabilia and actual photos of the attack.

Then, we watched a 23-minute documentary film featuring actual footage of the attack.

After the video we boarded the Navy shuttle boat that took us to the Arizona Memorial.

We stayed there a while and said a prayer for the dead and then we left the memorial and saw some of the damaged aircraft hangers at Hickam field that had been shot up by the Japanese fighter planes on December 7th. 1941.

The bullet holes from the Japanese fighter planes were still present in the walls.

Dick commented that

"If we could have waited one more week, we would have been at Pearl harbor on December 7th and the place would be full of veterans and tourists."

We were lucky that we were able to dodge that bullet.

Because, we would never had been able to do what we did.

What with all of the remembrance activities going on.

We had our fill of history and Dick and I decided that it was time to go and meet up with our ladies who were waiting on board ship for us to take them surfing.

We thought that they were joking because, neither Dick or I had ever been surfing before,

and we didn't know the first thing about the sport.

But, we decided that since the girls were patient with us while we checked out World War Two.

We'd better do something nice for them in return.

When we got back to the cruise ship we once again donned our Hawaiian print shirts, put on some swim trunks and pulled some cargo shorts over them.

Then we slipped on our huaraches sandals.

Dick put on a straw hat while I reached for my Aussie hat with the armadillo pin that held the left side of the brim up.

We grabbed our sun glasses and we made our way to the sundeck where we found our little dears sunning themselves and their mai tais in two lounge chairs.

"Show There you guys are… We've mished you!"

"Where the Hell have you been???"

They looked like they had been there for quite a while and their slurred speech was a dead giveaway that we'd need to get them to eat something soon.

Before even considering leaving the ship.

We noticed that lunch was just about to be served on the Sun Deck.

So, when the time came,

we got up and walked over to the buffet.

Ordered us some more drinks and then we filled some plates

with food.

And we took these items back to our grateful little ladies.

We all enjoyed a lunch of fish and shrimp and French fried potatoes, hush

puppies and Cole slaw.

We drank some tea to wash it all down.

Before long Dick and I decided to have some cocktails of our own after lunch.

It'd take us a while to catch up with the girls.

We never actually made it to the surfing competition.

We did however managed to make it to the nightly luau on the beach.

Where we were entertained by fire dancers and hula dancers.

And lots of singing and sitting and cuddling by the fire.

We heard stories about the fire God Pele.

And how the Hawaiian people used to sacrifice their young male children to

their hungry fire god.

We had more than enough to drink and the air was getting a little chilly.

Rhonda and Candace and Dick were raving on and on about how good

Hawaiian Kona coffee was.

Oh how rich and strong and flavorful the beans were.

Since the air had a bit of a chill to it.

How nice it was for them to have a hot beverage in their hands.

(I'd settle for some hot coco but there was none to be had).

I do not drink coffee so I was feeling left out.

But, I wouldn't begrudge my friends enjoying themselves over a nice hot" cup O' Joe".

Instead I fixated on how good fresh pineapple tasted.

Not like the semi dried pineapples that we tend to have back home.

This stuff was grown on the farm a few miles and another island over from where we were.

I also got to enjoy fresh coconuts for the first time.

And man the milk and coconut meat was terrific

We left the fireside and headed back to the docks to return to the ship.

We got to the docks and got back on the motor launch and rode back to the ship.

We could hear the sounds of Charles Dickens "A Christmas Carol

(the Movie of course) coming from the main theater.

It was the first Christmas movie of the trip.

And that reminded us that it was not summer.

But, getting close to Christmas.

You could almost forget the holidays here in Hawaii.

The Hybrid Diaries
Chapter Forty Eight
December 2nd 1990

We woke to news of the Super Typhoon Owen.

And how it was causing major damage all across the sub Pacific region.

Many ships had altered course.

But, were still caught up in the storms wake.

The Island Chain of the Caroline's was experiencing high winds and heavy rains.

We prayed for the safety of our comrades and the cargo that they kept watch over.

I was thankful that with the gaming systems and the TV and all of the game cartridges that we bought them.

that at least they would have something to do to keep them safely indoors and occupied.

We however were still under clear skies.

Our ship was now a couple of days behind schedule

But, otherwise we were on course for the Philippines.

The ship and its crew was doing their best to help us keep our holiday perspective.

Every night in the theater they would show a different Christmas movie.

Everything from Natalie Wood in Miracle on 34th Street to Bill Murray in "Scrooged" and every day more and more holiday decorations went up all around the ship.

While more and more Christmas songs were being pumped in over the ships intercom.

But, still it didn't feel right to most of the passengers.

Some of them preferred the sunny warm weather of the Pacific Ocean.

Compared to the snowy backdrop of sub zero weather.

There was plenty of sleet and snow in their home towns.

I guess that's why they booked their cruises so late in the season.

We simply had no other choice.

I have to keep reminding myself that we are probably the only people except for the crew.

Who are on this ship for business reasons.

Most people would have just flown to their destinations instead of taking the scenic route.

But, we had to travel exceptionally heavy.

And flying would not have been the best way for us to handle all of our traveling needs.

We were due to land in the Philippines on the sixth.

So, It would be another three days of eating and drinking and body massages and running around the sun deck.

Working out in the weight closet, and dancing in the ship's disco bedroom.

Really, this "Disco was a room no bigger than 10 feet by 10 feet.

With a DJ booth suspended from the ceiling to conserve floor space.

It was just about at the same height as the disco mirror ball that reflected light all over the walls.

There was a small bar in the disco room.

But, it was more of a why bother bar.

Because you had to pick your way through all the other guests just to get an drink.

And the bartender couldn't hear your order over all of the loud 70s disco music.

Blasting out of the speakers suspended from the ceiling overhead.

When you'd go in you could watch other people dance.

When you wanted to dance.

Everyone would have to get out of your way.

They in turn would have to watch you dance.

It was practically assholes and elbows.

So, we spent a lot of time in the ships Casino hallway.

The hallway was just that. A Hallway on the ship.

At night and when we are at sea,

The crew would pull open glass partitions.

Then wheel out the slot machines and the few gaming tables that were stowed onboard.

You could play craps, and roulette (which is where I spent most of my time) and poker

(which is where Dick spent most of his time).

The girls were stationed at one of the five slot machines.

Occasionally getting into it with some little old ladies who wanted their turns at the one armed bandits.

During the day the theater was transformed into a bingo hall.

Where we played bingo for prizes.

That's were a lot of the ships mostly elderly passengers were holed up.

A lot of the ladies on board were widows.

"out on the prowl for husband number two or three" was what David our Ships steward told us with a bit of a snicker and a thick English accent.

The Hybrid Diaries
Chapter Forty Nine
December 5th and 6th
Crossing the International Dateline and losing a day.

How does it feel to just lose a day?

If you are from The United States you pretty much already know what it feels like to lose an hour.

We do that every year when we switch over to daylight savings time.

We diligently "spring forward and then we fall back"

Yes it sounds corny, we "spring" forward."

What that means to the rest of the world is nothing.

The rest of the world doesn't acknowledge this practice originally thought up by the great Benjamin Franklin to conserve candles and maximize the available sunlight for maximum harvesting efficiency.

Spring forward means that we set the clocks an hour forward around 2:00 A.M. to 3:00AM in the spring time shortly after the Vernal (or Spring) Equinox.

And fall is for falling back or setting your clocks back an hour at around the Autumnal Equinox

Somewhere out in the sort of middle of the Pacific Ocean halfway between Hawaii and the Fiji Islands we crossed the imaginary International Date line.

Traveling across the International Date Line can be confusing.

Especially for people taking a short trip, such as from Fiji to Hawaii.

According to the clock,

A traveler would end up arriving in Hawaii before he or she had left Fiji.

Because Fiji is a day ahead of Hawaii, and two hours behind .

When it is noon in Hawaii, it is 10:00 AM in Fiji on the next day.

Confusions of days and schedules do sometimes lead to mishaps.

But most airlines and travel agencies keep the line in mind.

When informing travelers about schedules, expected arrival times, and itineraries.

But, I have to tell you.

We had a cocktail party that started one evening and by the calendar it went on for two days.

Even though the party only lasted a few hours.

It sure does confuse me when we crossed that date line.

I will admit it.

And I wasn't the only one.

Dick and Candace went to bed unusually early (for them) and they woke up two days later.

The ship had printed up scrolls for all of the passengers.

Proclaiming us all as members of the "Imperial Order of the Golden Dragon" just for crossing the International date line.

Just had to throw that little tidbit in there.

The Hybrid Diaries
Chapter Fifty

December 7th, 1990
Guess who didn't go to the Philippines?

Ok, We've been at sea for four (or should I say five) days now and we should have been getting ready to dock at Manila Port.

But the Super Typhoon Owen was dying out southeast in the Philippine Sea. North of the Caroline Island chain.

So, the Captain Decided that it would be in the best interest of the Passengers and Crew of the "Little Princess" to avoid making landfall.

While there were still dangerous weather conditions in the southern Pacific.

That could produce further storms that could easily turn into typhoons.

In fact the local weather services were already projecting a new Super typhoon forming from the Remnants of Owen.

They were calling this on Typhoon "Russ" and it was projected to me more Destructive than Owen.

Nothing like a ship full of sea sick old ladies riding the rails.

While the Ship bobbed up and down like a cork in a Jacuzzi.

Sure, it sounds funny, but, it's something I wouldn't care to experience.

So, We'll spend another couple of days at sea to avoid the nasty weather.

And err on the side of good sense and caution.

I just wonder how the guys are doing on the container ship.

If I knew that this sort of thing would have happened.

I would never have sent them on such a perilous voyage.

Well, back to the breakfast bar I guess.

And then more fun and games.

And tonight, I heard over the loud speaker that the Movie for tonight was Frank Capra's classic "It's a Wonderful Life,

Starring: Jimmy Stewart and Donna Reed.

Ah, Donna Reed always reminds me of Rhonda.

She's got that same kind of spirit.

She can always make me feel like the luckiest man in the world.

I wouldn't want to go on living on this world without her smiling face to look at.

My favorite scene is when they had a run on the building and loan and George gave up all of their honeymoon money to save the family business.

And Ernie the cab driver takes George to the Waldorf Hotel.

(aka 320 Sycamore, The old Granville House) where Mary has dinner prepared and a bedroom set up for them.

And she says to George as he opens the door.

"Welcome home Mr. Bailey!"

That's how Rhonda makes me feel like every single day that I am with her.

I couldn't have done all the things I have done.

If it wasn't for her encouragement.

I stare around the house, and I see the rotisserie with chickens roasting over a fire in the fireplace.

Being rotated by the phonograph with a piece of string and a thimble.

And I marvel at her ingenuity with a dumbfounded expression of wide eyed amazement.

I can't tell you how many times I have thought about that scene when I think about my girl.

I know that I don't say it near enough.

But, I do so love that woman.

The Crew had formed a Choral singing group who were walking through the corridors of the Ship singing Christmas Carols.

Some of them I was familiar with others were more English in the heritage.

But, they were all well done.

Just another friendly reminder of the upcoming Christmas holiday.

But, this time I actually enjoyed. It.

The Hybrid Diaries
Chapter Fifty One

December 11th 1990

It was three more days since we avoided the Philippines.

And we could begin to see signs of land

fall looming over the horizon.

Sea gulls filled the air and the unmistakable smell of ocean water crashing

onto rocky shores and breakers.

That distinctly fishy smell that reminds the nose and taste buds,

That "Yes you are smelling and or eating seafood.

Our ship's itinerary says (If you skipped past the Philippines section and go

right to Australia)

that the ship would hug the coastline for another day before we make land fall

in Sidney.

So, We had one more day at sea before we'd dock in Sidney Harbor.

Our grand ocean was just about at its halfway point.

David our cabin steward was sad that we would not be riding back

home on the little princess.

(honestly, I was growing tired of the monotony of ship life).

After the first few days of being on board the ship,

You've pretty much have done everything on board that the ship has to offer.

It's not a shortcoming with the ship or the crew.

In fact the crew and the entertainers do their very best to try and not do the

same things

over and over again.

The kitchen and the wait staff however are a slightly different story.

Every night in the main dining room (which was staffed by Italian waiters)

would come out and bring your dinners and then the head waiter would come

to our table and he'd proudly proclaim in broken English

"Ah, Tonight, I'll maka something special for you!

(and we'd all ooh and aahhh.)

"Tonight, I'ama makin…

"Fettuccini Alfredo"

And we'd all Clap and politely cheer like it was the very first time that he had ever done this.

After the second week,

I was pretty much snockered when I would enter the main dining room.

Because, I would know about the upcoming treat of a life time.

It was like an gastronomical déjà vu.

I did my very best not to mock him.

Because, he was trying to do his best to keep us entertained.

With his limited amount of English words at his disposal.

And after all we were his captive audience.

We did however get to try new thing to eat.

I had my first caviar (yuck) and my first escargot

(snails, again, yuck) and my first calamari (squid or Octopus, I really couldn't

tell one way or the other, they both had tentacles and once again yuck.)

The Hybrid Diaries
Chapter Fifty Two
December 13th 1990
Coming into Sidney and the brave new world

OK we got to Australia and we docked in Sidney. Is was and still is a big beautiful harbor.

We could see the world famous Sidney Opera house from the ship.

Fun fact here.

Did you know that the roof of the Sidney Opera House is made of a Canvas type material to make it look like the sails of a tall sailing ship coming into the harbor?

Me neither. I Just wanted to throw that one in there.

The cargo ship had docked in the harbor of Woolloomooloo about a week before we arrived in Sidney.

And the guys had already had everything off loaded onto the docks.

They had unpacked the cargo containers.

And had everything loaded into a couple of trailers that were to be towed by two Mack trucks.

They look similar to the Mack trucks that we have here in the states complete with "Mack the Bulldog" on the hoods.

Except they looked a little more robust than the trucks we drive in the USA.

They have massive cow catchers on the front that reminded me of old style steam locomotives

It kind of reminded me of the Movie " The Road Warrior" with it's right hand drive. (the steering wheel on the right side of the cab)

Just like all the other Holden's and Fords and Dodges and BMW's.

Their automobiles also have all the familiar auto names like the Ford Falcon and Caprice

And except for driving on the wrong side of the road.

The roads had a look that kind of reminded me of rural Texas from the late

1970's .

Lots of two lane roads.

Most of the roads are without any side shoulders to speak of.

Because we were from the "Other side of the lake" and being a bunch of dumb Yanks,

we decided it would be best if we hired drivers who could get us safely to Woolloomooloo, New South Wales without us getting killed on the way (for driving on the wrong side of the road.)

Woolloomooloo is a beautiful town set on the coast.

Just a few short miles down the road from Sidney.

Our trucks and trailers looked strange to the Aussies, who were not used to seeing an old 1936 Ford Stake bed Pickup truck painted in Aqua colored paint with many layers of clear coating over the Beautiful aqua finish.

And to them.

Left hand driving cars are a bit of a rarity there in Australia.

Much more rare when you throw in all of the logos for Reptex

Beer and the New Pauluxy Brewery and the long horn skull with the Texas flag. emblazoned right on the hood,

It kind of stood out amongst all of the Holden's and Fords.

So needless to say, we had made a statement.

We headed out to Woolloomooloo.

Where we were supposed to meet up with the race sponsors.

And a gentleman named James Kingston.

The trip there only took a few minutes.

Since Woolloomooloo is a suburb of Sidney.

Just slightly less than two miles from the actual dock yards.

It looked to me like we were driving through a rundown part of the city.

I was expecting the local roads to look like West Texas.

With lots of flat country and two lane blacktops.

With no shoulders on the side of the roads.

The road course on the other hand.

Should have plenty of that "Very rural".

Out in the middle of nowhere feel to it.

And with such great distances between stops.

I was thinking to myself

"No wonder these people wanted to promote alternative forms of energy and transportation".

Gasoline or Petrol as they call it was expensive.

Australia is a big country.

They need LOTS of it just to get around.

Just like in Texas.

These people needed to have new and better ways to get around.

When we got to our destination.

The bar at The Tilbury Hotel.

We met up with Mr. Kingston.

And the four other entries.

(Yes, I said four. As in "ONLY four.")

That had managed to make the trip.

There was the team that came from Japan.

And then there was the team from Denmark.

I was happy when I saw that the team from that college in Massachusetts that starts with the letter M and ends with the letter T.

Had also made the long journey to the outback.

(It was that team in particular that I had devoted all of my energies to defeating.

They represented to me

"The college elite."

My warped mind interpreted them as the team that believed that they are smarter than everyone else on the planet.

Now I may be wrong.

But, it was this type of thinking that got me to this point.

So, in a way I guess I owe them a debt of gratitude for inspiring me.

And setting the bar so high.

I want to beat them. I want to beat them BADLY.

They are the team I wanted to beat in the first place.

They are the standard bearers that I designed DeGuello for.

Deguello's purpose is to eat their race car for lunch.)

And there was also a cool looking solar racing car.

That was developed by and sponsored by General Motors.

I promised Rhonda that "

"I would do my very best to be civil and polite to all of the other teams "

"That is what I set off to do."

All of these futuristic vehicles were now in the parking lot of the the Tilbury Hotel.

A rather quaint old hotel that judging by the furniture in the bar.

It looks like it had been in business since the early 1960's.

The outside facade looks like it has been in business for a much longer period.

When we first pulled into the parking lot.

And decided to off load just one of our cars.

So that the competition could get a look at what we brung to the party.

We chose to off load DeGuello, because it looks the most ominous.

As she came rolling down the ramps in the back of the trailer.

Everyone ooo'd and ahhh'd they saw the fighter like canopy.

Then they saw her sinister sleek black carbon fiber body.

Glistening in the Aborigine sun.

They just stood there with their mouths hanging open.

As they gazed at Deguello in utter amazement.

Watching all of the sunlight reflecting off her indigo blue solar panels.

We unfolded them and I activated the photovoltaic sensor.

So that they could watch the solar panels track the movement of the sun.

And move all by themselves.

Greedily following the arc of the sun in the sky for maximum charging potential.

None of the other entrants had anything like that on their cars.

Back to Mr. Kingston.

We had all finally found Mr. Kingston in the hotel bar.

He was hammering down drink after drink.

As if he were desperately trying to forestall something bad.

By drinking his problems away.

That by itself should have told me something was wrong.

Mr. Kingston was kind enough to inform us that the race had been cancelled.

Due to a lack of entries and interest.

Needless to say, we all hit the roof.

All of our hard work, everything we had to endure .

All of the personal suffering.

Plus, all of the expense we shelled out was for nothing.

And it seems like they made little to no attempt to notify any of us.

About that little tidbit of information.

They said that they tried calling us at home.

But, got no answer.

And they said that they also sent us a letter to our home address

But, because we were already out to sea.

We must have just missed it

It seemed that the only one who would actually make anything out of the trip was Dick,

Who through his contacts was able to get import licenses.

And the distribution rights for his beer in Australia.

It was a major success for him, Candace and the New Paluxy Brewery.

Rhonda and I are very happy for their accomplishments

"Way to go Wigglers!"

It was a big open market for people who seem to love things from the Lone Star State.

Meanwhile every single one of the race teams

All FOUR of them.

Well five if you include us.

Was mad as hell that the race had been called off due to a lack of interest.

Where was the Government Sponsorship?

Where were the adoring crowds of fans?

(No one really cared about the prize money).

We have all come to Australia to compete.

So, after meeting with the other teams.

We all decided that we were going to "Go it alone."

And that since we were already there,

And since no one was going to win the prize money anyway.

 (Since there was no prize money to be won.)

We would race each other for in the spirit of competition and camaraderie.

And to advance the technology and promote world peace.

We all agreed that we would race on the original projected course.

That was outlined in the rule book.

It was roughly 1400 miles.

Basically 100 miles per day.

With check points along the way So the teams can make emergency repairs.

And to log our miles per hour.

And the miles per gallon.

And to monitor the vehicles efficiency.

We would all do our best to keep ourselves together.

And try to race in a sort of convoy for greater safety.

(Just like we did on our run down to Houston.)

It was going to be grueling ordeal.

It is summer down under in January.

And the heat gets to be around 45 degrees Celsius.

(about 120 degrees in the shade,)

Just about the same as a typical hot Texas summer.

Hot enough to bake the brains out of a horny toad.

Mr. Kingston kept apologizing for the lack of local support.

They didn't even have a local race team from Australia.

For the home fans to rally behind.

That was sad because it might have drummed up a little bit more interests.

At least on a local level.

He kept saying that he was sorry,

"That there would be no prize money."

"Because, there no large sponsors."

"And locally there was just not enough interest."

"And not enough co sponsors to give us a winner's purse."

We had already explained to him.

That we didn't care about the money.

And that was not our motivation for entering and wanting to compete in this challenge.

Dammit, we had just crossed the Pacific Ocean.

And we had spent untold thousands of dollars on each of our respective vehicles.

And each team had all worked their butts off.

I could relate to their struggles because of all the struggles that we had endured as well.

We had decided that we were going to have a race.

Even if it was for bragging rights only.

So, we all agreed to follow the original course plotted out by the "sponsors" of the race.

And we decided to set out for our trek out into the outback.

On the same day that Mr. Kingston had laid out originally.

Although, there would be no "Official local support" for us.

And no throngs of cheering crowds wishing us well.

And no prize money.

Because, the Australian people just didn't seem to care enough.

We would still be ready to put on a decent show.

And burn as little fossil fuels as possible.

In the process.

At least there might be a local television station there.

To cover us setting out into the desert.

There would be plenty of sun.

Two days before the big race the guys opened the containers.

They had unpacked everything from the containers .

And due to the typhoon saturating the inside of our containers,

we discovered that the sea water had seeped into the storage areas inside

of the containers.

The seawater was everywhere and the damage it had done had been extensive.

While the batteries were wrapped in plastic.

(And because of that they were still serviceable.)

(Thank you, Earl.)

The sea water had corroded the inner workings of our electric wheels.

None of them worked.

It's a good thing that we didn't have to drive our vehicle to the Tilbury Hotel.

We would not have been able to do that.

We would need to rebuild them.

Or else replace the corrupted components with new parts.

(Which were also damaged by the sea water)

Unfortunately, without Toby, it would take us too long for just me and the

guys to ferret out replacement parts locally.

We need Toby Caliente!!! I said.

So, I got on the phone and I called him back in Dallas,

Totally forgetting about the time delay.

It was six in the evening here.

And about six AM yesterday back at home,

So, while I didn't exactly wake him up out of a sound sleep.

I did however disturb his normal morning routine.

"Toby, It's me Andy." I began

"We need your help."

"The containers that we used to ship the vehicles were not water tight"

And the typhoon that the Container ship had encountered had exposed the wheels to salt water.

And now they don't work"

"We need a couple of new sets of wheels

for our racers. A.S.A.P.(As soon as possible)." I pleaded

"I'll go to the airport in Sidney and I'll book passage on a Quantas 747 for you and for your wife."

" Please, bring us some wheels so we can race."

"I know its short notice and you have things on your plate, "

But, Dog gone it we need you."

Toby, this is going to be your finest hour!"

Instead of him putting up a huge fuss.

And him hemming and hawing over the circumstances.

All Toby said was "OK, I'm yours."

So, Toby and Isabel went to D/FW airport and picked up their tickets at the Quantas

Terminal for the 747 that would whisk them high into the sky.

And across the ocean to Australia.

He had thoughtfully packed everything that he would need to build us the new wheels for DeGuello and for Spirit.

We met them at the airport the next day.

He had shipped the new parts and wheels on the same plane.

All we had to wait for customs to clear our crates of desperately needed Supplies.

Once we got the crates picked up.

We headed back to the hotel and we began to disassemble and rebuild the wheels.

Time was even shorter now.

We would need to have them running by tomorrow.

If we were going to compete in the race.

We pulled an all nighter and when the sun was rising,

we had just finished rebuilding the wheels for Deguello.

We had no time to get Spirit ready for the race.

Besides "our entry" was for only one vehicle.

Even though this was no longer an official race.

(so the rules of the race did not actually apply to any of us anymore,)

I would still have wanted to race the Spirit along with DeGuello.

Oh well.

Fate had conspired to make Spirit a rolling parts car.

Instead of the racer I had originally envisioned.

The Hybrid Diaries
Chapter Fifty Three
January 8th, 1991
The Race begins

So, the big day had finally come.

Everything that we had worked so hard for over the last few months would now come down to a Few grueling days in the intense Australian sun,

(Fourteen days to be exact).

All of the vehicles were lined up on the starting line.

And with the sound of the Starters gun and the Judge yelling,

" Ready Steady Go!" (I was expecting "On your Mark Get set Go, but, hey I guess they do things a little different down here).

Was easy to hear the starter.

Without the deafening, roaring sounds of internal combustion engines.

There was practically no sound at all.

Except for the small fuel efficient engine of the Japanese entry.

Blam! We all took off towards history.

The whine of electric motors filled the air

As the crowd of maybe forty people.

(Including the chase crews)

Cheered and blew whistles and honked air horns at the beginning of the race.

Our race cars dink dinked down the roadway

Without all of that sound of V-8 muscle

And all of the sounds of thunder that accompanies any NASCAR event.

The people were cheering.

And their cheers were much louder than the humming sounds our race cars made.

As we spooled up our motors to our vehicles top speeds.

None of our cars could break the century mark (100 miles per hour).

But, we were getting pretty close to eighty miles per hour.

Our testing indicated that our top speed was between seventy five and seventy eight miles per hour on a flat surface.

The wind blowing in through the car's fresh air vents felt like it was coming out of an electric hair dryer.

And the dry heat had almost no humidity in the air.

This was going to be rough.

Over the next few days, we started out strong,

Willy and Dylan and Christopher piloted our vehicle through the intense heat.

The camel back proved to be a life saver keeping the drivers hydrated.

In spite of the heat.

The Japanese were not quite so lucky.

Losing one of their drivers to heat prostration.

After only a couple of days behind the wheel of their sun

Tour racer.

Their design was powered by an electric motor, solar cells and a highly fuel efficient

Ethanol and gas mixture motorcycle engine.

Poor ventilation and the heat from the engine overwhelmed the drivers of their car in the dry and Intense heat of the Australian sun.

The Danes were also having their fair share of troubles.

Not enough vent holes to allow fresh air into their cramped cockpit.

And their bubble canopy had them being baked like ants under a magnifying glass.

Under the intense heat of the sun.

They had also been having communication problems with their radios.

We had yelled out a warning over our radios about an upcoming road train a few miles

Behind us and coming up unusually fast.

We did our best to alert everyone in the race.

To get the Hell off of the road. A.S.A.P.

Unfortunately, the Danish team had either ignored our warning or else failed to receive The alert messages.

In any case they were in the lead during this leg of the race.

And they did not notice us flash our lights to try and get their attention.

They stayed on the roadway,

The big Mack truck with a huge plate steel cow catcher on the front came barreling down the road at about 100 miles an hour.

The driver of the big rig saw our tiny race cars spread out over a mile of road. That did little to impress him.

He took no action other than continuing down the road at his blistering pace.

He did manage to blast his air horns at the little Danish solar powered car,

I am sure that it must have scared the living shit out of that poor little driver.

I imagined that he must have done his best to take evasive action.

And try to drive to the side of the road.

But, since there are no shoulders on the side of the road.

His little car drove off the roadway and into the brush.

Only to be caught up in the tremendous vortices generated by almost 150 feet of tractor trailer and sixty huge wheels.

We've been taught our lessons by practicing on an airport runway.

"So, we had experience with huge things generating terrific vortices when passing our cars."

We knew that the best course of action was to get the hell off of the roadway as fast as possible.

And to hold on tight just in case a vacuum caught our cars up in their wakes.

The Danish car was literally swept into the air,

It spun around on its rear wheels.

And then the solar panel was ripped from its hinges.

The car was thrown violently to the ground.

The driver was thrown clear of the vehicle and he landed face down in the brush.

We all stopped to render aid.

The driver was lucky that he was wearing a helmet.

Because, that saved his life.

The helmet took the brunt of the accident and was almost broken in two.

We had all agreed to stop at the nearest hospital.

And one of our chase trucks drove the Poor, battered Danish driver directly to the nearest hospital emergency ward (some twenty miles away) for treatment.

The road train truck didn't even stop.

He just kept on rolling at 100 plus miles per hour off into the distance.

Apparently unconcerned that he might have just killed somebody.

The guys did what field repairs we could to the damaged Danish racer.

We were able to Re-attach the solar panel,

And the Danes made a driver substitution and they started back

On with the race.

And with significant damage to their car.

One of their wheels was warped by the sheer force of the impact.

They could not run at the vehicles top speed anymore.

The vibrations must have been really bad.

Their car was shaking and almost hopping down the road.

They would need to either replace it with a spare wheel when we reached the next check point.

Or else have it trued so that they could continue racing.

My personal nemesis team had their own problems to contend with.

We had to pull over and help the team from M.I.T.

When one of their batteries caught fire and threatened to destroy the entire vehicle and the driver of the car.

Fortunately, their radios were working and they heard Christopher yelling at them to pull over because smoke was beginning to pour out of the back of the car.

Their chase truck had heard the warning as well.

And they came driving up to the solar powered race car.

And forced their car off of the roadway.

Just in time to put the fire out

And save their driver and their entry.

De Quello was humming right along,

Since the Houston run and all the problems that we encountered along the way had been addressed.

We had had adequate time to prepare for what may occur.

We learned that we should use the human powered flywheel generator sparingly

(Mainly when there were cameras filming us).

The electric wheels having been tested and retested and rebuilt.

Even though we had done our best to slap a working set together.

After the disastrous trip across the high seas.

The wheels had shown none of the weaknesses that we uncovered during the Shakedown ride from Dallas.

Toby had done his homework.

And we had finally had a set of secret weapons that no one

Had seen ever!

We all finally made it to the next check point and then we headed off to the hospital Several miles away.

By the time we got there the Danish driver had already been treated

For minor cuts and bruises and a mild concussion.

And although "Lars" would no longer be in any condition to compete

(relegated to the chase truck)

He would live to race another day.

The hardest thing for us to do was keeping the other teams from looking under our hood

So to speak.

The other teams and the press alike were all dying to get a look or a story on

What we had developed.

They all wanted to know everything that we had designed and developed on our own.

And they would have sold their grandmothers if we would show them.

Our blue prints and designs for the technology that we developed.

They were all dying to take a look at the technology that we had brought to the race.

We did our best to keep the curious away from our vehicles.

Sometimes having to be downright rude to these people

(This unfortunately is not the way we are under normal circumstances)

When they got a little too curious about what was under the hood.

After seeing the other racers up close.

We had suddenly realized why everyone was being so nosey.

We had basically brought a gun to a knife fight.

And yes, we were kicking butt and taking names.

Sometimes pulling away and leaving our competition far, far away in our

Dusty trails.

Our top speed of seventy eight miles per hour had put us right there with the GM

Sponsored team and the college team from MIT. (there I said it.M.I.T.)

I was so proud of everyone on my team.

And yes, I was proud of myself.

I did what I had set out to do.

I showed them all.

That one person could make a difference

Without the backing of a major university.

Even though we had to do it all for nothing more than a few lines in several local newspapers and a couple of spots on the local news channels.

Few people outside of Australia would even know what we were doing in the Outback.

We slept in our trucks and vehicles.

And we ate whatever we could find locally (mostly fruit)

And during the race day we drank lots of bottled water.

At day's end we drank Dick's beer and wine.

And on one night we had sake provided by the Japanese team.

We lived like gypsies for the duration of the race.

As was originally promised.

At the end of the race.

Mr. Kingston had managed to prepare some trophies for our efforts.

There were no losers.

Everyone who competed won.

We "won our race division for being the only human powered electric solar hybrid Vehicle in the race.

The Danes won their division.

Even though their car had been blown off the road course by a road train.

And had suffered major damage to their drive train when they flew off the course.

And into the underbrush.

Naturally the team from M.I.T. won for being the fastest pure solar vehicle.

And the GM team won for innovative design.

The Japanese won for most fuel efficient design.

Also as "Best in Show"

We all picked up our toys and then we packed everything up.

And we got the Hell out of Woolloomooloo, New South Wales and all of the lovely people there.

Thank you all for your hospitality.

Some of you DID manage to come out and see the future of cars.

And they cheered us on.

We did feel like champions when we crossed the finish line.

There were several hundred people there all hooping and hollering.

As each car crossed over the finish line.

I hope we put on an interesting enough show for you.

We all met up back at Sidney for one last get together.

We agreed that we would meet at The Sidney Opera house.

And start crawling from one bar to another.

So, we could take in as much local flavor as possible.

Our favorite local flavor came in the form of several quart sized cans of Foster's Lager Beer.

It comes in big blue cans that remind me of quart sized motor oil cans from the 1960's

It was a shame that we didn't get to take in all of the touristy sights of this great Country.

The Country of Australia has a lot of interesting sights to offer.

Although we were in another country, we had little time To ourselves.

Just to kick back and enjoy the sights.

I hope that they will welcome us back.

I know I keep saying this.

I was really blown away by the world famous Opera house.

It stood perched high upon a bluff in Sidney Harbor.

I had no idea that the roof of the Sidney Opera House was made out of canvas.

I thought it was made out of metal or fiberglass or something like that.

I know I already mentioned that but, man I still can't believe that it's canvas.

It looks so rigid.

I also didn't know that it was supposed to look like a sailing ship entering into the harbor.

That thought had never occurred to me.

Architecture is not my strong suit.

We made several toasts to each other and to our respective countries and to the future of Electric and hybrid vehicles in general.

Together we all crawled into several bars, got horribly over served.

Then we all had to somehow manage to crawl back to our hotels

For a good long hot shower.

A nice meal.

And a fresh change of clothes and underwear.

And a chance to sleep in a comfortable bed.

This would be our first time in over two weeks.

The Hybrid Diaries
The last Chapter.
January 22nd 1991
Sorry Sir, It's time to go home

Like a ninja thrusting a Samurai sword into our eye sockets.

Only hot coffee and iced tea and breakfast could abate the sheer feeling of agony that resided within our skulls.

And after breakfast.

We had to pack up all of our belongings and beat a hasty trip to the harbor.

So we could board our ship for the long trek back to America.

Then Texas and finally home.

Candace had booked us passage on a Mizzenmast barefoot Cruise ship.

We would ride home aboard the good ship " The SS. Idle wild".

These ships are tall three masted sailing vessels with auxiliary diesel engines on board

that not only generate power for all of the ships conveniences but also power the ship's propellers.

Just in case the ship gets becalmed and there is no wind.

It can still make twenty knots in a total calm.

A ship that is set up for total luxury and comfort.

Talk about the "Romance of sea travel."

Toby and Isabel would be flying back in a couple of days after doing a little sightseeing on their own.

Bill and the guys had the duty of packing the containers.

This time there would be extensive use of plastic to protect the electrical components from salt water damage.

We had the vehicles and all the spare parts loaded aboard another

Container ship with similar accommodations for Dylan and Christopher Willy and Bill

And the rest of the team.

It was the adventure of a lifetime.

Rhonda and I enjoyed it immensely.

We departed the great Island and the adventure seemed more and more like a dream.

As time and new distractions began to occupy a place in our minds and hearts.

And in about three weeks we would all be back at home and begin packing our belongings for our move to Austin, Texas.

So, I can be much closer to my job.

It is going to be hard moving all of our stuff with only an RX-7.

I found Dick sitting under the shade of the main sail on deck.

I found a deck chair and I ordered myself a cold one.

Dick was sitting on a deck chair relaxing with a tall

Cold drink in his hand.

When I sat down he asked me.

"So, you got any more bright ideas?"

"Yeah, As a matter of fact I do." I said

"Would you like to see my plans for my solar powered submarine?"

The Hybrid Diaries
Epilogue

Ironically the containers that held everything never showed back up at

Galveston.

Both vehicles had "somehow" disappeared in transit.

In spite all of our best efforts to keep an eye on our containers.

They were never off loaded at the Houston ship channel docks.

Yeah the vehicles and the contents of the containers were insured for a million dollars

That was not what concerned me.

All of our hard work and all of the technology that we as a team had developed had somehow magically disappeared off a huge nine hundred foot container ship.

Fortunately, they could not steal my ideas or our achievements.

But, dammit they stole my truck!

And my new trailer and both of my cars.

That just added even more insult to our losses.

The guys felt terrible about the missing containers.

They felt like they had let me down.

I never wanted them to feel bad about the missing containers or about letting me down,

I love those guys.

It was the fault of the Container ship's crew.

That somehow off loaded the wrong containers at the wrong port.

The original plans were safe back at home.

And Bill had all of the schematics for the carbon fiber bodies back at his shop in Grand Prairie.

So, we could always build another DeGuello.

but, the original vehicles.

The stuff that my dreams were made of were somehow taken away from us.

Dick Candice and Rhonda tried to console me.

I was having nothing of it.

I wanted to sue the container ship line that had somehow misplaced our cargo containers for negligence on their part.

I was told that this sort of thing happens all of the time.

And that most companies are able to write it off as a cost of doing business.

But, we were not some company conducting business overseas.

This was our prototypes for a better tomorrow.

Man I was upset.

But, since we filed our insurance claim.

Suing the Container ship Company was not going to be possible.

Somewhere out there are my prized human electric hybrid vehicles.

The only two of their kind in the whole wide world.

It was like Emilia Earhart's Lockheed Vega.

Both lost to history.

I wanted answers but, none were forthcoming.

The Hybrid Diaries
Epilogue to The Epilogue

It's been almost twenty three years since those fateful days.

When I first built my racers.

And now-a-days the technology that I invented has progressed by leaps and bounds.

The internet has been online for many years now.

And the free exchange of ideas goes around the planet with ease.

Just log into www.youtube.com and look up "e trykes" and

You will see just how far my ideas have come since that primitive time when it was just me in a welding shop.

Building a frame for my first human electric hybrid vehicle out of old bicycle parts.

People from all over the world have now stood upon my shoulders and the shoulders of many more people like myself to further develop and redevelop my technology that I had pioneered.

Although the original vehicles have still not been recovered.

It has come to my attention that the Chinese have more or less taken over the lion's share

of Electric hub motor development and lithium battery design.

And developments to such an extent that I am almost forced to conclude that my vehicles.

May be in someone's private garage in Shang-hi, China.

When we were in Woolloomooloo.

We were unaware that we had all kinds of reporters on hand.

From every major news outlet from all around the world.

Covering the race on a global scale,

Yes, and some of them were probably from China.

Either way,

The technology has been exported to other third world countries and to

Europe.

And in places like Malaysia and the Philippines.

The technology is producing many electric vehicles which are now helping to industrialize these

Countries which are short on fossil fuels.

It seems that even though I received little to no credit.

And the long distance race in the Outback of Australia has for the most part been relegated to the Dusty pages of history.

I am happy to think that somehow.

And in some small way.

I believe that I made a difference.

And that someone out there was watching us with great interest.

I only hope that my early visions.

And all of my hard work will end up back in the States

And that some people will embrace the products of my fertile imagination.

And that soon.

These three wheeled battery powered human electric hybrid vehicles will no longer be the oddities that they were so many years ago.

I wish that they would become a more main stream form of personal transportation for the masses.

Just think of the amount of fuel we would be saving.

Not to mention all of the pollution that we would no longer be dumping into our atmosphere.

Or all of the money we shell out on gasoline to power our cars.

From one short point to the other.

I see a time when the future will be brighter and more prosperous for everyone.

To that end.

I am happy that my story has been told.

www.ingramcontent.com/pod-product-compliance
Lightning Source LLC
Chambersburg PA
CBHW071840200526
45167CB00016B/2